# A·N·N·U·A·L E·DITIONS

# Geography
*Eighteenth Edition*

**03/04**

**EDITOR**

**Gerald R. Pitzl**
*Macalester College (Retired)*

Gerald R. Pitzl received his bachelor's degree in secondary social science education from the University of Minnesota in 1964 and his M.A. (1971) and Ph.D. (1974) in geography from the same institution. Dr. Pitzl has taught a wide array of geography courses, and he is the author of a number of articles on geography, the developing world, and the use of the Harvard case method.

*McGraw-Hill/Dushkin*
530 Old Whitfield Street, Guilford, Connecticut 06437

Visit us on the Internet
*http://www.dushkin.com*

# Credits

1. **Geography in a Changing World**
   Unit photo—© 2003 by PhotoDisc, Inc.
2. **Human-Environment Relationships**
   Unit photo—United Nations photo.
3. **The Region**
   Unit photo—United Nations photo.
4. **Spatial Interaction and Mapping**
   Unit photo—Courtesy of Digital Stock.
5. **Population, Resources, and Socioeconomic Development**
   Unit photo—Courtesy of United States Department of Agriculture.

# Copyright

Cataloging in Publication Data
Main entry under title: Annual Editions: Geography. 2003/2004.
1. Geography—Periodicals.  2. Anthropo-geography—Periodicals  3. Natural Resources—Periodicals
I. Pitzl, Gerald R., *comp.* II. Title: Geography.
ISBN 0–07–283817–5        658'.05        ISSN 1091–9937

© 2003 by McGraw-Hill/Dushkin, Guilford, CT 06437, A Division of The McGraw-Hill Companies.

Copyright law prohibits the reproduction, storage, or transmission in any form by any means of any portion of this publication without the express written permission of McGraw-Hill/Dushkin, and of the copyright holder (if different) of the part of the publication to be reproduced. The Guidelines for Classroom Copying endorsed by Congress explicitly state that unauthorized copying may not be used to create, to replace, or to substitute for anthologies, compilations, or collective works.

Annual Editions® is a Registered Trademark of McGraw-Hill/Dushkin, A Division of The McGraw-Hill Companies.

Eighteenth Edition

Cover image © 2003 PhotoDisc, Inc.
Printed in the United States of America    1234567890BAHBAH543    Printed on Recycled Paper

# Editors/Advisory Board

Members of the Advisory Board are instrumental in the final selection of articles for each edition of ANNUAL EDITIONS. Their review of articles for content, level, currentness, and appropriateness provides critical direction to the editor and staff. We think that you will find their careful consideration well reflected in this volume.

## EDITOR

**Gerald R. Pitzl (Retired)**
Macalester College

## ADVISORY BOARD

**Sarah Witham Bednarz**
Texas A & M University

**Robert S. Bednarz**
Texas A & M University

**Roger Crawford**
San Francisco State University

**James Fryman**
University of Northern Iowa

**Allison C. Gilmore**
Mercer University

**J. Michael Hammett**
Azusa Pacific University

**Miriam Helen Hill**
Jacksonville State University

**Lance F. Howard**
Clemson University

**Artimus Keiffer**
Wittenberg University

**Vandana Kohli**
California State University - Bakersfield

**David J. Larson**
California State University, Hayward

**Tom L. Martinson**
Auburn University - Main

**John T. Metzger**
Michigan State University

**Peter O. Muller**
University of Miami

**Robert E. Nunley**
University of Kansas

**Ray Oldakowski**
Jacksonville University

**Thomas R. Paradise**
University of Arkansas

**James D. Proctor**
University of California - Santa Barbara

**Donald M. Spano**
University of Southern Colorado

**Wayne G. Strickland**
Roanoke College

**John A. Vargas**
Pennsylvania State University - DuBois

**Daniel Weiner**
West Virginia University

**Randy W. Widdis**
University of Regina

# Staff

**Jeffrey L. Hahn,** Vice President/Publisher

## EDITORIAL STAFF

**Theodore Knight, Ph.D.,** Managing Editor
**Roberta Monaco,** Managing Developmental Editor
**Dorothy Fink,** Associate Developmental Editor
**Addie Raucci,** Senior Administrative Editor
**Robin Zarnetske,** Permissions Editor
**Marie Lazauskas,** Permissions Assistant
**Lisa Holmes-Doebrick,** Senior Program Coordinator

## TECHNOLOGY STAFF

**Richard Tietjen,** Senior Publishing Technologist
**Jonathan Stowe,** Executive Director of eContent
**Marcuss Oslander,** Sponsoring Editor of eContent
**Christopher Santos,** Senior eContent Developer
**Janice Ward,** Software Support Analyst
**Angela Mule,** eContent Developer
**Michael McConnel,** eContent Developer
**Ciro Parente,** Editorial Assistant
**Joe Offredi,** Technology Developmental Editor

## PRODUCTION STAFF

**Brenda S. Filley,** Director of Production
**Charles Vitelli,** Designer
**Mike Campbell,** Production Coordinator
**Eldis Lima,** Graphics
**Juliana Arbo,** Typesetting Supervisor
**Julie Marsh,** Project Editor
**Jocelyn Proto,** Typesetter
**Cynthia Powers,** Typesetter

# To the Reader

In publishing ANNUAL EDITIONS we recognize the enormous role played by the magazines, newspapers, and journals of the public press in providing current, first-rate educational information in a broad spectrum of interest areas. Many of these articles are appropriate for students, researchers, and professionals seeking accurate, current material to help bridge the gap between principles and theories and the real world. These articles, however, become more useful for study when those of lasting value are carefully collected, organized, indexed, and reproduced in a low-cost format, which provides easy and permanent access when the material is needed. That is the role played by ANNUAL EDITIONS.

The articles in this eighteenth edition of *Annual Editions: Geography* represent the wide range of topics associated with the discipline of geography. The major themes of spatial relationships, regional development, the population explosion, and socioeconomic inequalities exemplify the diversity of research areas within geography.

The book is organized into five units, each containing articles relating to geographical themes. Selections address the conceptual nature of geography and the global and regional problems in the world today. This latter theme reflects the geographer's concern with finding solutions to these serious issues. Regional problems, such as food shortages in the Sahel and the greenhouse effect, concern not only geographers but also researchers from other disciplines.

The association of geography with other fields is important, because expertise from related research will be necessary in finding solutions to some difficult problems. Input from the focus of geography is vital in our common search for solutions. This discipline has always been integrative. That is, geography uses evidence from many sources to answer the basic questions, "Where is it?" "Why is it there?" and "What is its relevance?" The first group of articles emphasizes the interconnectedness not only of places and regions in the world but of efforts toward solutions to problems as well. No single discipline can have all of the answers to the problems facing us today; the complexity of the issues is simply too great.

The writings in unit 1 discuss particular aspects of geography as a discipline and provide examples of the topics presented in the remaining four sections. Units 2, 3, and 4 represent major themes in geography. Unit 5 addresses important problems faced by geographers and others.

*Annual Editions: Geography 03/04* will be useful to both teachers and students in their study of geography. The anthology is designed to provide detail and case study material to supplement the standard textbook treatment of geography. The goals of this anthology are to introduce students to the richness and diversity of topics relating to places and regions on the Earth's surface, to pay heed to the serious problems facing humankind, and to stimulate the search for more information on topics of interest.

All members of the Macalester College community were saddened by the untimely passing of Barbara Wells-Howe, who was the administrative assistant through all of the seventeen editions of this work. Her typing, organization of materials, and many helpful suggestions were greatly appreciated. Without her diligence and professional efforts, these undertakings could not have been completed. Barbara's task has been taken over by Devi Benjamin, and I would like to express my gratitude to her for her invaluable assistance in preparing the eighteenth edition. Special thanks are also extended to McGraw-Hill/Dushkin's editorial staff for coordinating the production of the reader. A word of thanks must go as well to all those who recommended articles for inclusion in this volume and who commented on its overall organization. Artimus Keiffer, Peter O. Muller, and Wayne J. Strickland were especially helpful in that regard.

In order to improve the next edition of *Annual Editions: Geography*, we need your help. Please share your opinions by filling out and returning to us the postage-paid *article rating form* on the last page of this book. We will give serious consideration to all your comments.

Gerald R. Pitzl
*Editor*

# Contents

To the Reader                                                iv
Topic Guide                                                  xi
Selected World Wide Web Sites                                xiii

World Map                                                    xvi

## UNIT 1
## Geography in a Changing World

Eight articles discuss the discipline of geography and the extremely varied and wide-ranging themes that define geography today.

### Unit Overview                                            xviii

1. **The Big Questions in Geography,** Susan L. Cutter, Reginald Golledge, and William L. Graf, *The Professional Geographer,* August 2002

    The authors have taken the challenge of science correspondent John Noble Wilford to articulate the big questions in *geography.* Wilford's concern is that research by geographers is not being reported and that geographers may be missing the important questions in their research.                2

2. **Rediscovering the Importance of Geography,** Alexander B. Murphy, *The Chronicle of Higher Education,* October 30, 1998

    *Geography's* renaissance in U.S. education is the key theme of this piece. The author insists that geography be recognized not as an exercise in place names, but because it addresses physical and human processes and sheds light on the nature and meaning of changing *spatial arrangements* and *landscapes.*     11

3. **Birth of the African Union,** Sarah Coleman and Julius Dawu, *World Press Review Online,* July 24, 2002

    Amid both hope and skepticism, a new *regional* organization, the African Union, was launched on July 9, 2002. A major outcome of this group's work is *economic growth* to close the gap between African countries and the developed world.    13

4. **The Four Traditions of Geography,** William D. Pattison, *Journal of Geography,* September/October 1990

    This key article, originally published in 1964, was reprinted, with the author's later comments, in the 75-year retrospective of the *Journal of Geography.* It is a classic in the *history of geography.* William Pattison discusses the four main themes that have been the focus of work in the discipline of geography for centuries—the spatial concept, area studies, land-human relationships, and earth science.    15

5. **Restoring an Ecosystem Torn Asunder by a Dam,** Sandra Blakeslee, *New York Times.com,* June 11, 2002

    The installation of the Glen Canyon Dam on the Colorado River profoundly altered the *environment* of the area. Now, efforts are under way to use adaptive management techniques to address problems of sand accumulations and alterations in fish populations within the Colorado River *ecosystem.*    19

6. **Sculpting the Earth From Inside Out,** Michael Gurnis, *Scientific American,* March 2001

    How has Earth's surface been sculpted? Through the use of new and more powerful computer models, insights are gained into the complexity of the *geological processes* involved in creating the overall *physical geography* of the planet.    22

The concepts in bold italics are developed in the article. For further expansion, please refer to the Topic Guide and the Index.

v

7. **Nuclear Power: A Renaissance That May Not Come,** *The Economist,* May 19, 2001

    The recent **energy** crisis in California and a growing **global demand for energy** have spurred new interest in expanding nuclear power generation in the United States. Although it has been some time since a serious nuclear accident has occurred, doubts still remain about expansion.   **29**

8. **Filling the Void,** Cathleen McGuigan, *Newsweek,* September 9, 2002

    An unprecedented **urban** renewal project is unfolding in New York City as plans are considered for structures to rise at the site of the World Trade Center twin towers. Interest in this renewal project is both intense and **global.**   **33**

# UNIT 2
# Human-Environment Relationships

Eleven articles examine the relationship between humans and the land on which we live. Topics include global warming, water management, urban sprawl, pollution, and the effects of human society on the global environment.

**Unit Overview**   **36**

9. **Global Warming: The Contrarian View,** William K. Stevens, *New York Times,* February 29, 2000

    William Stevens's article focuses on a debate about the causes of **global warming.** The contrarian view suggests that the effects of **greenhouse gas** accumulations in the atmosphere are overestimated.   **38**

10. **The Pollution Puzzle,** Tom Arrandale, *Governing,* August 2002

    State governments from New Hampshire to California are taking the lead in the fight against **pollution** and **environmental** degradation. In 2000, the National Academy of Public Administrators (NAPA) insisted that the new leader of the Environmental Protection Agency begin new campaigns to curb **greenhouse gases,** reduce smog, and keep fertilizers out of rivers and streams.   **43**

11. **Human Modification of the Geomorphically Unstable Salt River in Metropolitan Phoenix,** Martin Roberge, *The Professional Geographer,* May 2002

    Proposals have been made to alter the physical restrictions on the Salt River and return the waterway to its original braided form. The power of the channeled river has caused **environmental** disruptions in the **urban** area of Phoenix.   **48**

12. **Texas and Water: Pay Up or Dry Up,** *The Economist,* May 26, 2001

    **Water** availability continues to be a problem of **regional** concern in the American Southwest. Many **urban** areas along the course of the Rio Grande, for instance, may literally run out of water within 20 years if conservation measures are not enacted.   **60**

13. **Beyond the Valley of the Dammed,** Bruce Barcott, *Utne Reader,* May/June 1999

    A number of **dams** in the United States are coming down. Bruce Barcott reports on recent **water** management thinking, resulting in the removal of some of the 75,000 existing dams. What is known now is that sediments deposited in the reservoirs become polluted with chemicals and toxic waste, which is damaging to the **environment.**   **62**

14. **Environmental Enemy No. 1,** *The Economist,* July 6, 2002

    **Coal** is a widely used **energy** source and a prime contributor to **carbon dioxide** in the atmosphere, the major cause of **global warming.** Carbon sequestration is seen as a positive move to reduce carbon dioxide accumulations, an intermediate step toward the development of environmentally friendly hydrogen fuel cells.   **67**

The concepts in bold italics are developed in the article. For further expansion, please refer to the Topic Guide and the Index.

15. **Carbon Sequestration: Fired Up With Ideas,** *The Economist,* July 6, 2002

    Scientists are proposing novel ideas for reducing the amount of carbon dioxide in the atmosphere in order to deter further **global warming.** One process involves capturing carbon dioxide and storing it within ocean waves, in depleted oil and gas reservoirs, and in coal seams. Carbon sequestration could prove to be a positive action to deter **environmental** degradation.   **68**

16. **Past and Present Land Use and Land Cover in the USA,** William B. Meyer, *Consequences,* Spring 1995

    Cities, towns, and villages now cover slightly more than 4 out of every 100 acres of the land in the continental United States, and every day the percentage grows. William Meyer looks at both the positive and negative **consequences of changes in land use.**   **70**

17. **Operation Desert Sprawl,** William Fulton and Paul Shigley, *Governing,* August 1999

    **Population** growth in Las Vegas has been staggering (Clark County has grown from 790,000 in 1990 to over 1.2 million in 1998). Las Vegas is one of the fastest-growing **urban** places in the United States. Its **economic** base of tourism has not been damaged, but strains are showing in its **transportation** and **water delivery systems.**   **78**

18. **A Modest Proposal to Stop Global Warming,** Ross Gelbspan, *Sierra,* May/June 2001

    Environmental damage caused by **global warming** can be stopped, the author contends, by calling an end to the Carbon Age. **Alternative energy sources** and a reasonable approach to a post–**Kyoto Protocol** era is proposed.   **82**

19. **A Greener, or Browner, Mexico?,** *The Economist,* August 7, 1999

    The North American Free Trade Agreement (NAFTA) was claimed to be the world's first **environmentally friendly trade treaty.** Critics suggest that NAFTA has resulted in a more **polluted Mexico.** Employment in Mexico's maquiladoras has doubled since 1991, but so have the levels of **toxic waste.**   **86**

# UNIT 3
## The Region

Ten selections review the importance of the region as a concept in geography and as an organizing framework for research. A number of world regional trends, as well as the patterns of area relationships, are examined.

### Unit Overview   88

20. **The Rise of the Region State,** Kenichi Ohmae, *Foreign Affairs,* Spring 1993

    The **global** market has dictated a new, more significant set of **region states** that reflect the real flow of economic activity. The old idea of the **nation-state** is considered an anachronism, an unnatural and dysfunctional unit for organizing either human activity or emerging economic systems.   **90**

21. **Continental Divide,** Torsten Wohlert, *World Press Review,* January 2000

    Torsten Wohlert reports that the **geopolitical** picture in Europe may change greatly as the European Union (EU) considers applications for membership from Central European countries. One of the controversies is the EU's insistence on intensive **agricultural** practices, a process not beneficial to the Central European **region.**   **94**

22. **A Dragon With Core Values,** *The Economist,* March 30, 2002

    A **regional** competition has pitted Hong Kong and Shanghai, two major **urban** centers in China, against each other for leadership in the emerging **economic growth** of China.   **96**

The concepts in bold italics are developed in the article. For further expansion, please refer to the Topic Guide and the Index.

vii

23. **The Late Great Wall,** Melinda Liu, *Newsweek,* July 29, 2002
   The Great Wall of China, one of the most recognizable ***regional*** structures in the world, is in need of extensive repairs. Some of the problems are ***environmental:*** Gobi Desert sands cover more stretches of the wall than before and sections made of an adobe-like substance are crumbling to dust.   **97**

24. **A Continent in Peril,** *Time,* February 12, 2001
   This ***map*** of Africa illustrates the ***regional*** devastation of ***AIDS*** on the continent. In 2000 there were 36 million adults in the world living with AIDS; 70 percent of them were in sub-Saharan Africa.   **100**

25. **AIDS Has Arrived in India and China,** Ann Hwang, *World Watch,* January/February 2001
   The number of ***deaths from AIDS*** will soon surpass the number of deaths associated with the Black Death epidemic in the fourteenth century. These numbers will increase in years to come now that AIDS has made significant inroads in India and China. It was hoped that the Asian ***region*** would be spared.   **102**

26. **Greenville: From Back Country to Forefront,** Eugene A. Kennedy, *Focus,* Spring 1998
   Eugene Kennedy traces the rise of Greenville, South Carolina, from a back-country ***region*** in its early days, to "Textile Center of the World" in the 1920s, to a supremely successful, diversified ***economic*** stronghold in the 1990s. Greenville's location within a thriving area midway between Charlotte and Atlanta gives it tremendous ***growth potential.***   **109**

27. **Death of a Small Town,** Dirk Johnson, *Newsweek,* September 10, 2001
   Small towns in the ***region*** of the Great Plains are suffering from ***population*** declines and the loss of viable ***economic*** activities. Many of the small towns in this area lack accessibility to service centers and viable health care facilities.   **116**

28. **The Río Grande: Beloved River Faces Rough Waters Ahead,** Steve Larese, *New Mexico,* June 2000
   As Steve Larese reports, the Rio Grande is in danger as the American Southwest ***region*** comes into a protracted period of ***drought.*** Natural occurrences and the ***damming*** of the river have constricted the flow of ***water*** in Albuquerque and areas farther downstream.   **118**

29. **"You Call That Damn Thing a Boat?",** Charles W. Ebeling, *American Heritage of Invention & Technology,* Fall 2001
   For decades in the nineteenth and early twentieth centuries, metal tankers called "whalebacks" were the primary carriers of bulk commodities such as coal and wheat on the Great Lakes. Whalebacks proved to be invaluable in the complex ***transportation system*** in the ***region.***   **124**

# UNIT 4
## Spatial Interaction and Mapping

Eight articles discuss the key theme in geographical analysis: place-to-place spatial interaction. Human diffusion, transportation system, and cartography are some of the themes examined.

**Unit Overview**   **126**

30. **Transportation and Urban Growth: The Shaping of the American Metropolis,** Peter O. Muller, *Focus,* Summer 1986
   Peter Muller reviews the importance of ***transportation*** in the growth of American ***urban places.*** The highly compact urban form of the mid-nineteenth century was transformed in successive eras by electric streetcars, highways, and expressways. The city of the future may rely more on ***communication*** than on the actual movement of people.   **128**

The concepts in bold italics are developed in the article. For further expansion, please refer to the Topic Guide and the Index.

31. **Internet GIS: Power to the People!,** Bernardita Calinao and Candace Brennan, *GEO World,* June 2002

    A ***GIS***-based Web site allowed citizens of Erie, Pennsylvania, to help choose which airport runway extension alternatives would work best. The Web site featured environmental ***maps,*** which proved useful in making suggestions for modifying this ***transportation system.*** GIS on the Internet revolutionizes how ***environmental assessments*** are conducted. **137**

32. **ORNL and the Geographic Information Systems Revolution,** Jerome E. Dobson and Richard C. Durfee, *ORNL Review,* September 1, 2002

    The work of the Oak Ridge National Laboratory in ***Geographic Information Systems (GIS)*** from the 1960s to the present is reviewed in this essay. Its work includes GIS applications in coastal change analysis, ***environmental*** restoration, and ***transportation*** modeling and analysis. **140**

33. **Mapping the Outcrop,** J. Douglas Walker and Ross A. Black, *Geotimes,* November 2000

    Digitized ***topographic maps*** and the application of principles from the powerful field of ***Geographic Information Systems (GIS)*** technology, mainstays in geography, are revolutionizing geology instruction and its practice. **158**

34. **Gaining Perspective,** Molly O'Meara Sheehan, *World Watch,* March/April 2000

    This in-depth article discusses important uses for satellite imagery to monitor ***weather*** patterns, ***flooding,*** and problems in ***agriculture.*** Molly Sheehan describes how ***mapping*** has been greatly enhanced through satellite technology. **160**

35. **Counties With Cash,** John Fetto, *American Demographics,* April 2000

    This county-based ***choropleth*** map illustrates the great range of per capita income in the United States. A table of counties with the highest dollar values include ***metropolitan*** and rural counties. **168**

36. **A City of 2 Million Without a Map,** Oakland Ross, *World Press Review,* July 2002

    Managua, Nicaragua, an ***urban*** place of over two million people, was struck by an enormous earthquake in 1972 that significantly disrupted the grid network of the place. The old ***maps*** were no longer useful and the city has yet to be totally rebuilt. Novel ways of giving directions from place to place have emerged and ***accessibility*** within Managua has become difficult. **171**

37. **China Journal I,** Henry Petroski, *American Scientist,* May/June 2001

    The Three Gorges project in China will create the world's largest hydroelectric generating ***dam.*** However, the ***environmental*** costs will be staggering. There is also a ***spatial interaction*** outcome: the dam will extend the reach of ocean-going vessels to Chongqing, nearly 2100 kilometers inland. **172**

# UNIT 5
## Population, Resources, and Socioeconomic Development

Eight articles examine the effects of population growth on natural resources and the resulting socioeconomic level of development.

**Unit Overview** **178**

38. **Before the Next Doubling,** Jennifer D. Mitchell, *World Watch,* January/February 1998

    The article warns of the possibility of another doubling of world ***population*** in the next century. Food shortages, inadequate supplies of fresh water, and greatly strained ***economic*** systems will result. Family planning efforts need to be increased, and a change in the desire for large families should be addressed. **180**

The concepts in bold italics are developed in the article. For further expansion, please refer to the Topic Guide and the Index.

39. **A Turning-Point for AIDS?,** *The Economist,* July 15, 2000

The spread of AIDS, particularly devastating in the sub-Saharan *region,* is addressed. A picture of its effects on Botswana is shown in superimposed *population* pyramids for 2020, one showing the country's projected population without the AIDS epidemic and one with AIDS. **187**

40. **AIDS Scourge in Rural China Leaves a Generation of Orphans,** Elisabeth Rosenthal, *New York Times.com,* August 25, 2002

East Asia is the last major *region* to experience the devastation of the *AIDS* pandemic. Recent estimates suggest over one million AIDS cases in China alone. **191**

41. **The Next Oil Frontier,** *Business Week,* May 27, 2002

U.S. interest in petroleum extraction is mounting as *economic development* in the old USSR's Central Asian Republics is creating a complex *geopolitical* situation. The United States is planning the installation of an oil pipeline from the Caspian Sea in Kazakhstan west to the Black Sea. **194**

42. **Gray Dawn: The Global Aging Crisis,** Peter G. Peterson, *Foreign Affairs,* January/February 1999

Countries in the *developed* world will experience a decided increase in the number of elderly in the *population.* This *demographic transition* will have dramatic *economic* impact. By 2020, one in four people in the developed world will be 65 years old or older. **197**

43. **A Rare and Precious Resource,** Houria Tazi Sadeq, *The UNESCO Courier,* February 1999

Fresh *water* is fast becoming a scarce *resource.* Water, necessary for every human being, is extensively used in *agriculture* and industry. The availability of fresh water is a *regional* problem. As *population* increases, more regions are facing water shortages. **203**

44. **In Race to Tap the Euphrates, the Upper Hand Is Upstream,** Douglas Jehl, *New York Times.com,* August 25, 2002

Dwindling fresh *water* supplies *globally* will be a major factor slowing *economic growth* in future decades. *Drought* conditions in many world *regions* are already causing serious water shortages. **207**

45. **The Coming Water Crisis,** Marianne Lavelle and Joshua Kurlantzick, *U.S. News & World Report,* August 12, 2002

Environmental Protection Agency head Christine Whitman calls *water* quality and quantity "the biggest *environmental* issue that we face in the 21st century." Infrastructure degradation, *regional* supply imbalances, and higher cost are just a few of the reasons why fresh water is a key issue in the United States and elsewhere. **210**

**Index** 217
**Test Your Knowledge Form** 220
**Article Rating Form** 221

The concepts in bold italics are developed in the article. For further expansion, please refer to the Topic Guide and the Index.

# Topic Guide

This topic guide suggests how the selections in this book relate to the subjects covered in your course. You may want to use the topics listed on these pages to search the Web more easily.

On the following pages a number of Web sites have been gathered specifically for this book. They are arranged to reflect the units of this *Annual Edition*. You can link to these sites by going to the DUSHKIN ONLINE support site at *http://www.dushkin.com/online/*.

**ALL THE ARTICLES THAT RELATE TO EACH TOPIC ARE LISTED BELOW THE BOLD-FACED TERM.**

### Accessibility
36. A City of 2 Million Without a Map

### Africa
39. A Turning-Point for AIDS?

### Agriculture
21. Continental Divide
34. Gaining Perspective
43. A Rare and Precious Resource

### AIDS
24. A Continent in Peril
25. AIDS Has Arrived in India and China
39. A Turning-Point for AIDS?

### Cartography
34. Gaining Perspective
35. Counties With Cash

### Climate change
9. Global Warming: The Contrarian View

### Communication
30. Transportation and Urban Growth: The Shaping of the American Metropolis

### Dams
12. Texas and Water: Pay Up or Dry Up
13. Beyond the Valley of the Dammed
28. The Rio Grande: Beloved River Faces Rough Waters Ahead
37. China Journal I

### Demographic transition
38. Before the Next Doubling
42. Gray Dawn: The Global Aging Crisis

### Development
38. Before the Next Doubling
42. Gray Dawn: The Global Aging Crisis

### Drought
28. The Rio Grande: Beloved River Faces Rough Waters Ahead
44. In Race to Tap the Euphrates, the Upper Hand Is Upstream

### Economic growth
3. Birth of the African Union
22. A Dragon With Core Values
30. Transportation and Urban Growth: The Shaping of the American Metropolis
41. The Next Oil Frontier
44. In Race to Tap the Euphrates, the Upper Hand Is Upstream

### Economic issues
14. Environmental Enemy No. 1
17. Operation Desert Sprawl
23. The Late Great Wall
26. Greenville: From Back Country to Forefront
27. Death of a Small Town
38. Before the Next Doubling
42. Gray Dawn: The Global Aging Crisis

### Ecosystem
5. Restoring an Ecosystem Torn Asunder by a Dam

### Energy
7. Nuclear Power: A Renaissance That May Not Come
18. A Modest Proposal to Stop Global Warming

### Environment
5. Restoring an Ecosystem Torn Asunder by a Dam
10. The Pollution Puzzle
11. Human Modification of the Geomorphically Unstable Salt River in Metropolitan Phoenix
13. Beyond the Valley of the Dammed
14. Environmental Enemy No. 1
15. Carbon Sequestration: Fired Up With Ideas
19. A Greener, or Browner, Mexico?
31. Internet GIS: Power to the People!
32. ORNL and the Geographic Information Systems Revolution
37. China Journal I
45. The Coming Water Crisis

### Flooding
34. Gaining Perspective

### Geographic information systems (GIS)
31. Internet GIS: Power to the People!
32. ORNL and the Geographic Information Systems Revolution
33. Mapping the Outcrop

### Geography
1. The Big Questions in Geography
2. Rediscovering the Importance of Geography

### Geography, history of
4. The Four Traditions of Geography

### Geological processes
6. Sculpting the Earth From Inside Out

### Geopolitics
21. Continental Divide
41. The Next Oil Frontier

### Global issues
7. Nuclear Power: A Renaissance That May Not Come
8. Filling the Void
9. Global Warming: The Contrarian View
15. Carbon Sequestration: Fired Up With Ideas
18. A Modest Proposal to Stop Global Warming
20. The Rise of the Region State
44. In Race to Tap the Euphrates, the Upper Hand Is Upstream

### Greenhouse effect
- 9. Global Warming: The Contrarian View
- 10. The Pollution Puzzle

### Kyoto Protocol
- 18. A Modest Proposal to Stop Global Warming

### Landscape
- 2. Rediscovering the Importance of Geography
- 16. Past and Present Land Use and Land Cover in the USA

### Maps
- 24. A Continent in Peril
- 31. Internet GIS: Power to the People!
- 33. Mapping the Outcrop
- 34. Gaining Perspective
- 35. Counties With Cash
- 36. A City of 2 Million Without a Map

### Natural resources
- 43. A Rare and Precious Resource

### Physical geography
- 6. Sculpting the Earth From Inside Out

### Pollution
- 10. The Pollution Puzzle
- 19. A Greener, or Browner, Mexico?

### Population
- 17. Operation Desert Sprawl
- 27. Death of a Small Town
- 33. Mapping the Outcrop
- 38. Before the Next Doubling
- 39. A Turning-Point for AIDS?
- 42. Gray Dawn: The Global Aging Crisis
- 43. A Rare and Precious Resource

### Region
- 3. Birth of the African Union
- 12. Texas and Water: Pay Up or Dry Up
- 21. Continental Divide
- 22. A Dragon With Core Values
- 24. A Continent in Peril
- 25. AIDS Has Arrived in India and China
- 26. Greenville: From Back Country to Forefront
- 27. Death of a Small Town
- 28. The Rio Grande: Beloved River Faces Rough Waters Ahead
- 29. "You Call That Damn Thing a Boat?"
- 39. A Turning-Point for AIDS?
- 40. AIDS Scourge in Rural China Leaves a Generation of Orphans
- 43. A Rare and Precious Resource
- 44. In Race to Tap the Euphrates, the Upper Hand Is Upstream
- 45. The Coming Water Crisis

### Region-state
- 20. The Rise of the Region State

### Resource management
- 43. A Rare and Precious Resource

### Spatial arrangements
- 2. Rediscovering the Importance of Geography
- 37. China Journal I

### Spatial geography
- 4. The Four Traditions of Geography

### Transportation systems
- 17. Operation Desert Sprawl
- 29. "You Call That Damn Thing a Boat?"
- 30. Transportation and Urban Growth: The Shaping of the American Metropolis
- 31. Internet GIS: Power to the People!
- 32. ORNL and the Geographic Information Systems Revolution

### Urban areas
- 8. Filling the Void
- 11. Human Modification of the Geomorphically Unstable Salt River in Metropolitan Phoenix
- 12. Texas and Water: Pay Up or Dry Up
- 16. Past and Present Land Use and Land Cover in the USA
- 17. Operation Desert Sprawl
- 22. A Dragon With Core Values
- 30. Transportation and Urban Growth: The Shaping of the American Metropolis
- 35. Counties With Cash

### Water
- 12. Texas and Water: Pay Up or Dry Up
- 13. Beyond the Valley of the Dammed
- 17. Operation Desert Sprawl
- 28. The Rio Grande: Beloved River Faces Rough Waters Ahead
- 43. A Rare and Precious Resource
- 44. In Race to Tap the Euphrates, the Upper Hand Is Upstream
- 45. The Coming Water Crisis

### Weather
- 34. Gaining Perspective

# World Wide Web Sites

The following World Wide Web sites have been carefully researched and selected to support the articles found in this reader. The easiest way to access these selected sites is to go to our DUSHKIN ONLINE support site at http://www.dushkin.com/online/.

# AE: Geography 03/04

**The following sites were available at the time of publication. Visit our Web site—we update DUSHKIN ONLINE regularly to reflect any changes.**

## General Sources

**About: Geography**
http://geography.about.com
This Web site, created by the About network, contains hyperlinks to many specific areas of geography, including cartography, population, country facts, historic maps, physical geography, topographic maps, and many others.

**The Association of American Geographers (AAG)**
http://www.aag.org
Surf this site of the Association of American Geographers to learn about AAG projects and publications, careers in geography, and information about related organizations.

**Geography Network**
http://www.geographynetwork.com
The Geography Network is an online resource to discover and access geographical content, including live maps and data, from many of the world's leading providers.

**National Geographic Society**
http://www.nationalgeographic.com
This site provides links to National Geographic's huge archive of maps, articles, and other documents. Search the site for information about worldwide expeditions of interest to geographers.

**The New York Times**
http://www.nytimes.com
Browsing through the archives of the *New York Times* will provide you with a wide array of articles and information related to the different subfields of geography.

**Social Science Internet Resources**
http://www.wcsu.ctstateu.edu/library/ss_geography.html
This site is a definitive source for geography-related links to universities, browsers, cartography, associations, and discussion groups.

**U.S. Geological Survey (USGS)**
http://www.usgs.gov
This site and its many links are replete with information and resources for geographers, from explanations of El Niño, to mapping, to geography education, to water resources. No geographer's resource list would be complete without frequent mention of the USGS.

## UNIT 1: Geography in a Changing World

**Alternative Energy Institute (AEI)**
http://www.altenergy.org
The AEI will continue to monitor the transition from today's energy forms to the future in a "surprising journey of twists and turns." This site is the beginning of an incredible journey.

**Mission to Planet Earth**
http://www.earth.nasa.gov
This site will direct you to information about NASA's Mission to Planet Earth program and its Science of the Earth System. Surf here to learn about satellites, El Niño, and even "strategic visions" of interest to geographers.

**Nuclear Power Introduction**
http://library.thinkquest.org/17658/nuc/nucintroht.html?tqskip1=1&tqtime=0125
Here you will find information regarding alternative energy forms. There is a brief introduction to nuclear power and a link to the geography of nuclear power—maps that show where nuclear power plants exist.

**Poverty Mapping**
http://www.povertymap.net
Poverty maps can quickly provide information on the spatial distribution of poverty. Here you will find maps, graphics, data, publications, news, and links that provide the public with poverty mapping from the global to the subnational level.

**Santa Fe Institute**
http://acoma.santafe.edu/sfi/research/
This home page of the Santa Fe Institute—a nonprofit, multidisciplinary research and education center—will lead you to a plethora of valuable links related to its primary goal: to create a new kind of scientific research community, pursuing emerging science. Such links as Evolution of Language, Ecology, and Local Rules for Global Problems are offered.

**Solstice: Documents and Databases**
http://solstice.crest.org/docndata.shtml
In this online source for sustainable energy information, the Center for Renewable Energy and Sustainable Technology (CREST) offers documents and databases on renewable energy, energy efficiency, and sustainable living. The site also offers related Web sites, case studies, and policy issues. Solstice also connects to CREST's Web presence.

## UNIT 2: Human-Environment Relationships

**Alliance for Global Sustainability (AGS)**
http://www.global-sustainability.org
The AGS is a cooperative venture seeking solutions to today's urgent and complex environmental problems. Research teams from four research universities study large-scale, multidisciplinary environmental problems that are faced by the world's ecosystems, economies, and societies.

**Environment News Service: Global Warming Could Make Water a Scarce Resource**
http://ens.lycos.com/ens/dec2000/2000L-12-15-06.html
This article, by Cat Lazaroff, makes interesting reading. Lazaroff reports the results of a 2-year study of the potential impacts of climate change on the nation's fresh and salt water systems. One of the conclusions: "Humans are changing the climate."

xiii

# www.dushkin.com/online/

### Human Geography
*http://www.geog.le.ac.uk/cti/hum.html*

The CTI Centre for Geography, Geology, and Meteorology provides this site, which contains links to human geography in relation to agriculture, anthropology, archaeology, development geography, economic geography, geography of gender, and many others.

### The North-South Institute
*http://www.nsi-ins.ca/ensi/index.html*

Searching this site of the North-South Institute—which works to strengthen international development cooperation and enhance gender and social equity—will help you find information on a variety of development issues.

### United Nations Environment Programme (UNEP)
*http://www.unep.ch*

Consult this home page of UNEP for links to critical topics of concern to geographers, including desertification and the impact of trade on the environment. The site will direct you to useful databases and global resource information.

### World Health Organization
*http://www.who.int*

This home page of the World Health Organization will provide you with links to a wealth of statistical and analytical information about health in the developing world.

## UNIT 3: The Region

### AS at UVA Yellow Pages: Regional Studies
*http://xroads.virginia.edu/~YP/regional.html*

Those interested in American regional studies will find this site a gold mine. Links to periodicals and other informational resources about the Midwest/Central, Northeast, South, and West regions are provided here.

### Can Cities Save the Future?
*http://www.huduser.org/publications/econdev/habitat/prep2.html*

This press release about the second session of the Preparatory Committee for Habitat II is an excellent discussion of the question of global urbanization.

### IISDnet
*http://iisd.ca*

The International Institute for Sustainable Development, a Canadian organization, presents information through gateways entitled Business and Sustainable Development, Developing Ideas, and Hot Topics. Linkages is its multimedia resource for environment and development policymakers.

### NewsPage
*http://www.individual.com*

Individual, Inc., maintains this business-oriented Web site. Geographers will find links to much valuable information about such fields as energy, environmental services, media and communications, and health care.

### Treaty on Urbanization
*http://www.geocities.com/atlas/urb/tretyurb.html*

The original 1992 Treaty on Urbanization is available at this site. Its goal is just, democratic, and sustainable cities, town, and villages.

### Virtual Seminar in Global Political Economy/Global Cities & Social Movements
*http://csf.colorado.edu/gpe/gpe95b/resources.html*

This Web site is rich in links to subjects of interest in regional studies, such as sustainable cities, megacities, and urban planning. Links to many international nongovernmental organizations are included.

### World Regions & Nation States
*http://www.worldcapitalforum.com/worregstat.html*

This site provides strategic and competitive intelligence on regions and individual states, geopolitical analyses, geopolitical factors of globalization factors, geopolitics of production, and much more.

## UNIT 4: Spatial Interaction and Mapping

### Edinburgh Geographical Information Systems
*http://www.geo.ed.ac.uk/home/gishome.html*

This valuable site, hosted by the Department of Geography at the University of Edinburgh, provides information on all aspects of Geographic Information Systems and provides links to other servers worldwide. A GIS reference database as well as a major GIS bibliography is included.

### Geography for GIS
*http://www.ncgia.ucsb.edu/cctp/units/geog_for_GIS/GC_index.html*

This hyperlinked table of contents was created by Robert Slobodian of Malaspina University. Here you will find information regarding GIS technology.

### GIS Frequently Asked Questions and General Information
*http://www.census.gov/ftp/pub/geo/www/faq-index.html*

Browse through this site to get answers to FAQs about Geographic Information Systems. It can direct you to general information about GIS as well as guidelines on such specific questions as how to order U.S. Geological Survey maps. Other sources of information are also noted.

### International Map Trade Association
*http://www.maptrade.org*

The International Map Trade Association offers this site for those interested in information on maps, geography, and mapping technology. Lists of map retailers and publishers as well as upcoming IMTA conferences and trade shows are noted.

### PSC Publications
*http://www.psc.isr.umich.edu*

Use this site and its links from the Population Studies Center of the University of Michigan for spatial patterns of immigration and discussion of white and black flight from high immigration metropolitan areas in the United States.

### U.S. Geological Survey
*http://www.usgs.gov/research/gis/title.html*

This site discusses the uses for Geographic Information Systems and explains how GIS works, addressing such topics as data integration, data modeling, and relating information from different sources.

## UNIT 5: Population, Resources, and Socioeconomic Development

### African Studies WWW (U.Penn)
*http://www.sas.upenn.edu/African_Studies/AS.html*

Access to rich and varied resources that cover such topics as demographics, migration, family planning, and health and nutrition is available at this site.

### Geography and Socioeconomic Development
*http://www.ksg.harvard.edu/cid/andes/Documents/Background%20Papers/Geography&Socioeconomic%20Development.pdf*

John L. Gallup wrote this 19-page background paper examining the state of the Andean region. He explains the strong and

# www.dushkin.com/online/

pervasive effects geography has on economic and social development.

**Human Rights and Humanitarian Assistance**
*http://www.etown.edu/vl/humrts.html*
　　Through this site, part of the World Wide Web Virtual Library, you can conduct research into a number of human-rights topics in order to gain a greater understanding of the issues affecting indigenous peoples in the modern era.

**Hypertext and Ethnography**
*http://www.umanitoba.ca/faculties/arts/anthropology/tutor/aaa_presentation.new.html*
　　This site, presented by Brian Schwimmer of the University of Manitoba, will be of great value to people who are interested in culture and communication. He addresses such topics as multivocality and complex symbolization, among many others.

**Research and Reference (Library of Congress)**
*http://lcweb.loc.gov/rr/*
　　This research and reference site of the Library of Congress will lead you to invaluable information on different countries. It provides links to numerous publications, bibliographies, and guides in area studies that can be of great help to geographers.

**Space Research Institute**
*http://arc.iki.rssi.ru/eng/*
　　Browse through this home page of Russia's Space Research Institute for information on its Environment Monitoring Information Systems, the IKI Satellite Situation Center, and its Data Archive.

**World Population and Demographic Data**
*http://geography.about.com/cs/worldpopulation/*
　　On this site you will find information about world population and additional demographic data for all the countries of the world.

We highly recommend that you review our Web site for expanded information and our other product lines. We are continually updating and adding links to our Web site in order to offer you the most usable and useful information that will support and expand the value of your Annual Editions. You can reach us at: *http://www.dushkin.com/annualeditions/*.

# World Map

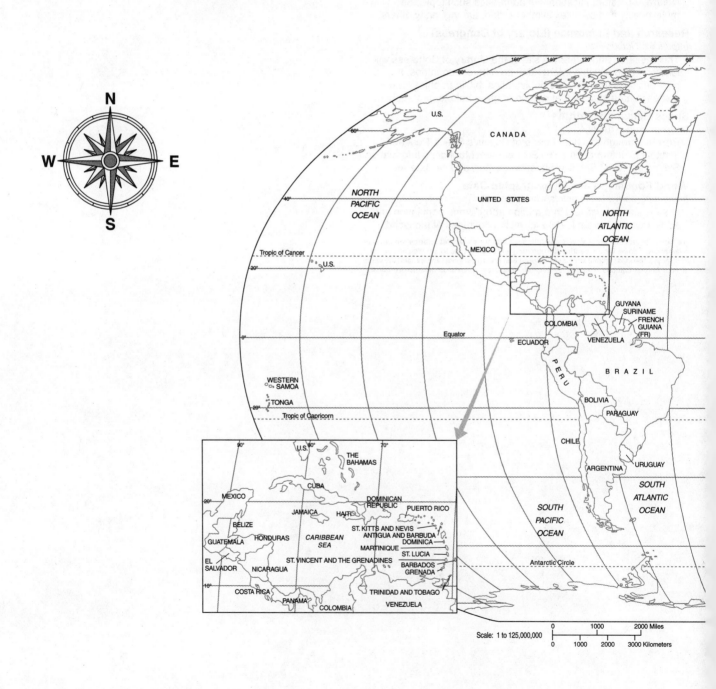

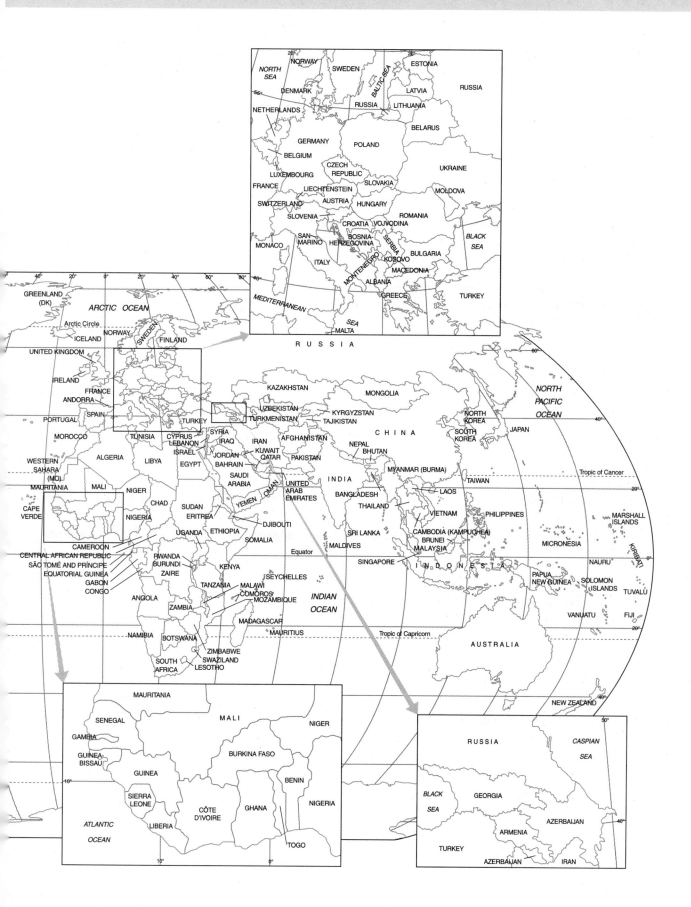

# UNIT 1

# Geography in a Changing World

## Unit Selections

1. **The Big Questions in Geography**, Susan L. Cutter, Reginald Golledge, and William L. Graf
2. **Rediscovering the Importance of Geography**, Alexander B. Murphy
3. **Birth of the African Union**, Sarah Coleman and Julius Dawu
4. **The Four Traditions of Geography**, William D. Pattison
5. **Restoring an Ecosystem Torn Asunder by a Dam**, Sandra Blakeslee
6. **Sculpting the Earth From Inside Out**, Michael Gurnis
7. **Nuclear Power: A Renaissance That May Not Come**, *The Economist*
8. **Filling the Void**, Cathleen McGuigan

## Key Points to Consider

- Why is geography called an integrating discipline?
- How is geography related to earth science? Give some examples of these relationships.
- What are area studies? Why is the spatial concept so important in geography? What is your definition of geography?
- Discuss whether or not change is a good thing. Why is it important to anticipate change?
- Will nuclear power generation increase?
- What does interconnectedness mean in terms of places? Give examples of how you as an individual interact with people in other places. How are you "connected" to the rest of the world?
- What will the world be like in the year 2010? Tell why you are pessimistic or optimistic about the future. What, if anything, can you do about the future?

 **Links: www.dushkin.com/online/**
These sites are annotated in the World Wide Web pages.

**Alternative Energy Institute (AEI)**
  http://www.altenergy.org
**Mission to Planet Earth**
  http://www.earth.nasa.gov
**Nuclear Power Introduction**
  http://library.thinkquest.org/17658/nuc/nucintroht.html?tqskip1=1&tqtime=0125
**Poverty Mapping**
  http://www.povertymap.net
**Santa Fe Institute**
  http://acoma.santafe.edu/sfi/research/
**Solstice: Documents and Databases**
  http://solstice.crest.org/docndata.shtml

What is geography? This question has been asked innumerable times, but it has not elicited a universally accepted answer, even from those who are considered to be members of the geography profession. The reason lies in the very nature of geography as it has evolved through time. Geography is an extremely wide-ranging discipline, one that examines appropriate sets of events or circumstances occurring at specific places. Its goal is to answer certain basic questions.

The first question—Where is it?—establishes the location of the subject under investigation. The concept of location is very important in geography, and its meaning extends beyond the common notion of a specific address or the determination of the latitude and longitude of a place. Geographers are more concerned with the relative location of a place and how that place interacts with other places both far and near. Spatial interaction and the determination of the connections between places are important themes in geography.

Once a place is "located," in the geographer's sense of the word, the next question is, Why is it here? For example, why are people concentrated in high numbers on the North China plain, in the Ganges River Valley in India, and along the eastern seaboard in the United States? Conversely, why are there so few people in the Amazon basin and the Central Siberian lowlands? Generally, the geographer wants to find out why particular distribution patterns occur and why these patterns change over time.

The element of time is another extremely important ingredient in the geographical mix. Geography is most concerned with the activities of human beings, and human beings bring about change. As changes occur, new adjustments and modifications are made in the distribution patterns previously established. Patterns change, for instance, as new technology brings about new forms of communication and transportation and as once-desirable locations decline in favor of new ones. For example, people migrate from once-productive regions such as the Sahel when a disaster such as drought visits the land. Geography, then, is greatly concerned with discovering the underlying processes that can explain the transformation of distribution patterns and interaction forms over time. Geography itself is dynamic, adjusting as a discipline to handle new situations in a changing world.

Geography is truly an integrating discipline. The geographer assembles evidence from many sources in order to explain a particular pattern or ongoing process of change. Some of this evidence may even be in the form of concepts or theories borrowed from other disciplines. The first article of this unit suggests the big questions in geography and the second stresses the importance of geography as a discipline and proclaims its "rediscovery." The next article announces the birth of the African Union.

Throughout its history, four main themes have been the focus of research work in geography. These themes or traditions, according to William Pattison in "The Four Traditions of Geography," link geography with earth science, establish it as a field that studies land-human relationships, engage it in area studies, and give it a spatial focus. Although Pattison's article first appeared over 30 years ago, it is still referred to and cited frequently today. Much of the geographical research and analysis engaged in today would fall within one or more of Pattison's traditional areas, but new areas are also opening for geographers. The Glen Canyon Dam on the Colorado River is the subject of the next article. The surface of the Earth is subject to geological processes, the focus of the next article. The question of nuclear power plant expansion in the United States is raised next. The last article considers plans for rebuilding on the site of the World Trade Center following the terrorist attacks of September 11, 2001.

# Article 1

# The Big Questions in Geography

In noting his fondness for geography, John Noble Wilford, science correspondent for *The New York Times*, nevertheless challenged the discipline to articulate those big questions in our field, ones that would generate public interest, media attention, and the respect of policymakers. This article presents our collective judgments on those signi.cant issues that warrant disciplinary research. We phrase these as a series of ten questions in the hopes of stimulating a dialogue and collective research agenda for the future and the next generation of geographic professionals. **Key Words: geographic ideas, geographic research, geographic thought.**

**Susan L. Cutter**
*University of South Carolina*

**Reginald Golledge**
*University of California, Santa Barbara*

**William L. Graf**
*University of South Carolina*

## Introduction

At the 2001 national meeting of the Association of American Geographers (AAG) in New York City, the opening session featured an address by John Noble Wilford, science correspondent for *The New York Times*. In very candid language, Wilford challenged the discipline to articulate the big questions in our field—questions that would capture the attention of the public, the media, and policymakers (Abler 2001). The major questions posed by Wilford's remarks include the following: Are geographers missing big questions in their research? Why is the research by geographers on big issues not being reported? And what role can the AAG play in improving geographic contributions to address big issues?

First, geographers are doing research on some major issues facing modern society, but not all of them. Geographic thinking is a primary component of the investigation of global warming, for example. Products of that research seen by decision makers and the public often take the form of maps and remote sensing images that explain the geographic outcomes of climate change. Geographic approaches are at the heart of much of the analysis addressing natural and technological hazards, with public interaction taking place through the mapping media. Earthquake, volcanic, coastal, and riverine hazards are all subject to spatial analysis that has become familiar to the public. The terrorist attacks of 11 September 2001 have stimulated new interest in geographic information systems that can be used in response to hazardous events and as guidance in emergency preparedness and response (Figure 1).

In addition to these recent challenges, however, there are major issues that geographers are not addressing adequately at the present time, as illustrated by the accounting that follows in this article. A primary reason for the disconnect between capability to help solve problems and the application of those skills for many major issues is the sociology of the discipline of geography. The majority of AAG members, for example, are academicians, and their agendas and reward structures are targeted at specialized research deeply buried in paradigms that are obscure to decision makers and the public. Additionally, this social structure tends to lead geographic researchers into investigations on small problems that can be solved quickly, produce professional publications, and support a drive for promotion and tenure, rather than investigating more complex, bigger problems that are not easily or quickly solved and do not necessarily lead to academic publications of a type the genre usually demands.

With few exceptions, those geographers outside the university setting are scattered and work individually, in small groups, or as members of larger interdisciplinary teams for governmental agencies, businesses, or private organizations. Because there are few true "institutes" of geographic research, it is difficult to focus geographic energy on big problems. Many geographers in these settings are responding to immediate and short-term demands on their time and talents, rather than leading the larger-scale investigations.

The work in which many geographers engage to address major problems is not reported for two reasons: it does not fit the classic mold for the research journals where geographers get their greatest career awards, and work related to policy often emerges without attribution to the researchers of origin. A significant example that illustrates this point is the work of the Committee on Geography of the Board on Earth Sciences and Resources in the National Research Council (NRC). The committee oversees study committees, which produce geographic studies and reports to guide the federal government in a wide variety of issues that qualify as big questions. Recent work, contributed to primarily by geographers and accomplished from a geographic perspective, includes advice to the U.S. Geological Survey on

# Article 1. The Big Questions in Geography

**Figure 1** Manhattan, New York, before the terrorist attacks of 11 September 2001 (left), and after (right). Photos by S. L. Cutter.

reformulating its research programs to address geographic issues entangled in urban expansion, hazards, and mapping. In other cases, one study committee is producing direction for the federal government on what decision makers and the general public need to know about the world of Islam, while another is investigating transportation issues related to urban congestion and the development of livability indicators. Other geographers participate in the Water Science and Technology Board of the NRC, with recent contributions including the use of the watershed concepts in ecosystem management and the role of dams in the security of public water supplies. Another example involves a geographer-led multidisciplinary group to investigate spatial thinking, and another geographer led a major effort in global mapping. In all of these cases, geographers play a central role, but the product of their work is ascribed only to an organization (the NRC), and individuals are recognized only in lists of contributors. If the reports successfully influence policy, the decision makers who actuate that policy take credit for the process, rather than the original investigators who made the recommendations.

The AAG plays a role in stimulating the research that addresses big questions of importance to modern society by recognizing such work and publicizing it. It may be that individual researchers will be more willing to undertake such research if their work is recognized by their colleagues in the discipline as being important and worthy of praise. The AAG can influence the National Science Foundation, the National Endowment for the Humanities, the National Institutes of Health, the National Geographic Society, and other funding sources to channel attention and resources to individuals or teams examining the big questions. Individual geographers are not likely to be able to exert much influence, except when they serve on review panels for these organizations, but the AAG can exert its influence from its steady and visible presence in Washington.

In trying to identify those issues that might qualify as big questions (Table 1), we have included wide-ranging concepts that encompass some conceptual issues (such as scale), but also point out specific topical areas that seem to demand particular attention at the moment. We argue that such diverse big ideas belong in this accounting because, in the end, they are related to each other and mutually supportive. Some of these big questions may be obscure to the public, but most of them are familiar to researchers and policymakers alike, who have already begun to address them. There is little hope that any collection of big questions can identify problems of equal "bigness," but the ones we have identified all seem to warrant teams of researchers and significant funding rather than following the discipline's usual mode of a single or small group of investigators with funding limited to one or two years in duration. The communication of geographic research findings to the public in thoughtful, useful ways represents a major challenge. This challenge, by itself, can also be regarded as yet another big problem facing the discipline. With these introductory comments in mind, we now turn to those questions that we feel are important for the geographic community to address.

## What Makes Places and Landscapes Different from One Another and Why Is This Important?

This first question goes to the core of the discipline the relevance of similarities and differences among people, places, and regions. What is the nature of uneven economic development and what can geography contribute to understanding this phenomenon? More specifically, how can national and global policies be implemented in a world that is increasingly fragmented politically, socially, culturally, and environmentally?

To elaborate on this question, we accept an assumption that the human mind is not constructed to handle large-scale continuous

## Table 1  Big Questions in Geography

1. What makes places and landscapes different from one another, and why is this important?
2. Is there a deeply held human need to organize space by creating arbitrary borders, boundaries, and districts?
3. How do we delineate space?
4. Why do people, resources, and ideas move?
5. How has the earth been transformed by human action?
6. What role will virtual systems play in learning about the world?
7. How do we measure the unmeasurable?
8. What role has geographical skill played in the evolution of human civilization, and what role can it play in predicting the future?
9. How and why do sustainability and vulnerability change from place to place and over time?
10. What is the nature of spatial thinking, reasoning, and abilities?

chaos. Nor does it function optimally when dealing with large-scale perfect uniformity. Between these two extremes there is variability, which is the dominant characteristic of both the natural world and the human world. To understand the nature of physical and human existence, we need to examine the occurrence and distribution of variability in various domains. For geographers, this examination has involved exploring the nature of spatial distributions, patterns, and associations, examining the effects of scale, and developing modes of representation that best communicate the outcomes of these explorations. In the course of this search for understanding of the essentials of spatial variation, geographers have attempted to comprehend the interaction between physical and human environments, how people adapt to different environments, and how knowledge about human-environment relations can be communicated through appropriate representational media.

Even in the absence of humans, the earth and the phenomena found on this planet are incredibly diverse. Variability is widespread; uniformity is geographically restricted. Determining the nature and occurrence of variability and uniformity are at the heart of the discipline of geography. No other area of inquiry has, as its primary goal, discovering, representing, and explaining the nature of spatial variability in natural and human environments at scales beyond the microscopic and the figural (body space) such as vista, environmental, or gigantic and beyond (Montello 1993). Most geography has been focused on vista, environmental or gigantic scales, but some (e.g., cognitive behavioral) emphasizes figural scale. Finding patterns or trends towards regularity at some definable scale amidst this variability provides the means for generalizing, modeling, and transferring knowledge from one spatial domain to another. Law-like and theoretical statements can be made, and confidence in the relevance of decisions and policies designed to cope with existence can be determined.

Among other things, geographers have repeatedly found, at some scales, spatial regularity in distributions of occurrences that seem random or indeed chaotic at other scales. Sometimes this results from selecting an appropriate scale and format for summarizing and representing information. Examples include using very detailed environmental-scale data to discover the topologic properties of stream networks, or establishing the regular and random components of human settlement patterns in different environments.

Realizing the spatial variability in all phenomena is a part of the naïve understanding of the world. Being able to explain the nature of variability is the academic challenge that drives the discipline of geography. Like other scientists, geographers examine variability in their search for knowledge and understanding of the world we live in, particularly in the human environment relations and interactions that are a necessity for our continued existence.

## Is There a Deeply Held Human Need to Organize Space by Creating Arbitrary Borders, Boundaries, and Districts?

Humans, by their very nature, are territorial. As human civilizations grew from hunter gatherers to more sedentary occupations, physical manifestation of the demarcation of space ensued. Hadrian's Wall kept the Scots and Picts out, the Great Wall of China protected the Ming Dynasties from the Mongols, and the early walled cities of Europe protected those places from barbarians and other acquisitive sociocultural groups.

At a more limited scale, internal spaces in cities were also divided, often based on occupation and/or class. As civilizations grew, space was organized and reorganized into districts that supported certain economic activities. City-states begat nation-states, and eventually most of the world was carved up into political spaces. Nation-states required borders and boundaries (all involving geography), as land and the oceans (and the resources contained within) were carved up into non-equal units. Within nations, land partitioning has been a factor in the decline of environmental quality. For example, the erection of barbed-wire fencing on the Great Plains to separate farming and ranching homesteads from each other did more to hasten the decline of indigenous species and landscape degradation than any other invention at the time (Worster 1979, 1993).

The modern equivalent of the human need and desire for delineating space is the notion of private property. Suburban homes with tall fences between neighbors, for example, help foster the ideal of separation from neighbors and disengagement from the community, both predicated on the need to protect "what's mine" (and of course the ubiquitous property value) as well as providing a basic need for privacy. The tendency for the rich to get richer and the poor to get poorer also applies to the values of these divided properties. The diffusion of democratically controlled, market based economies to much of the globe increases the significance of research that explores why we divide space. Pressing research questions include, for example: Are ghettos bursting with poverty-level inhabitants an inevitable consequence of democratic capitalist societies? Are such societies amenable to concerns for social justice? And how would such concerns influence the patterns and distributions of living activities?

We also lack some of the basic understanding of how the physical delineation of space affects our perception of it. Furthermore, we need better knowledge of how perceptions of physical space alter social, physical, and environmental processes. Finally, has globalization changed our view of the social construction of space? Does physical space still support spatial relations and spatial interactions, or are they becoming somewhat independent, as may be the case in social space, intellectual space, and cyberspace? How will the interactions between people, places, and regions change as our view of space (and time, for that matter) changes?

## How Do We Delineate Space?

Once we understand *why* we partition space, we face a closely related issue: *how* do we do it? The definition of regions by drawing boundaries is deceptively simple. The criteria by which we delineate space have far-reaching consequences, because the resulting divisions of space play a large role in determining how we perceive the world. A map of the United States showing the borders of the states, for example, evokes a very different perception of the nation than a similar scale map showing the borders of the major river basins. A further difference in perception is created if the map shows major rivers as networks rather than as basins, and the resulting difference between perception of networks and perception of regions can direct knowledge and its application in divergent ways (National Research Council 1999, x). For example, should we conduct pollution oriented research on rivers or on watersheds, or on the state administrative units that potentially might control pollution? What are the implications of our choice of geographic framework?

The logical, rational delineation of spaces on the globe depends on the criteria to be used, but geographic research offers few established, widely accepted rules about what these criteria should be or how they might be employed. The designation of political boundaries without respect to ethnic cultures has wrought havoc in much of post colonial Africa and central Europe, for example, but geographers have not yet offered workable alternatives that account for the complexities of multicultural populations. In natural-science research and management, a major issue is the establishment of meaningful regions that can be aggregated together to scale up, or that can be disaggregated to scale down. Natural scientists also experience significant difficulty in designing compatible regions across topical subjects. For example, the blending of watersheds, ecosystems, and ranges of particular species poses significant problems in environmental management. Adding to the complexity from a management and policy perspective is the tangle of administrative regions, whose boundaries are often derived from political boundaries rather than natural ones. Recognition of these problems is easy but offering thoughtful geographic solutions to them is not.

Geographers have much to contribute to the delineation of space by developing new knowledge and techniques for defining subdivisions of earth space based on specific criteria, including economic efficiency, compatibility across applications, ease of aggregation and disaggregation, repeatability, and universality of application. Geographers need to develop methods for delineating space that either resist change over time or accommodate temporal changes smoothly.

A continuing example of delineating space that has important political implications is the process of defining American congressional districts once each decade based on the population census. The need for fair representation, relative uniformity in population numbers in each district, recognition of traditional communities, and accommodation of changing population distributions comprise some of the criteria that need not equate to partisan politics in constructing at least the first approximation of redrawn district boundaries (Monmonier 2001). Some states have nonpartisan commissions to delineate the districts, yet geography provides very little substantive advice on the subject to guide such groups.

## Why do People, Resources, and Ideas Move?

One of the fundamental concepts in geography is the understanding that goods, services, people, energy, materials, money, and even ideas flow through networks and across space from place to place. Although geography faces questions about all these movements, one of the most pressing questions concerns the movement of people. We have some knowledge about the behavior of people who move their residences from one place to another, and we can observe obvious economic forces leading to the migration of people toward locations of relative economic prosperity. However, we have much less understanding about the episodic movements of people in cities. In most developed countries, the congestion of vehicular traffic has become a significant negative feature in assessing the quality of life, and in lesser-developed countries the increasing number of vehicles used in the context of inadequate road networks results in frustrating delays. Geography can and should address fundamental issues such as the environmental consequences of the decision to undertake laborious journeys to work (e.g., contributions of vehicle exhaust to air pollution, the possible environmental changes induced by telecommuting, and the need for alternative-fuel, low-pollution vehicles). In addition to understanding the environmental consequences of daily moves, the discipline has much to offer in describing, explaining, and predicting the sociocultural consequences of these decisions.

The flow of vehicles on roadways involves obvious physical networks, but there are other flows demanding attention that operate through more abstract spaces. The diffusion of culture—particularly "Western" culture, with its emphasis on materialism and individualism—is one of the leading edges of globalization of the world economy. Geographers must begin to address how these social, cultural, and economic forces operate together to diffuse, from a few limited sources, an extensive array of ideas and attitudes that are accepted by a diverse set of receiving populations. Even if such diffusion takes place through digital space, it probably does so in a distinctive geography that we should understand if we are to explain and predict the world in the twenty-first century.

The electrical energy crisis of 2001 made us aware, quite vividly, of the finiteness of nonrenewable resources such as oil and gas and of the difficulties in their distributions. We have already consumed more than 50 percent of the world's known reserves of these resources. Historically, as one energy source has replaced another (as when coal power replaced water power), there have been changes in the locational patterns, growth, importance of settlements, and significance of regions. Examples include the decline of heavy industrial areas into "rust belts" and their replacement with service- and information-based centers that have more locational flexibility. As current energy sources change, what will happen to urban location and growth? Will the geopolitical power structure of the world change markedly? For example, will the countries that are part of the Organization of the Petroleum Exporting Countries (OPEC) retain their global economic power and political strength? Will existing populations and settlements decline, or relocate to alternative sources of energy? What will be the geographic configuration of the economic and political power that goes with such changes?

Finally, the more physically oriented flows, such as those of energy and materials, present a demanding set of questions for geographers. While geochemists are deriving the magnitudes of elemental fluxes of such substances as carbon and nitrogen, for example, it is incumbent on geographers to point out that these fluxes do not take place in aspatial abstract ways, but rather in a physically and socially defined landscape that has important locational characteristics. In other words, although there may very well be an understanding of the amounts of nitrogen circulating from earth to oceans to atmosphere, that circulation is not everywhere equal. How does human management affect the nitrogen and other elemental cycles? What explains its geographic vari-

**Figure 2** A local example of transformations brought about in the natural world by humans. The lower Sandy River of western Oregon appears to be a pristine river, but it has radically altered water, sediment, and biological systems because of upstream dams. Photo by W. L. Graf.

ability? How does that variability change in response to controls not related to human intervention? This leads to our next big question.

## How Has the Earth Been Transformed by Human Action?

Humans have altered the earth, its atmosphere, and its water on scales ranging from local to global (Thomas 1956; Turner et al. 1990). At the local scale, many cities and agricultural landscapes represent nearly complete artificiality in a drive to create comfortable places in which to live and work, and to maximize agricultural production for human benefit. The transformations have also had negative effects at local scales, such as altering the chemical characteristics of air and water, converting them into media that are toxic for humans as well as other species. At regional scales, human activities have resulted in wholesale changes in ecosystems, such as the deforestation of northwest Europe over the past several centuries, a process that seems to be being replicated in many tropical regions today. At a global scale, the introduction of industrial gases into the atmosphere plays a still emerging role in global climate change. Taken together, these transformations have had a geographically variable effect that geographers must better define and explain. Dilsaver and Colten (1992, 9) succinctly outlined the basic questions almost a decade ago: How have human pursuits transformed the environment, and how have human social organizations exerted their control over environments? Graf (2001) recently asked how we can undo some of our previous efforts at environmental change and control.

In many instances, this explanation of variation might emphasize the physical aspects of changes, or understanding the underlying dynamics of why the changes occur (Dilsaver, Wyckoff, and Preston 2000). Wide-ranging assessments of river basins, for example, must rely on a plethora of controlling factors ranging from land use to water, sediment, and contaminant movements. Geographers must employ more complicated and insightful approaches, however, to truly understand why transformations vary from place to place, largely in response to the connection between the biophysical environment and the human society that occupies it. Understanding this delicate interplay between nature and society as a two-way connection can lead us to new knowledge about social and environmental landscapes, but it can also help us make better decisions on how to achieve future landscapes that are more often transformed in nondestructive ways.

One of the primary issues facing many societies in their relationship with their supporting environments is how much of the biophysical world should be left unchanged, or at least changed to the minimal degree possible. The amount of remaining "natural" landscape in many nations is small—probably less than 5 percent of the total surface—so time is growing short to decide what areas should be set aside and preserved (Figure 2). Not only do these preservation decisions affect land and water surfaces; they also profoundly affect nonhuman species that use the surfaces for habitat. If human experience is enriched by diverse ecosystems, then the decline in biodiversity impoverishes humanity as well. Which areas should be preserved and why? How should preserved areas be linked with one other? How can public and private property productively coexist with nearby preserved areas?

## What Role Will Virtual Systems Play in Learning about the World?

Stated another way, what will virtual systems allow us to do in the future that we cannot do now? What new problems can be pursued (Golledge forthcoming)? Providing an answer opens a Pandora's box of questions concerning the geographic impacts of new technologies (Goodchild 2000). What new multimodal interfaces for interpreting visualized onscreen data need to be developed in order to overcome current technological constraints of geographic data visualization? Can we produce a virtual geography? Do we really want to?

One serious problem that deserves immediate attention is the examination of the geographic implications of the development of economies and societies based on information technology. In particular, the sociospatial implications of an increasing division between the digital haves and have-nots demand attention. Pursuing such a problem will require answering questions about the geographic consequences of employment in cyberspace and its implication for human movements such as migration, intraurban mobility, commuting, and activity-space restructuring. The current extensive demand for and use of transportation for business purposes may need to be re-examined. It may be argued that, in the world of business communication, geographic distance is a decreasingly important factor, because both digital and visual interaction can take place at the click of a mouse button without the need for person-to-person confrontation. If this is so, what are the longer-term impacts for living and lifestyles, and how could the inhabitation and use of geographic environments be affected? If this is true, why is it that we see dramatic concentrations of cyber-businesses in a few areas, similar to the locational behavior of pre-digital industries? Are Silicon Valley in California and Route 128 in Massachusetts simply the "rust belts" of the future?

Research has shown that the most effective way of learning about an environment is by directly experiencing it, so that all sensory modalities are activated during that experience (Figure 3) (Gale 1984; Lloyd and Heivly 1987; MacEachren 1992). However, many places are distant or inaccessible to most people. The interior of the Amazon rainforest, the arctic tundra of northern Siberia, Himalayan peaks, the interior of the Sahara desert, Antarctica and the South Pole, the barrios of Rio de Janeiro, and the Bosnian highlands can become much closer to us. Satellite imagery provides detailed digitized imagery of these

**Figure 3** *Exploring immersive virtual worlds with equipment developed between 1992 and 2001, showing the original and the miniaturized versions of a GPS-driven auditory virtual environment at the University of California, Santa Barbara. The more cumbersome 1992 version is shown on the left, with the reduced 2001 version on the right. Psychologist Jack Loomis and associates developed the system, demonstrated here by author Reginald Golledge. Photo courtesy of R. Golledge.*

places. A problem awaiting solution is how to use this extensive digital database to build virtual systems that will allow immersive experiences with such environments. Problems of motion-sickness experienced by some people in immersive systems need to be solved; assuming this will be achieved, virtual reality could become the laboratory of the future for experiencing different places and regions around the world.

Discovering how best to deal with problematic futures, on earth or on other planets, is definitely one of the big problems facing current and future geographers. Many land use planning, transportation, and social policies are made on an "if _____ then _____" basis. Because we are unable to change the world experimentally, we need to investigate other ways of observing environmental events and changes. Examples include changing a street for vehicles to a pedestrian mall to explore human movement behavior, or experiencing the action and consequences of snow or mud avalanches in tourist-dependent alpine environments. What more can we learn by building and manipulating virtual environments? In a virtual system, we can raise local pollution levels, accelerate global warming, change sea levels by melting ice caps, or simulate the impacts of strictly enforcing land conversion policies at the rural-urban fringes of large cities. In the face of an increasingly international economy and globalization of environmental issues, there is a need to develop a way to explore possible scenarios before implementing policies theoretically designed to deal with global (or more local) problems.

## How Do We Measure the Unmeasurable?

Geography is normally practiced at local to national scales at which we can get a clear sense of the existence or development of patterns and processes. People, landscapes, and resources are not evenly distributed on the earth's surface, so we begin with a palette that is diverse. How can we accommodate such diversity in policies to avoid winners and losers? Economists, for example, assume away all spatial variability in their economic models. What happens to general models when space is introduced? How can we transform from the local to the global and vice versa? The question of scale transformation, especially the calibration of large-scale global circulation models or the development of climate-impact models globally with local or regional applications is a major area in which geography can contribute and is playing a leading role (AAG GCLP Group forthcoming).

We need to develop more compatible databases that have an explicit geographic component, with geocoded data that permit us to scale up and scale down as the need arises. Data collection, archiving, and dissemination all are areas that require our expertise, be it demographic data, environmental data, or land-use data. The large question is, how do we maintain a global information system that goes beyond the petty tyrannies of nation-states (and the need to protect information for "security" reasons), yet protects individuals' right to privacy? The selective use of remote-sensing techniques to monitor environmental conditions has been helpful in understanding the linkages between local activities and global impacts. However, can we use advanced technology to support demographic data collection and analyses and still maintain safeguards on privacy protection (Liverman et al. 1998)? For example, recent Supreme Court decisions have placed important legal protections on the use of thermal infrared sensors in public safety.

Another series of issues involves the aggregation of human behaviors. How can we geographically aggregate data along a set of common dimensions to insure its representation of reality and get around the thorny issues of averaging and the mean-areal-center or modified areal-unit problems? We often use techniques to handle aggregated populations and areas that in fact, depart from reality, creating a type of artificial environment. Unfortunately, public policies all too often are based on these constructed realities, thus further exacerbating the distribution of goods, services, and resources. What new spatial statistical tools do we need to address this concern?

Lastly, in a post-11-September world, how do we measure the geography of fear? Does the restriction of geographic data (presumably for national security reasons) attenuate or amplify fear of the unknown? The discipline requires the open access to information and data about the world and the people who live there. Data access will be one of the key issues for our community to address in the coming years.

## What Role Has Geographical Skill Played in the Evolution of Human Civilization, and What Role Can It Play in Predicting the Future?

Is there a necessary geographic base to human history? If so, how can we improve our ability to predict spatial events and events that have spatial consequences that will fundamentally shape the future? Can we develop the geographic equivalent of leading economic indicators?

From the early cradles of civilization in Africa and Asia, humankind gradually colonized the earth. This process of redistributing people in space (migration) was caused by population growth, resource exhaustion, attractive untapped resources, environmental change, environmental hazard, disease, or invasion and succession by other human groups. But what skills and abilities were required to ensure success in relocation movements? Were the movements random or consciously directed? If they were directed, then what skills and/or abilities were required by explorers, leaders, and followers to ensure success? What criteria had to be satisfied before resettlement was possible? What new geographic skills and abilities have been developed throughout human history, and which ones have deteriorated or disappeared? Have geographic skills and abilities been maintained equally in males and females? If not, what developments in the evolution of human civilizations have mediated such losses or changes?

While we know much about human history, we know little about the geographical basis of world history, and we know little of the extent to which the presence or absence of geographic knowledge played a significant part in historical development. For example, would Napoleon's invasion of Russia been more successful had skilled and knowledgeable geographers counseled him on the route chosen and the appropriate season for movement? Historians often tell us that understanding the past is the key to knowing the present and to successfully predicting the future. We cannot fully understand the past if we ignore or diminish the importance of environmental diversity and knowledge about those variations that are the result of spatial and geographic thinking and reasoning. A similar argument can be made for predicting future events and behaviors. What geographic knowledge is likely to be important in prediction? Must we rely on assumptions about uniform environments, population characteristics, tastes and preferences, customs, beliefs, and values? Such a procedure is precarious at best. However, we do not currently know how to incorporate geographic variability into our models, or indeed what variables should be incorporated into predictive models. Achieving such a goal is a necessary part of increasing our very limited predictive capabilities.

## How and Why Do Sustainability and Vulnerability Change from Place to Place and over Time?

Historically, geography was an integrative science with a particular focus on regions. It then switched from breadth to depth, with improvements in theory, methods, and techniques. We are now returning to that earlier perspective as we look for common ground in the interactions between human systems and physical systems. Increasing population pressures, the regional depletion or total exhaustion of resources, environmental degradation, and rampant development are processes that affect the sustainability of natural systems and constructed environments. There is a movement toward the integration of many different social and natural science perspectives into a field called sustainability science (Kates et al. 2001). Understanding what constrains and enhances sustainable environments will be an important research theme in the future. How can we maintain and improve the quality of urban environments for general living (social, economic, and environmental conditions)? How long can the processes of urban and suburban growth continue without deleterious and fundamental changes in the landscape and the escalation in costs of environmental restoration? Suburban sprawl is already a major policy issue. What is the long-term impact on human survival of the constant usurpation of agricultural land by the built environment? How long can we continue slash and burn agriculture in many parts of the tropical world? What triggers the environmental insecurity of nations, and how does this lead to armed conflicts and mass migrations of people? How have these processes varied in time and space? What are the greatest threats to the sustainability of human settlements, agriculture, energy use, for example and how can we mitigate or reduce those threats (NRC 2000)?

Nonsustainable environments enhance the effect of risks and hazards and ultimately increase both biophysical and social vulnerability, often resulting in disasters of one kind or another. When societies or ecosystems lack the ability to stop decay or decline and they do not have the adequate means to defend against such changes, there can be potentially catastrophic results. Examples include the environmental degradation of the Aral Sea, the increasing AIDS pandemic, and the human and environmental costs of coastal living (Heinz Center 2000). Vulnerability can be thought of as a continuum of processes, ranging from the initial susceptibility to harm to resilience (the ability to recover) to longer-term adaptations in response to large-scale environmental changes (Cutter, Mitchell, and Scott 2000). These processes manifest themselves at different geographic scales, ranging from the local to the global. What is the threshold when vulnerability ceases to become something we can deal with and becomes something we cannot? At what point does the built environment or ecosystem extend beyond its own ability to recover from natural or social forces?

## What Is the Nature of Spatial Thinking, Reasoning, and Abilities?

Geographic knowledge is the product of spatial thinking and reasoning (Golledge 2002b). These processes require the ability to comprehend scale changes; transformations of phenomena, or representations among one, two, and three spatial dimensions. They also require understanding of: the effect of distance, direction, and orientation on developing spatial knowledge; the nature of reference frames for identifying locations, distributions and patterns; the nature of spatial hierarchies; the nature of forms by extrapolating from cross sections; the significance of adjacency and nearest neighbor concepts; the spatial properties of density, distance, and density decay; and the configurations of patterns and shapes in various dimensions and with differing degrees of completeness. It also requires knowing the implications of spatial association and understanding other concepts not yet adequately articulated or understood. What geography currently lacks is an elaboration of the fundamental geographic concepts and skills that are necessary for the production and communication of spatial and geographic information. In the long run, this will be needed before geography can develop a well-articulated knowledge base of a type similar to other human and physical sciences.

# Conclusion

In the American Declaration of Independence, Thomas Jefferson wrote that among the most basic of human rights are life, liberty, and the pursuit of happiness. Each of these rights is played out upon a geographic stage, has geographic properties, and operates as a geographical process. Geography, as a field of knowledge and as a perspective on the world, has paid too little attention to these grand ideas, and they are fertile ground for the seeds of new geographic research. How and why does the opportunity for the pursuit of happiness vary from one place to another, and does the very nature of that pursuit change geographically?

In pursuit of answers to the big questions articulated above, we will inevitably need to think about doing research on problems such as:

- What are the spatial constraints on pursuing goals of life, liberty, and the pursuit of happiness?
- What are our future resource needs, and where will we find the new resources that have not, at this stage, been adequately explored?
- When does geography start and finish? Does it matter?
- What are likely to be the major problems in doing the geography of other planets?
- Will cities of the future remain bound to the land surface, or will they move to what we now consider unlikely or exotic locations (under water or floating in space)?

The big questions posed here are not all encompassing. They represent our collective judgments (and biases) on what issues are significant for the discipline, and those that should provide a focus for our considerable intellectual capital. Not everyone will agree with us, nor should they. We view this article as the beginning of a dialogue within the discipline as to what are the probable big questions for the next generation of geographers.

# Literature Cited

Abler, R. F. 2001. From the meridian—Wilford's "science writer's view of geography." *AAG Newsletter* 36 (4): 2, 9.

Association of American Geographers (AAG) Global Change in Local Places (GCLP) Research Group. Forthcoming. *Global change and local places: Estimating, understanding, and reducing greenhouse gases.* Cambridge, U.K.: Cambridge University Press.

Cutter, S. L., J. T. Mitchell, and M. S. Scott. 2000. Revealing the vulnerability of people and places: A case study of Georgetown County, South Carolina. *Annals of the Association of American Geographers* 90:713–37.

Dilsaver, L. M., and C. E. Colten, eds. 1992. *The American environment: Interpretations of past geographies.* Lanham, MD: Rowan and Littlefield Publishers.

Dilsaver, L. M., W. Wyckoff, and W. L. Preston. 2000. Fifteen events that have shaped California's human landscape. *The California Geographer* XL: 1–76.

Gale, N. D. 1984. Route learning by children in real and simulated environments. Ph.D. diss., Department of Geography, University of California, Santa Barbara.

Golledge, R. G. 2002. The nature of geographic knowledge. *Annals of the Association of American Geographers* 92 (1): 1–14.

———. Forthcoming. *Spatial cognition and converging technologies.* Paper presented at the Workshop on Converging Technology (NBIC) for Improving Human Performance, sponsored by the National Science Foundation. Washington, D.C. In press.

Goodchild, M. F. 2000. Communicating geographic information in a digital age. *Annals of the Association of American Geographers* 90:344–55.

Graf, W. L. 2001. Dam age control: Restoring the physical integrity of America's rivers. *Annals of the Association of American Geographers* 91:1–27.

Heinz Center. 2000. *The hidden costs of coastal erosion.* Washington, D.C.: The H. John Heinz III Center for Science, Economics and the Environment.

Kates, R. W., W. C. Clark, R. Corell, J. M. Hall, C. C. Jaeger, I. Lowe, J. J. McCarthy, H. J. Schnellnhuber, B. Bolin, N. M. Dickson, S. Faucheux, G. C. Gallopin, A. Grubler, B. Huntley, J. Jager, N. S. Jodha, R. E. Kasperson, A. Mabogunje, P. Matson, H. Mooney, B. Moore III, T. O'Riodan, and U. Svedin. 2001. Sustainability science. *Science* 292:641–42.

Liverman, D., E. F. Moran, R. R. Rindfuss, and P. C. Stern, eds. 1998. *People and pixels: Linking remote sensing and social science.* Washington, D.C.: National Academy Press.

Lloyd, R. E., and C. Heivly. 1987. Systematic distortion in urban cognitive maps. *Annals of the Association of American Geographers* 77:191–207.

MacEachren, A. M. 1992. Application of environmental learning theory to spatial knowledge acquisition from maps. *Annals of the Association of American Geographers* 82 (2): 245–74.

Monmonier, M. S. 2001. *Bushmanders and Bullwinkles: How politicians manipulate electronic maps and census data to win elections.* Chicago: University of Chicago Press.

Montello, D. R. 1993. Scale and multiple psychologies of space. In *Spatial information theory: A theoretical basis for GIS. Lecture notes in computer science 716. Proceedings, European Conference, COSIT '93. Marciana Marina, Elba Island, Italy, September*, ed. A. U. Frank and I. Campari. 312–21. New York: Springer-Verlag.

National Research Council (NRC). 1999. *New strategies for America's watersheds.* Washington, D.C.: National Academy Press.

———. 2000. *Our common journey: A transition toward sustainability.* Washington, D.C.: National Academy Press.

Thomas, W. L., Jr., ed. 1956. *Man's role in changing the face of the earth.* Chicago: The University of Chicago Press.

Turner, B. L. II, W. C. Clark, R. W. Kates, J. F. Richards, J. T. Mathews, and W. Meyer, eds. 1990. *The earth as transformed by human action: Global and regional changes in the biosphere over the past 300 years.* Cambridge, U.K.: University of Cambridge Press.

Worster, D. E. 1979. *Dust bowl: The Southern plains in the 1930s.* Oxford: Oxford University Press.

———. 1993. *The wealth of nature: Environmental history and the ecological imagination.* New York: Oxford University Press.

**ANNUAL EDITIONS**

SUSAN L. CUTTER is Carolina Distinguished Professor, Department of Geography, University of South Carolina, Columbia, SC 29208. E-mail: scutter@sc.edu. She served as president of the Association of American Geographers from 2000-2001, and is a fellow of the American Association for the Advancement of Science (AAAS). Her research interests are vulnerability science, and environmental hazards policy and management.

REGINALD GOLLEDGE is a Professor of Geography at the University of California, Santa Barbara, Santa Barbara, CA 93106. E-mail: golledge@geog.ucsb.edu and served as AAG president from 1999 to 2000. His research interests include various aspects of behavioral geography (spatial cognition, cognitive mapping, spatial thinking), the geography of disability (particularly the blind), and the development of technology (guidance systems and computer interfaces) for blind users.

WILLIAM L. GRAF is Education Foundation University Professor and Professor of Geography at the University of South Carolina, Columbia, SC 29208. E-mail: graf@sc.du. He served as AAG president from 1998–1999, and is a National Associate of the National Academy of Science. His specialties are fluvial geomorphology and policy for public land and water.

From *The Professional Geographer,* August 2002, pp. 305-317. © 2002 by Blackwell Publishing, Inc.

# POINT OF VIEW

# Rediscovering the Importance of Geography

*By Alexander B. Murphy*

As AMERICANS STRUGGLE to understand their place in a world characterized by instant global communications, shifting geopolitical relationships, and growing evidence of environmental change, it is not surprising that the venerable discipline of geography is experiencing a renaissance in the United States. More elementary and secondary schools now require courses in geography, and the College Board is adding the subject to its Advanced Placement program. In higher education, students are enrolling in geography courses in unprecedented numbers. Between 1985–86 and 1994–95, the number of bachelor's degrees awarded in geography increased from 3,056 to 4,295. Not coincidentally, more businesses are looking for employees with expertise in geographical analysis, to help them analyze possible new markets or environmental issues.

In light of these developments, institutions of higher education cannot afford simply to ignore geography, as some of them have, or to assume that existing programs are adequate. College administrators should recognize the academic and practical advantages of enhancing their offerings in geography, particularly if they are going to meet the demand for more and better geography instruction in primary and secondary schools. We cannot afford to know so little about the other countries and peoples with which we now interact with such frequency, or about the dramatic environmental changes unfolding around us.

From the 1960s through the 1980s, most academics in the United States considered geography a marginal discipline, although it remained a core subject in most other countries. The familiar academic divide in the United States between the physical sciences, on one hand, and the social sciences and humanities, on the other, left little room for a discipline concerned with how things are organized and relate to one another on the surface of the earth—a concern that necessarily bridges the physical and cultural spheres. Moreover, beginning in the 1960s, the U.S. social-science agenda came to be dominated by pursuit of more-scientific explanations for human phenomena, based on assumptions about global similarities in human institutions, motivations, and actions. Accordingly, regional differences often were seen as idiosyncrasies of declining significance.

Although academic administrators and scholars in other disciplines might have marginalized geography, they could not kill it, for any attempt to make sense of the world must be based on some understanding of the changing human and physical patterns that shape its evolution—be they shifting vegetation zones or expanding economic contacts across international boundaries. Hence, some U.S. colleges and universities continued to teach geography, and the discipline was often in the background of many policy issues—for example, the need to assess the risks associated with foreign investment in various parts of the world.

By the late 1980s, Americans' general ignorance of geography had become too widespread to ignore. Newspapers regularly published reports of surveys demonstrating that many Americans could not identify major countries or oceans on a map. The real problem, of course, was not the inability to answer simple questions that might be asked on *Jeopardy!*; instead, it was what that inability demonstrated about our collective understanding of the globe.

Geography's renaissance in the United States is due to the growing recognition that physical and human processes such as soil erosion and ethnic unrest are inextricably tied to their geographical context. To understand modern Iraq, it is not enough to know who is in power and how the political system functions. We also need to know something about the country's ethnic groups and their settlement patterns, the different physical environments and resources within the country, and its ties to surrounding countries and trading partners.

Those matters are sometimes addressed by practitioners of other disciplines, of course, but they are rarely central to the analysis. Instead, generalizations are often made at the level of the state, and little attention is given to spatial patterns and practices that play out on local levels or across international boundaries. Such preoccupations help to explain why many scholars were caught off guard by the explosion of ethnic unrest in Eastern Europe following the fall of the Iron Curtain.

Similarly, comprehending the dynamics of El Niño requires more than knowledge of the behavior of ocean and air currents; it is also important to understand how those currents are situated with respect to land masses and how they relate to other climatic patterns, some of which have been altered by the burning of fossil fuels and other human activities. And any attempt to understand the nature and extent of humans' impact on the environment requires consideration of the relationship between human and physical contributions to environmental change. The factories and cars in a city produce smog, but surrounding mountains may trap it, increasing air pollution significantly.

TODAY, academics in fields including history, economics, and conservation biology are turning to geographers for help with some of their concerns. Paul Krugman, a noted economist at the Massachusetts Institute of Technology, for example, has turned conventional wisdom on its head by pointing out the role of historically rooted regional inequities in how international trade is structured.

Geographers work on issues ranging from climate change to ethnic conflict to urban sprawl. What unites their work is its focus on the shifting organization and character of the earth's surface. Geographers examine changing patterns of vegetation to study global warming; they analyze where ethnic groups live in Bosnia to help understand the pros and cons of competing administrative solutions to the civil war there; they map AIDS cases in Africa to learn how to reduce the spread of the disease.

Geography is reclaiming attention because it addresses such questions in their relevant spatial and environmental contexts. A growing number of scholars in other disciplines are realizing that it is a mistake to treat all places as if they were essentially the same (think of the assumptions in most economic models), or to undertake research on the environment that does not include consideration of the relationships between human and physical processes in particular regions.

Still, the challenges to the discipline are great. Only a small number of primary- and secondary-school teachers have enough training in geography to offer students an exciting introduction to the subject. At the college level, many geography departments are small; they are absent altogether at some high-profile universities.

Perhaps the greatest challenge is to overcome the public's view of geography as a simple exercise in place-name recognition. Much of geography's power lies in the insights it sheds on the nature and meaning of the evolving spatial arrangements and landscapes that make up our world. The importance of those insights should not be underestimated at a time of changing political boundaries, accelerated human alteration of the environment, and rapidly shifting patterns of human interaction.

*Alexander B. Murphy is a professor and head of the geography department at the University of Oregon, and a vice-president of the American Geographical Society.*

---

Originally appeared in *The Chronicle of Higher Education*, October 30, 1998, p. 54. © 1998 by Alexander B. Murphy. Reprinted by permission.

# Article 3

## Can a New Union Bring Africa Together?

# Birth of the African Union

Sarah Coleman in New York and Julius Dawu in Harare

July 24, 2002

It was, by all accounts, a great party. As more than 25,000 people watched in Durban's ABSA rugby stadium, a fleet of parachutists descended from the sky, and a colorful parade marched by that included everything from bare-breasted Zulu maidens to two national soccer teams. The occasion was the launch of the African Union (AU) on July 9, an event that marshaled a record attendance of 43 heads of state from across the continent.

The AU replaces the Organization of African Unity (OAU), which was founded in 1963 to address the issue of decolonization, but had devolved into a bloated and debt-ridden bureaucracy. Though it memorably fought against apartheid in South Africa, the OAU was often criticized for failing to take a stand against the continent's home-grown despots. Over the years, it turned a blind eye to human rights abuses by the likes of Uganda's Idi Amin, Zaire's Mobutu Sese Seko, and the Central African Republic's Jean-Bedel Bokassa.

The AU has pledged to do better. Unlike the OAU, which respected the sovereign powers of its member countries, the AU has been planned as a pan-Africanist organization. Modeled on the European Union, it will have its own Parliament, central bank, and court of justice. Perhaps most significantly, its Peace and Security Council has a mandate to intervene in regional conflicts involving war crimes, genocide, and crimes against humanity, giving it powers that far exceed those of the OAU.

Amid the razzmatazz of its opening ceremonies, the advent of the AU was greeted with both optimism and skepticism in the African press. For Naseem Mahatey, writing in Johannesburg's black community-oriented *City Press* (June 30), the AU was "the new continental vehicle purpose-designed to take on the challenges of the 21st century and to realize Africa's renaissance."

But an editorial in Gabarone's independent weekly *Mmegi* (July 5-11) called it "a giant paper tiger" and worried that "the continent's diversity of religions, political structures, economic advancement, [and] caliber of leadership can only show the AU to be pie in the sky. Though the idea of the AU is noble, the situation on the ground points to the fact that its birth might be premature. Africa needs to put its house in order first and totally overhaul its act for the AU baby to thrive into mature adulthood."

Banjul's independent biweekly *The Independent* (July 13) warned the AU to guard against colonialism. "While the whole continent basks in the euphoria, amidst the pomp and pageantry of the birth of the AU, we nevertheless call for cautious optimism, dogged determination, and stoic reserve to resist every attempt by organized global vultures in dove's clothing to thwart the lofty project by sheer subtlety," *The Independent* said in an editorial.

The presence at the launch of such leaders as Zimbabwe's Robert Mugabe and Libya's Muammar Qaddafi led many commentators to wonder whether the AU would live up to its promise of imposing standards of good governance on its member countries. "How can rulers who are themselves clearly the problem be part of the solution?" asked Shyaka Kanuma in Johannesburg's liberal *Mail & Guardian* (July 12). "These men are not troubled by niceties such as respect for human rights, concern for their populations' material welfare, or consensual decision-making."

In the week preceding the launch, such questions dogged the nascent organization as the memberships of Swaziland and Madagascar were challenged. After a fierce debate, Madagascar—whose current leader, Marc Ravalomanana, took power after a disputed election in December—was barred from the union, much to the disgust of Swazi opposition groups, who wanted their own country banned.

"The criteria used to deal with different countries has been riddled with inconsistency," wrote Bongani Masuku, secretary-general of the Swaziland Solidarity Network, a network of Swazi opposition groups, in a letter to the AU quoted by Durban's liberal *Daily News* (July 5). "The barring of Madagascar from the AU while accepting the Swazi despot, [King] Mswati [III], indicates the deep-seated nature of this problem."

Zimbabwe's pro-government *Herald* took a different view, saying that the AU's decision to bar Madagascar demonstrated the new organization's independence from

Western influence: "The continent's principled position on Madagascar has… exposed the double standards of the United States, Britain, the European Union, International Monetary Fund and World Bank on issues to do with good governance, democracy, and constitutionality," wrote the *Herald* in an editorial comment (July 14).

Another controversial element was the New Partnership for Africa's Development (NEPAD), the divisive proposal that has been adopted as the AU's economic policy. NEPAD, which is the brainchild of the presidents of South Africa, Nigeria, Senegal, and Algeria, proposes that member countries will pledge good governance in return for foreign aid. It has been widely criticized for being a presidential initiative that puts Africa's needs at the whim of First World countries.

"Put politely, NEPAD… is nothing other than an earnest appeal to the West's conscience for help," wrote Kanuma in the *Mail & Guardian* (July 12). But an editorial [in] Kampala's government-owned *New Vision* (July 11) put a positive spin on the fact that a recently concluded G-8 summit had failed to commit a package of aid to NEPAD: "The fact that the expected check from the G-8 bounced… helped to clear misconceptions and renewed consensus that whether it is NEPAD or AU, we have to do things ourselves… not because somebody in Europe or America likes them or not."

Despite all the criticisms, the launch of the AU did inspire optimism in some quarters. London's bimonthly newsmagazine *New African* praised the fact that commitment to human rights had been written into the AU treaty: "The old days, when those rights could be overlooked because of one-sided concern for government authority, have definitely gone," it wrote (July/August).

Others expressed hope that the AU—under the leadership of South Africa's President Thabo Mbeki—would have the political muscle to close the gap between Africa and the developed world. "Enlightened African leaders are trying to harness the continent's bargaining power in the face of globalization," read an editorial in Nairobi's independent *East African Standard* (July 2). "We hope Africa [has] learned from the failures of the OAU to make [the] AU work."

---

From *World Press Review Online*, July 24, 2002. © 2002 by World Press Review. Reprinted by permission.

# The Four Traditions of Geography

*William D. Pattison*

Late Summer, 1990

To Readers of the *Journal of Geography:*

I am honored to be introducing, for a return to the pages of the *Journal* after more than 25 years, "The Four Traditions of Geography," an article which circulated widely, in this country and others, long after its initial appearance—in reprint, in xerographic copy, and in translation. A second round of life at a level of general interest even approaching that of the first may be too much to expect, but I want you to know in any event that I presented the paper in the beginning as my gift to the geographic community, not as a personal property, and that I re-offer it now in the same spirit.

In my judgment, the article continues to deserve serious attention—perhaps especially so, let me add, among persons aware of the specific problem it was intended to resolve. The background for the paper was my experience as first director of the High School Geography Project (1961–63)—not all of that experience but only the part that found me listening, during numerous conference sessions and associated interviews, to academic geographers as they responded to the project's invitation to locate "basic ideas" representative of them all. I came away with the conclusion that I had been witnessing not a search for consensus but rather a blind struggle for supremacy among honest persons of contrary intellectual commitment. In their dialogue, two or more different terms had been used, often unknowingly, with a single reference, and no less disturbingly, a single term had been used, again often unknowingly, with two or more different references. The article was my attempt to stabilize the discourse. I was proposing a basic nomenclature (with explicitly associated ideas) that would, I trusted, permit the development of mutual comprehension **and** confront all parties concerned with the pluralism inherent in geographic thought.

This intention alone could not have justified my turning to the NCGE as a forum, of course. The fact is that from the onset of my discomfiting realization I had looked forward to larger consequences of a kind consistent with NCGE goals. As finally formulated, my wish was that the article would serve "to greatly expedite the task of maintaining an alliance between professional geography and pedagogical geography and at the same time to promote communication with laymen" (see my fourth paragraph). I must tell you that I have doubts, in 1990, about the acceptability of my word choice, in saying "professional," "pedagogical," and "layman" in this context, but the message otherwise is as expressive of my hope now as it was then.

I can report to you that twice since its appearance in the *Journal*, my interpretation has received more or less official acceptance—both times, as it happens, at the expense of the earth science tradition. The first occasion was Edward Taaffe's delivery of his presidential address at the 1973 meeting of the Association of American Geographers (see *Annals AAG*, March 1974, pp. 1–16). Taaffe's working-through of aspects of an interrelation among the spatial, area studies, and man-land traditions is by far the most thoughtful and thorough of any of which I am aware. Rather than fault him for omission of the fourth tradition, I compliment him on the grace with which he set it aside in conformity to a meta-epistemology of the American university which decrees the integrity of the social sciences as a consortium in their own right. He was sacrificing such holistic claims as geography might be able to muster for a freedom to argue the case for geography as a social science.

The second occasion was the publication in 1984 of *Guidelines for Geographic Education: Elementary and Secondary Schools*, authored by a committee jointly representing the AAG and the NCGE. Thanks to a recently published letter (see *Journal of Geography*, March-April 1990, pp. 85–86), we know that, of five themes commended to teachers in this source,

> The committee lifted the human environmental interaction theme directly from Pattison. The themes of place and location are based on Pattison's spatial or geometric geography, and the theme of region comes from Pattison's area studies or regional geography.

Having thus drawn on my spatial, area studies, and man-land traditions for four of the five themes, the committee could have found the remaining theme, movement, there too—in the spatial tradition (see my sixth paragraph). However that may be, they did not avail themselves of the earth science tradition, their reasons being readily surmised. Peculiar to the elementary and secondary schools is a curriculum category framed as much by theory of citizenship as by theory of knowledge: the social studies. With admiration, I see already in the committee members' adoption of the theme idea a strategy for assimilation of their program to the established repertoire of social studies practice. I see in their exclusion of the earth science tradition an intelligent respect for social studies' purpose.

Here's to the future of education in geography: may it prosper as never before.

W. D. P., 1990

In 1905, one year after professional geography in this country achieved full social identity through the founding of the Association of American Geographers, William Morris Davis responded to a familiar suspicion that geography is simply an undisciplined "omnium-gatherum" by describing an approach that as he saw it imparts a "geographical quality" to some knowledge and accounts for the absence of the quality elsewhere.[1] Davis spoke as president of the AAG. He set an example that was followed by more than one president of that organization. An enduring official concern led the AAG to publish, in 1939 and in 1959, monographs exclusively devoted to a critical review of definitions and their implications.[2]

Every one of the well-known definitions of geography advanced since the founding of the AAG has had its measure of success. Tending to displace one another by turns, each definition has said something true of geography.[3] But from the vantage point of 1964, one can see that each one has also failed. All of them adopted in one way or another a monistic view, a singleness of preference, certain to omit if not to alienate numerous professionals who were in good conscience continuing to participate creatively in the broad geographic enterprise.

The thesis of the present paper is that the work of American geographers, although not conforming to the restrictions implied by any one of these definitions, has exhibited a broad consistency, and that this essential unity has been attributable to a small number of distinct but affiliated traditions, operant as binders in the minds of members of the profession. These traditions are all of great age and have passed into American geography as parts of a general legacy of Western thought. They are shared today by geographers of other nations.

There are four traditions whose identification provides an alternative to the competing monistic definitions that have been the geographer's lot. The resulting pluralistic basis for judgment promises, by full accommodation of what geographers do and by plain-spoken representation thereof, to greatly expedite the task of maintaining an alliance between professional geography and pedagogical geography and at the same time to promote communication with laymen. The following discussion treats the traditions in this order: (1) a spatial tradition, (2) an area studies tradition, (3) a man-land tradition and (4) an earth science tradition.

## Spatial Tradition

Entrenched in Western thought is a belief in the importance of spatial analysis, of the act of separating from the happenings of experience such aspects as distance, form, direction and position. It was not until the 17th century that philosophers concentrated attention on these aspects by asking whether or not they were properties of things-in-themselves. Later, when the 18th century writings of Immanuel Kant had become generally circulated, the notion of space as a category including all of these aspects came into widespread use. However, it is evident that particular spatial questions were the subject of highly organized answering attempts long before the time of any of these cogitations. To confirm this point, one need only be reminded of the compilation of elaborate records concerning the location of things in ancient Greece. These were records of sailing distances, of coastlines and of landmarks that grew until they formed the raw material for the great *Geographia* of Claudius Ptolemy in the 2nd century A.D.

A review of American professional geography from the time of its formal organization shows that the spatial tradition of thought had made a deep penetration from the very beginning. For Davis, for Henry Gannett and for most if not all of the 44 other men of the original AAG, the determination and display of spatial aspects of reality through mapping were of undoubted importance, whether contemporary definitions of geography happened to acknowledge this fact or not. One can go further and, by probing beneath the art of mapping, recognize in the behavior of geographers of that time an active interest in the true essentials of the spatial tradition—*geometry* and *movement*. One can trace a basic favoring of movement as a subject of study from the turn-of-the-century work of Emory R. Johnson, writing as professor of transportation at the University of Pennsylvania, through the highly influential theoretical and substantive work of Edward L. Ullman during the past 20 years and thence to an article by a younger geographer on railroad freight traffic in the U.S. and Canada in the *Annals* of the AAG for September 1963.[4]

One can trace a deep attachment to geometry, or positioning-and-layout, from articles on boundaries and population densities in early 20th century volumes of the *Bulletin of the American Geographical Society*, through a controversial pronouncement by Joseph Schaefer in 1953 that granted geographical legitimacy only to studies of spatial patterns[5] and so onward to a recent *Annals* report on electronic scanning of cropland patterns in Pennsylvania.[6]

One might inquire, is discussion of the spatial tradition, after the manner of the remarks just made, likely to bring people within geography closer to an understanding of one another and people outside geography closer to an understanding of geographers? There seem to be at least two reasons for being hopeful. First, an appreciation of this tradition allows one to see a bond of fellowship uniting the elementary school teacher, who attempts the most rudimentary instruction in directions and mapping, with the contemporary research geographer, who dedicates himself to an exploration of central-place theory. One cannot only open the eyes of many teachers to the potentialities of their own instruction, through proper exposition of the spatial tradition, but one can also "hang a bell" on research quantifiers in geography, who are often thought to have wandered so far in their intellectual adventures as to have become lost from the rest. Looking outside geography, one may anticipate benefits from the readiness of countless persons to associate the name "geography" with maps. Latent within this readiness is a willingness to recognize as geography, too, what maps are about—and that is the geometry of and the movement of what is mapped.

## Area Studies Tradition

The area studies tradition, like the spatial tradition, is quite strikingly represented in classical antiquity by a practitioner to whose surviving work we can point. He is Strabo, celebrated for his *Geography* which is a massive production addressed to the statesmen of Augustan Rome and intended to sum up and regularize knowledge not of the location of places and associated cartographic facts, as in the somewhat later case of Ptolemy, but of the nature of places, their character and their differentiation. Strabo exhibits interesting attributes of the area-studies tradition that can hardly be overemphasized. They are a pronounced tendency toward subscription primarily to literary standards, an almost omnivorous appetite for information and a self-conscious companionship with history.

It is an extreme good fortune to have in the ranks of modern American geography the scholar Richard Hartshorne, who has pondered the meaning of the area-studies tradition with a legal acuteness that few persons would challenge. In his *Nature of Geography*, his 1939 monograph already cited,[7] he scrutinizes exhaustively the implications of the "interesting attributes" identified in connection with Strabo, even though his concern is with quite other and much later authors, largely German. The major literary problem of unities or wholes he considers from every angle. The Gargantuan appetite for miscellaneous information he accepts and rationalizes. The companionship between area studies and history he clarifies by appraising the so-called idiographic content of both and by affirming the tie of both to what he and Sauer have called "naively given reality."

The area-studies tradition (otherwise known as the chorographic tradition) tended to be excluded from early American professional geography. Today it is beset by certain champions of the spatial tradition who would have one believe that somehow the area-studies way of organizing knowledge is only a subdepartment of spatialism. Still, area-studies as a method of presentation lives and prospers in its own right. One can turn today for reassurance on this score to practically any issue of the *Geographical Review*, just as earlier readers could turn at the opening of the century to that magazine's forerunner.

What is gained by singling out this tradition? It helps toward restoring the faith of many teachers who, being accustomed to administering learning in the area-studies style, have begun to wonder if by doing so they really were keeping in touch with professional geography. (Their doubts are owed all too much to the obscuring effect of technical words attributable to the very professionals who have been intent, ironically, upon protecting that tradition.) Among persons outside the classroom the geographer stands to gain greatly in intelligibility. The title "area-studies" itself carries an understood message in the United States today wherever there is contact with the usages of the academic community. The purpose of characterizing a place, be it neighborhood or nation-state, is readily grasped. Furthermore, recognition of the right of a geographer to be unspecialized may be expected to be forthcoming from people generally, if application for such recognition is made on the merits of this tradition, explicitly.

## Man-Land Tradition

That geographers are much given to exploring man-land questions is especially evident to anyone who examines geographic output, not only in this country but also abroad. O. H. K. Spate, taking an international view, has felt justified by his observations in nominating as the most significant ancient precursor of today's geography neither Ptolemy nor Strabo nor writers typified in their outlook by the geographies of either of these two men, but rather Hippocrates, Greek physician of the 5th century B.C. who left to posterity an extended essay, *On Airs, Waters and Places*.[8] In this work made up of reflections on human health and conditions of external nature, the questions asked are such as to confine thought almost altogether to presumed influence passing from the latter to the former, questions largely about the effects of winds, drinking water and seasonal changes upon man. Understandable though this uni-directional concern may have been for Hippocrates as medical commentator, and defensible as may be the attraction that this same approach held for students of the condition of man for many, many centuries thereafter, one can only regret that this narrowed version of the man-land tradition, combining all too easily with social Darwinism of the late 19th century, practically overpowered American professional geography in the first generation of its history.[9] The premises of this version governed scores of studies by American geographers in interpreting the rise and fall of nations, the strategy of battles and the construction of public improvements. Eventually this special bias, known as environmentalism, came to be confused with the whole of the man-land tradition in the minds of many people. One can see now, looking back to the years after the ascendancy of environmentalism, that although the spatial tradition was asserting itself with varying degrees of forwardness, and that although the area-studies tradition was also making itself felt, perhaps the most interesting chapters in the story of American professional geography were being written by academicians who were reacting against environmentalism while deliberately remaining within the broad man-land tradition. The rise of culture historians during the last 30 years has meant the dropping of a curtain of culture between land and man, through which it is asserted all influence must pass. Furthermore work of both culture historians and other geographers has exhibited a reversal of the direction of the effects in Hippocrates, man appearing as an independent agent, and the land as a sufferer from action. This trend as presented in published research has reached a high point in the collection of papers titled *Man's Role in Changing the Face of the Earth*. Finally, books and articles can be called to mind that have addressed themselves to the most difficult task of all, a balanced tracing out of interaction between man and environment. Some chapters in the book mentioned above undertake just this. In fact the separateness of this approach is discerned only with difficulty in many places; however, its significance as a general research design that rises above environmentalism, while refusing to abandon the man-land tradition, cannot be mistaken.

The NCGE seems to have associated itself with the man-land tradition, from the time of founding to the present day, more than with any other tradition, although all four of the traditions are amply represented in its official magazine, *The Journal of Geography* and in the proceedings of its annual meetings. This apparent preference on the part of the NCGE members *for defining geography in terms of the man-land tradition* is strong evidence of the appeal that man-land ideas, separately stated, have for persons whose main job is teaching. It should be noted, too, that this inclination reflects a proven acceptance by the general public of learning that centers on resource use and conservation.

## Earth Science Tradition

The earth science tradition, embracing study of the earth, the waters of the earth, the atmosphere surrounding the earth and the association between earth and sun, confronts one with a paradox. On the one hand one is assured by professional geographers that their participation in this tradition has declined precipitously in the course of the past few decades, while on the other one knows that college departments of geography across the nation rely substantially, for justification of their role in general education, upon curricular content springing directly from this tradition. From all the reasons that combine to account for this state of affairs, one may, by selecting only two, go far toward achieving an understanding of this tradition. First, there is the fact that American college geography, growing out of departments of geology in many crucial instances, was at one time greatly overweighted in favor of earth science, thus rendering the field unusually liable to a sense of loss as better balance came into being. (This one-time disproportion found reciprocate support for many years in the narrowed, environmentalistic interpretation of the man-land tradition.) Second, here alone in earth science does one encounter subject matter in the normal sense of the term as one reviews geographic traditions. The spatial tradition abstracts certain aspects of reality; area studies is distinguished by a point of view; the man-land tradition dwells upon relationships; but earth science is identifiable through concrete objects. Historians, sociologists and other academicians tend not only to accept but also to ask for help from this part of geography. They readily appreciate earth science as something physically associated with their subjects of study, yet generally beyond their competence to treat. From this appreciation comes strength for geography-as-earth-science in the curriculum.

Only by granting full stature to the earth science tradition can one make sense out of the oft-repeated addage, "Geography is the mother of sciences." This is the tradition that emerged in ancient Greece, most clearly in the work of Aristotle, as a wide-ranging study of natural processes in and near the surface of the earth. This is the tradition that was rejuvenated by Varenius in the 17th century as "Geographia Generalis." This is the tradition that has been subjected to subdivision as the development of science has approached the present day, yielding mineralogy, paleontology, glaciology, meterology and other specialized fields of learning.

Readers who are acquainted with American junior high schools may want to make a challenge at this point, being aware that a current revival of earth sciences is being sponsored in those schools by the field of geology. Belatedly, geography has joined in support of this revival.[10] It may be said that in this connection and in others, American professional geography may have faltered in its adherence to the earth science tradition but not given it up.

In describing geography, there would appear to be some advantages attached to isolating this final tradition. Separation improves the geographer's chances of successfully explaining to educators why geography has extreme difficulty in accommodating itself to social studies programs. Again, separate attention allows one to make understanding contact with members of the American public for whom surrounding nature is known as the geographic environment. And finally, specific reference to the geographer's earth science tradition brings into the open the basis of what is, almost without a doubt, morally the most significant concept in the entire geographic heritage, that of the earth as a unity, the single common habitat of man.

## An Overview

The four traditions though distinct in logic are joined in action. One can say of geography that it pursues concurrently all four of them. Taking the traditions in varying combinations, the geographer can explain the conventional divisions of the field. Human or cultural geography turns out to consist of the first three traditions applied to human societies; physical geography, it becomes evident, is the fourth tradition prosecuted under constraints from the first and second traditions. Going further, one can uncover the meanings of "systematic geography," "regional geography," "urban geography," "industrial geography," etc.

It is to be hoped that through a widened willingness to conceive of and discuss the field in terms of these traditions, geography will be better able to secure the inner unity and outer intelligibility to which reference was made at the opening of this paper, and that thereby the effectiveness of geography's contribution to American education and to the general American welfare will be appreciably increased.

## Notes

1. William Morris Davis, "An Inductive Study of the Content of Geography," *Bulletin of the American Geographical Society*, Vol. 38, No. 1 (1906), 71.
2. Richard Hartshorne, *The Nature of Geography*, Association of American Geographers (1939), and idem., *Perspective on the Nature of Geography*, Association of American Geographers (1959).
3. The essentials of several of these definitions appear in Barry N. Floyd, "Putting Geography in Its Place," *The Journal of Geography*, Vol. 62, No. 3 (March, 1963), 117–120.
4. William H. Wallace, "Freight Traffic Functions of Anglo-American Railroads," *Annals of the Association of American Geographers*, Vol. 53, No. 3 (September, 1963), 312–331.
5. Fred K. Schaefer, "Exceptionalism in Geography: A Methodological Examination," *Annals of the Association of American Geographers*, Vol. 43, No. 3 (September, 1953), 226–249.
6. James P. Latham, "Methodology for an Instrumental Geographic Analysis," *Annals of the Association of American Geographers*, Vol. 53, No. 2 (June, 1963), 194–209.
7. Hartshorne's 1959 monograph, *Perspective on the Nature of Geography*, was also cited earlier. In this later work, he responds to dissents from geographers whose preferred primary commitment lies outside the area studies tradition.
8. O. H. K. Spate, "Quantity and Quality in Geography," *Annals of the Association of American Geographers*, Vol. 50, No. 4 (December, 1960), 379.
9. Evidence of this dominance may be found in Davis's 1905 declaration: "Any statement is of geographical quality if it contains… some relation between an element of inorganic control and one of organic response" (Davis, *loc. cit.*).
10. Geography is represented on both the Steering Committee and Advisory Board of the Earth Science Curriculum Project, potentially the most influential organization acting on behalf of earth science in the schools.

---

From *Journal of Geography*, September/October 1990, pp. 202–206. © 1990 by the National Council for Geographic Education. Reprinted by permission.

# Restoring an Ecosystem Torn Asunder by a Dam

By SANDRA BLAKESLEE

FLAGSTAFF, Ariz.—Forty years after one generation dammed the Colorado River at the upper end of the Grand Canyon, a new generation of engineers and scientists is struggling to deal with the consequences: colossal loss of sand, shrinking beaches, an invasion of outside fish and plants, the extinction of native species, erosion of archaeological sites and the sudden appearance of an Asian tapeworm, to name a few.

To solve such problems, the engineers and scientists charged with managing the river have evolved a new approach: instead of trying to conquer nature, they are bold enough to think they can outfox it. Using a process called adaptive management, they hope to stave off or reverse damage done to the river and its life forms by manipulating flows from the dam.

Adaptive management, they say, rests on two core principles. First, complex systems are inherently unpredictable; it is impossible to know the consequences of various human actions. Second, the only way to manage complex problems is through a collaborative process in which everyone with a stake agrees to try new measures. When experiments fail, as some do, these stakeholders must stand ready to try something else based on common interests.

The technique is being tried in many ecosystems, including the Everglades, the Columbia River basin and the San Francisco Bay delta. But it is furthest along here at the Grand Canyon, said Dr. Carl Walters, a professor in the Fisheries Center at the University of British Columbia, who helped create the approach. "It's working there," he said, "because of a dedicated community of people willing to work together and a relatively simple institutional setting."

The Grand Canyon's adaptive managers—scientists and engineers who work for a variety of state and federal agencies—have their work cut out for them. The Colorado River ecosystem has changed radically since the Glen Canyon Dam was finished in 1963 with the aim of storing water for Colorado, New Mexico, Utah and Wyoming.

Until then, the river was notoriously erratic. In the spring, gathering snowmelt from its vast watershed could send 300,000 cubic feet of ice-cold water per second through narrow canyon walls—enough to cover a football field 10 feet deep every second. The flooding river carried enough sand to fill a football stadium to the rim every three hours. In summer, the water slowed to a trickle—perhaps 1,000 cubic feet per second—and warmed to 85 degrees. Great piles of driftwood accumulated along the river banks, along with tall sand dunes.

The 710-foot-high dam, at the northern edge of the Grand Canyon, is a plug holding back the world's longest reservoir—Lake Powell, extending 186 miles upriver. Water flowing through the dam is held to 31,000 cubic feet a second and never exceeds 48 degrees. It contains no sand, no driftwood and few nutrients.

Somehow no one realized four decades ago that the dam would profoundly alter the downstream ecosystem, said Dr. Barry Gold, chief of the Grand Canyon Monitoring and Research Center in Flagstaff, Ariz., where the river's adaptive management plans are coordinated. A lone civil engineer warned in 1971 that the river would eventually lose all its sand, he said, "but no one listened."

For a time, Dr. Gold said, the river seemed to hold up well. Eight species of native fish could still be found, their muscular bodies, leathery skin and tiny eyes adapted for life in the formerly erratic, silt-clouded river.

Wildlife officials stocked the river with nonnative rainbow trout, turning the 15-mile stretch below the dam into a blue-ribbon trout fishery. Rafting and camping exploded in popularity. Sandstorms disappeared. A nonnative tree called tamarisk gave shelter to endangered songbirds like the Southwestern flycatcher. The muddy waters turned into rippling sheets of emerald green with diamond facets. Peregrine falcons swooped down to eat trout.

And the dam, which powered generators, proved to be a big moneymaker for the federal government. Because power demands varied, so did the amount of water allowed to pass through the turbines: from 5,000 to 25,000 cubic feet per second, in a single day. The flow sent sinuous tides pulsing down the river.

But by 1990, the dam's negative effects began to show, said Randall Peterson, manager of the federal Bureau of Reclamation's adaptive management division in Salt Lake City. The huge daily fluctuations in water flow were eroding the river channel and sandbars, making rafting difficult. Native fish were in decline. Wind had blown away high banks of sand covering archaeological sites. Rocky debris from side canyons was accumulating in the main channel, making the rapids more dangerous.

Alarmed, the federal Department of the Interior, which oversees the dam, decided in 1991 to curtail daily fluctuations. From then on, the amount of water released downstream could not vary more than 8,000 cubic feet per second in any one day. Power revenue fell $30 million a year. The daily pulses of water flattened out, making the river safer for rafting and fishing. A year later Congress passed the Grand Canyon Protection Act. Dr. Gold's research center was formed in 1995.

"We were asked to figure out how to use the dam to protect the river and associated resources, and make recommendations to stakeholders," Dr. Gold said. His center convened 25 interested parties, representing state and federal agencies, Indian tribes, environmental groups, recreation interests and power contractors, to discuss proposed solutions.

The center found that conditions on the river were deteriorating alarmingly. Whole beaches had disappeared. Four species of native fish had become extinct. An Asian tapeworm appeared; it now infects most native fish that survive. Rainbow trout, now spawning naturally in the wild, increased their numbers sixfold, so that some parts of the river contained 17,000 trout per mile. Steadier flows apparently increased their survival rates but reduced their food resources, so they became smaller and thinner.

In the spring of 1996, Mr. Peterson said, researchers tried out their first big experiment using the Glen Canyon Dam. For one week, they released 45,000 cubic feet of water per second, using special spillways. They figured the high water would lift sand stored on the bottom of the river and deposit it onto beaches.

While the experiment looked like a huge success at first, it quickly went awry. A year later, most of the sand was gone. "We made a huge mistake," said Dr. Theodore Melis, a sediment expert at the research center. The sand that built the beaches, it turned out, had come not from the river bottom but from existing beaches and eddies. Then fluctuating flows continued to erode sand as before. Two different experiments in 1997 and 2000 also failed to make beaches or retain sand.

Meanwhile, the rainbow trout continue to proliferate, said Dr. Lew Coggins, a fisheries biologist at the center. As many as a million rainbow trout are now in the river, eating midges, plants and possibly a native fish called the humpback chub. Ten years ago, some 8,300 adult chub lived in the river; today there are only 2,100 large enough to spawn. Biologists worry that this may not be enough to sustain the population.

If each rainbow trout around the mouth of the Little Colorado River ate just one baby chub a day when the chub are most vulnerable, Dr. Coggins said, the chub might not survive. Moreover, a second introduced species, brown trout, is lurking in the area, and "they are voracious predators," he said.

To address these two problems—too many trout and not enough sand—scientists recently proposed a new experiment to the participating stakeholders.

"You can't make new beaches unless you have new sand to work with," said Dr. David Rubin, a sedimentologist at the United States Geological Survey in Santa Cruz, Calif.

In this area, the Colorado River's main source for new sand is the Paria River, 16 miles downstream from the dam. It can dump millions of tons of sediment into the bigger river during monsoon storms in the late summer and fall.

"Ideally we'd like to send a pulse of floodwater through the dam immediately after a monsoon rain," Dr. Rubin said, to transfer the new sand onto beaches. But dam regulations do not permit experimental floods until January, after most of the new sand is already lost. The alternative, he said, is to keep flows very low after a big storm to preserve as much new sand as possible and then to release a two-day flood in January to capture it along hundreds of miles of shoreline.

To reduce the rainbow trout, managers want to run high fluctuating flows through the Grand Canyon for three months. This would strand and dry out trout beds, killing eggs and some large adults, without harming most other fish, which tend to dwell in side channels. Scientists also plan to capture trout at the mouth of the Little Colorado and examine their stomachs for chub (the remains will be dumped downstream).

"This is a bold experiment that can be stopped after a few years if it turns out the chub population is not improving," said Dr. James Kitchell, a professor of zoology and director of the Center for Limnology at the

University of Wisconsin. "Regardless of that, a reduction in trout should improve the trout fishery by reducing crowding and producing more trophy fish in coming years."

"Of course," he went on, "the best way to disentangle the puzzle is to repeat the experiment with high numbers of trout and low numbers of trout. But in this case, we may not have the luxury of proper experimental design. Time may be short for the chub.

"We need to create a sense of urgency and pursue the most direct, immediate actions required to reverse the decline. That's the essence of adaptive management. You learn from experience, design new experiments, test a hypothesis and go from there."

Mr. Peterson says adaptive management works only if interested parties endorse the experiments. Now, a majority is behind the plan to reduce trout and use the monsoon flows to provide new sand to make beaches; the most vocal opposition comes from 25 licensed trout guides who generate $20 million a year taking 30,000 anglers from Lee's Ferry to good fishing spots below the dam.

"We disagree there are too many trout," said Terry Gunn, owner of the Lee's Ferry Fly Shop. "And there's no good evidence that rainbow trout eat chub."

The collaborative process can never please everyone, Mr. Peterson said. Dr. Gold agreed. "It's a rugby scrum," he said. But he added: "Just as the river channel, which is not seen from the surface, shapes the way the water flows, the leadership skills brought to an adaptive management program shape the science and policy that make up the program."

---

From the *New York Times.com,* June 11, 2002. © 2002 by The New York Times Company. Reprinted by permission.

# Sculpting the Earth from Inside Out

Powerful motions deep inside the planet do not merely shove fragments of the rocky shell horizontally around the globe—they also lift and lower entire continents

by Michael Gurnis

Credit for sculpting the earth's surface typically goes to violent collisions between tectonic plates, the mobile fragments of the planet's rocky outer shell. The mighty Himalayas shot up when India rammed into Asia, for instance, and the Andes grew as the Pacific Ocean floor plunged beneath South America. But even the awesome power of plate tectonics cannot fully explain some of the planet's most massive surface features.

Take southern Africa. This region boasts one of the world's most expansive plateaus, more than 1,000 miles across and almost a mile high. Geologic evidence shows that southern Africa, and the surrounding ocean floor, has been rising slowly for the past 100 million years, even though it has not experienced a tectonic collision for nearly 400 million years.

The African superswell, as this uplifted landmass is known, is just one example of dramatic vertical movement by a broad chunk of the earth's surface. In other cases from the distant past, vast stretches of Australia and North America bowed down thousands of feet—and then popped up again.

Scientists who specialize in studying the earth's interior have long suspected that activity deep inside the earth was behind such vertical changes at the surface. These geophysicists began searching for clues in the mantle—the middle layer of the planet. This region of scalding-hot rock lies just below the jigsaw configuration of tectonic plates and extends down more than 1,800 miles to the outer edge of the globe's iron core. Researchers learned that variations in the mantle's intense heat and pressure enable the solid rock to creep molassesslike over thousands of years. But they could not initially decipher how it could give rise to large vertical motions. Now, however, powerful computer models that combine snapshots of the mantle today with clues about how it might have behaved in the past are beginning to explain why parts of the earth's surface have undergone these astonishing ups and downs.

The mystery of the African superswell was among the easiest to decipher. Since the early half of the 20th century, geophysicists have understood that over the unceasing expanse of geologic time, the mantle not only creeps, it churns and roils like pot of thick soup about to boil. The relatively low density of the hottest rock makes that material buoyant, so it ascends slowly; in contrast, colder, denser rock sinks until heat escaping the molten core warms it enough to make it rise again. These three-dimensional motions, called convection, are known to enable the horizontal movement of tectonic plates, but it seemed unlikely that the forces they created could lift and lower the planet's surface. That skepticism about the might of the mantle began to fade away when researchers created the first blurry images of the earth's interior.

About 20 years ago scientists came up with a way to make three-dimensional snapshots of the mantle by measuring vibrations that are set in motion by earthquakes originating in the planet's outer shell. The velocities of these vibrations, or seismic waves, are determined by the chemical composition, temperature and pressure of the rocks they travel through. Waves become sluggish in hot, low-density rock, and they speed up in colder, denser regions. By recording the time it takes for seismic waves to travel from an earthquake's epicenter to a particular recording station at the surface, scientists can infer the temperatures and densities in a given segment of the interior. And by compiling a map of seismic velocities from thousands of earthquakes around the globe they can begin to map temperatures and densities throughout the mantle.

These seismic snapshots, which become increasingly more detailed as researches find more accurate ways to compile their measurements, have recently revealed some unexpectedly immense formations in the deepest parts of the mantle. The largest single structure turns out to lie directly below Africa's southern tip. About two years ago seismologists Jeroen Ritsema and Hendrik-Jan van Heijst of the California Institute of Technology calculated that this mushroom-shaped mass stretches some 900 miles upward from the core and spreads across several thousand miles [see "Mantle Map".]

The researchers immediately began to wonder whether this enormous blob could be shoving Africa skyward. Be-

### Article 6. Sculpting the Earth from Inside Out

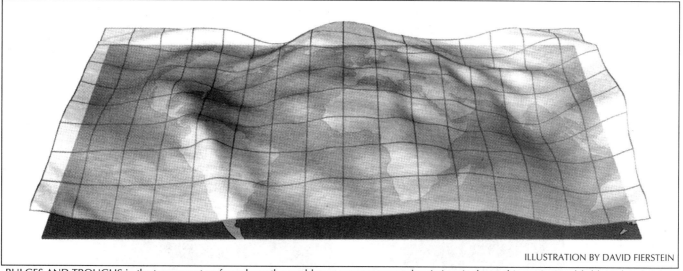

BULGES AND TROUGHS in the transparent surface above the world map represent natural variations in the earth's gravitational field. High points indicate stronger-than-normal gravity caused by a pocket of excess mass within the planet's interior; low areas occur above regions where a deficiency of mass produces a band of low gravity. Such differences in gravity hint at the location of oddities in the structure of the earth's mantle.

cause the blob is a region where seismic waves are sluggish, they assumed that it was hotter than the surrounding mantle. The basic physics of convection suggested that a hot blob was likely to be rising. But a seismic snapshot records only a single moment in time and thus only one position of a structure. If the blob were of a different composition than the surrounding rock, for instance, it could be hotter and still not rise. So another geophysicist, Jerry X. Mitrovica of the University of Toronto, and I decided to create a time-lapse picture of what might be happening. We plugged the blob's shape and estimated density, along with estimates of when southern Africa began rising, into a computer program that simulates mantle convection. By doing so, we found last year that the blob is indeed buoyant enough to rise slowly within the mantle—and strong enough to push Africa upward as it goes.

Seismic snapshots and computer models—the basic tools of geophysicists—were enough to solve the puzzle of the African superswell, but resolving the up-and-down movements of North America and Australia was more complicated and so was accomplished in a more circuitous way. Geophysicists who think only about what the mantle looks like today cannot fully explain how it sculpts the earth's surface. They must therefore borrow from the historical perspective of traditional geologists who think about the way the surface has changed over time.

## Ghosts from the Past

The insights that would help account for the bobbings of Australia and North America began to emerge with investigations of a seemingly unrelated topic: the influence of mantle density on the earth's gravitational field. The basic principles of physics led scientists in the 1960s to expect that gravity would be lowest above pockets of hot rock, which are less dense and thus have less mass. But when geophysicists first mapped the earth's gravitational variations, they found no evidence that gravity correlated with the cold and hot parts of the mantle—at least not in the expected fashion.

Indeed, in the late 1970s and early 1980s Clement G. Chase uncovered the opposite pattern. When Chase, now at the University of Arizona, considered geographic scales of more than 1,000 miles, he found that the pull of gravity is strongest not over cold mantle but over isolated volcanic regions called hot spots. Perhaps even more surprising was what Chase noticed about the position of a long band of low gravity that passes from Hudson Bay in Canada northward over the North Pole, across Siberia and India, and down into Antarctica. Relying on estimates of the ancient configuration of tectonic plates, he showed that this band of low gravity marked the location of a series of subduction zones—that is, the zones where tectonic plates carrying fragments of the seafloor plunge back into the mantle—from 125 million years ago. The ghosts of ancient subduction zones seemed to be diminishing the pull of gravity. But if cold, dense chunks of seafloor were still sinking through the mantle, it seemed that gravity would be high above these spots, not low, as Chase observed.

In the mid-1980s geophysicist Bradford H. Hager, now at the Massachusetts Institute of Technology, resolved this apparent paradox by proposing that factors other than temperature might create pockets of extra or deficient mass within the mantle. Hager developed his theory from the physics that describe moving fluids, whose behavior the mantle imitates over the long term. When a low-density fluid rises upward, as do the hottest parts of the mantle, the force of the flow pushes up the higher-density fluid above it. This gentle rise atop the upwelling itself creates an excess of mass (and hence stronger gravity) near the planet's surface. By the same token, gravity can be lower over cold, dense material: as this heavy matter sinks, it drags down

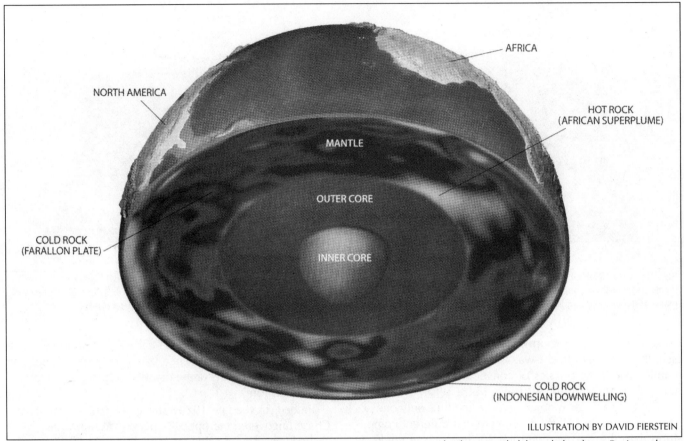

MANTLE MAP integrates measurements of thousands of earthquake vibrations, or seismic waves, that have traveled through the planet. Regions where waves moved quickly usually denote cold, dense rock. Regions where waves slowed down denote hot, less compact rock. Under southern Africa and the South Atlantic lies a pocket of sluggish velocities—a buoyant blob of hot rock called the African superplume. The map also reveals cold, sinking material that is tugging on North America and Indonesia.

mass that was once near the surface. This conception explained why the ghosts of subduction zones could generate a band of low gravity: some of that cold, subducted seafloor must still be sinking within the mantle—and towing the planet's surface downward in the process. If Hager's explanation was correct, it meant that the mantle did not merely creep horizontally near the planet's surface; whole segments of its up-and-down movements also reached the surface. Areas that surged upward would push the land above it skyward, and areas that sank would drag down the overlying continents as they descended.

## Bobbing Continents

At the same time that Chase and Hager were discovering a mechanism that could dramatically lift and lower the earth's surface, geologists were beginning to see evidence that continents might actually have experienced such dips and swells in the past. Geologic formations worldwide contain evidence that sea level fluctuates over time. Many geologists suspected that this fluctuation would affect all continents in the same way, but a few of them advanced convincing evidence that the most momentous changes in sea level stemmed from vertical motions of continents. As one continent moved, say, upward relative to other landmasses, the ocean surface around that continent would become lower while sea level around other landmasses would stay the same.

Most geologists, though, doubted the controversial notion that continents could move vertically—even when the first indications of the bizarre bobbing of Australia turned up in the early 1970s. Geologist John J. Veevers of Macquarie University in Sydney examined outcrops of ancient rock in eastern Australia and discovered that sometime in the early Cretaceous period (about 130 million years ago), a shallow sea rapidly covered that half of Australia while other continents flooded at a much more leisurely pace. Sea level climaxed around those landmasses by the late Cretaceous (about 70 million years ago), but by then the oceans were already retreating from Australia's shores. The eastern half of the continent must have sunk several thousand feet relative to other landmasses and then popped back up before global sea level began to fall.

Veevers's view of a bobbing continent turned out to be only part of Australia's enigmatic story. In 1978 geologist Gerard C. Bond, now at Columbia University's Lamont-Doherty Earth Observatory, discovered an even stranger turn of events while he was searching global history for examples of vertical continental motion. After Australia's dip and rise during the Cretaceous, it sank again, this time by 600 feet, between then

Article 6. Sculpting the Earth from Inside Out

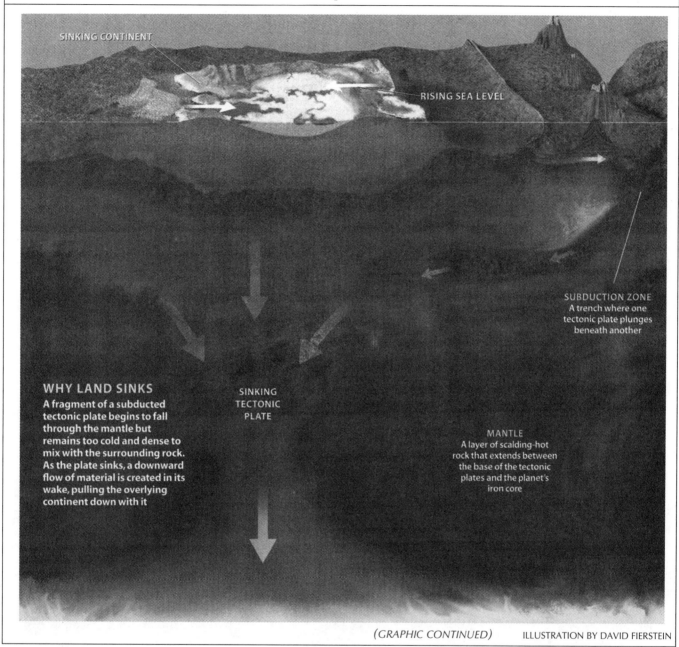

(GRAPHIC CONTINUED) ILLUSTRATION BY DAVID FIERSTEIN

and the present day. No reasonable interpretation based on plate tectonics alone could explain the widespread vertical motions that Bond and Veevers uncovered. Finding a satisfactory explanation would require scientists to link this information with another important clue: Hager's theory about how the mantle can change the shape of the planet's surface.

The first significant step in bringing these clues together was the close examination of another up-and-down example from Bond's global survey. In the late 1980s this work inspired Christopher Beaumont, a geologist at Dalhousie University in Nova Scotia, to tackle a baffling observation about Denver, Colo. Although the city's elevation is more than a mile above sea level, it sits atop flat, undeformed marine rocks created from sediments deposited on the floor of a shallow sea during the Cretaceous period. Vast seas covered much of the continents during that time, but sea level was no more than about 400 feet higher than it is today. This means that the ocean could never have reached as far inland as Denver's current position—unless this land was first pulled down several thousand feet to allow waters to flood inland.

Based on the position of North America's coastlines during the Cretaceous, Beaumont estimated that this bowing downward and subsequent uplift to today's elevation must have affected an area more than 600 miles across. This geographic scale was problematic for the prevailing view that plate tectonics alone molded the surface. The mechanism

ILLUSTRATION BY DAVID FIERSTEIN

of plate tectonics permits vertical motions within only 100 miles or so of plate edges, which are thin enough to bend like a stiff fishing pole, when forces act on them. But the motion of North America's interior happened several hundred miles inland—far from the influence of plate collisions. As entirely different mechanism had to be at fault.

Beaumont knew that subducted slabs of ancient seafloor might sit in the mantle below North America and that such slabs could theoretically drag down the center of a continent. To determine whether downward flow of the mantle could have caused the dip near Denver, Beaumont teamed up with Jerry Mitrovica, then a graduate student at the University of Toronto, and Gary T. Jarvis of York University in Toronto. They found that the sinking of North America during the Cretaceous could have been caused by a plate called the Farallon as it plunged into the mantle beneath the western coast of North America. Basing their conclusion on a computer model, the research team argued that the ancient plate thrust into the mantle nearly horizontally. As it began sinking, it created a downward flow in its wake that tugged North America low enough to allow the ocean to rush in. As the Farallon plate sank deeper, the power of its trailing wake decreased. The continent's tendency to float eventually won out, and North America resurfaced.

When the Canadian researchers advanced their theory in 1989, the Farallon plate had long since vanished into the

### Article 6. Sculpting the Earth from Inside Out

## Australia's Ups and Downs
A computer model reveals how the ghost of an ancient subduction zone dragged down a continent

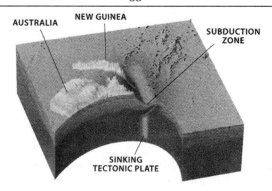

**130 Million Years Ago**
Australia is bordered by a subduction zone, a deep trench where the tectonic plate to the east plunges into the mantle. The sinking plate pulls the surrounding mantle and the eastern edge of Australia down with it. Later, subduction ceases and the continent begins to drift eastward.

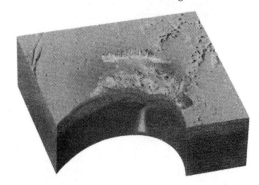

**90 Million Years Ago**
The entire eastern half of Australia sinks about 1,000 feet below sea level as the continent passes eastward over the sinking tectonic plate. About 20 million years later the plate's downward pull diminishes as it descends into the deeper mantle. As a result, the continent then pops up again.

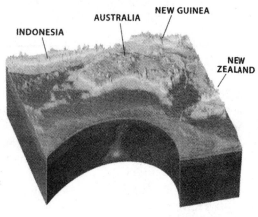

**Today**
Australia lies north of its former site, having been pushed there by activity in adjacent tectonic plates beginning about 45 million years ago. The entire continent has dropped relative to its greatest elevation as the result of a downward tug in the mantle under Indonesia—a landmass that is also sinking.

mantle, so its existence had only been inferred from geologic indications on the bottom of the Pacific Ocean. At that time, no seismic images were of high enough resolution to delineate a structure as small as a sinking fragment of the seafloor. Then, in 1996, new images of the mantle changed everything. Stephen P. Grand of the University of Texas at Austin and Robert D. van der Hilst of M.I.T., seismologists from separate research groups, presented two images based on entirely different sets of seismic measurements. Both pictures showed virtually identical structures, especially the cold-mantle down-wellings associated with sinking slabs of seafloor. The long-lost Farallon plate was prominent in the images as an arching slab 1,000 miles below the eastern coast of the U.S.

## Moving Down Under

Connecting the bobbing motion of North America to the subduction of the seafloor forged a convincing link between ancient sea-level change and goings-on in the mantle. It also became clear that the ancient Farallon slab sits within the band of low gravity that Chase had observed two decades earlier. I suspected that these ideas could also be applied to the most enigmatic of the continental bobbings, that of Australia during and since the Cretaceous. I had been simulating mantle convection with computer models for 15 years, and many of my results showed that the mantle was in fact able to lift the surface by thousands of feet—a difference easily great enough to cause an apparent drop in sea level. Like Chase, Veevers and other researchers before me, I looked at the known history of plate tectonics for clues about whether something in the mantle could have accounted for Australia's bouncing. During the Cretaceous period, Australia, South America, Africa, India, Antarctica and New Zealand were assembled into a vast supercontinent called Gondwana, which had existed for more than 400 million years before it fragmented into today's familiar landmasses. Surrounding Gondwana for most of this time was a huge subduction zone where cold oceanic plates plunged into the mantle.

I thought that somehow the subduction zone that surrounded Gondwana for hundreds of millions of years might have caused Australia's ups and downs. I became more convinced when I sketched the old subduction zones on maps of ancient plate configurations constructed by R. Dietmar Müller, a seagoing geophysicist at Sydney University. The sketches seemed to explain the Australian oddities. Australia would have passed directly over Gondwana's old subduction zone at the time it sank.

To understand how the cold slab would behave in the mantle as Gondwana broke apart over millions of years, Müller and I joined Louis Moresi of the Commonwealth Scientific and Industrial Research Organization in Perth to run a computer simulation depicting the mantle's influence on Australia over time. We knew the original position of the ancient subduction zone, the history of horizontal plate motions in the region and the estimated properties—such as viscosity—of the mantle below. Operating under these constraints, the computer played out a scenario for Australia that fit our hypotheses nearly perfectly [see "Australia's Ups and Downs"].

The computer model started 130 million years ago with ocean floor thrusting beneath eastern Australia. As Australia broke away from Gondwana, it passed over the cold, sinking slab, which sucked the Australian plate downward. The continent rose up again as it continued its eastward migration away from the slab.

Our model resolved the enigma of Australia's motion during the Cretaceous, originally observed by Veevers, but we were still puzzled by the later continentwide sinking of Australia that Bond discovered. With the help of another geophysicist, Carolina Lithgow-Bertelloni, now at the University of Michigan, we confirmed Bond's observation that as Australia moved northward toward Indonesia after the Cretaceous, it subsided by about 6500 feet. Lithgow-Bertelloni's global model of the mantle, which incorporated the history of subduction, suggested that Indonesia is sucked down more than any other region in the world because it lies at the intersection of enormous, present-day subduction systems in the Pacific and Indian oceans. And as Indonesia sinks, it pulls Australia down with it. Today Indonesia is a vast submerged continent—only its highest mountain peaks protrude above sea level.

Which brings us back to Africa. In a sense, Indonesia and Africa are opposites; Indonesia is being pulled down while Africa is being pushed up. These and other changes in the mantle that have unfolded over the past few hundred million years are intimately related to Gondwana. The huge band of low gravity that Chase discovered 30 years ago is created by the still-sinking plates of a giant subduction zone that once encircled the vast southern landmass. At the center of Gondwana was southern Africa, which means that the mantle below this region was isolated from the chilling effects of sinking tectonic plates at that time—and for the millions of years since. This long-term lack of cold, downward motion below southern Africa explains why a hot superplume is now erupting in the deep mantle there.

With all these discoveries, a vivid, dynamic picture of the motions of the mantle has come into focus. Researchers are beginning to see that these motions sculpt the surface in more ways than one. They help to drive the horizontal movement of tectonic plates, but they also lift and lower the continents. Perhaps the most intriguing discovery is that motion in the deep mantle lags behind the horizontal movement of tectonic plates. Positions of ancient plate boundaries can still have an effect on the way the surface is shaped many millions of years later.

Our ability to view the dynamics of mantle convection and plate tectonics will rapidly expand as new ways of observing the mantle and techniques for simulating its motion are introduced. When mantle convection changes, the gravitational field changes. Tracking variations in the earth's gravitational field is part of a joint U.S. and German space mission called GRACE, which is set for launch in June. Two spacecraft, one chasing the other in earth orbit, will map variations in gravity every two weeks and perhaps make it possible to infer the slow, vertical flow associated with convection in the mantle. Higher-resolution seismic images will also play a pivotal role in revealing what the mantle looks like today. Over the five- to 10-year duration of a project called USArray, 400 roving seismometers will provide a 50-mile-resolution view into the upper 800 miles of the mantle below the U.S.

Plans to make unprecedented images and measurements of the mantle in the coming decade, together with the use of ever more powerful supercomputers, foretell an exceptionally bright future for deciphering the dynamics of the earth's interior. Already, by considering the largest region of the planet—the mantle—as a chuck of rock with a geologic history, earth scientists have made extraordinary leaps in understanding the ultimate causes of geologic changes at the surface.

## Further Information

DYNAMICS OF CRETACEOUS VERTICAL MOTION OF AUSTRALIA AND THE AUSTRALIAN-ANTARCTIC DISCORDANCE. Michael Gurnis, R. Dietmar Müller and Louis Moresi in *Science*, Vol. 279, pages 1499–1504; March 6, 1998.

DYNAMIC EARTH: PLATES, PLUMES AND MANTLE CONVECTION. Geoffrey F. Davies. Cambridge University Press, 2000.

CONSTRAINING MANTLE DENSITY STRUCTURE USING GEOLOGICAL EVIDENCE OF SURFACE UPLIFT RATES: THE CASE OF THE AFRICAN SUPERPLUME. Michael Gurnis, Jerry X. Mitrovica, Jeroen Ritsema and Hendrik-Jan van Heijst in *Geochemistry, Geophysics, Geosystems*, Vol. 1, Paper No. 1999GC000035; 2000. Available online at http://146.201.254.53/publicationsfinal/articles/1999GC000035/fs1999G000035.html

Gurnis's Computational Geodynamics Research Group Web site: www.gps.caltech.edu/~gurnis/geodynamics.html

---

MICHAEL GURNIS is a geophysicist who is interested in the dynamics of plate tectonics and the earth's interior. These physical processes, which govern the history of the planet, have intrigued him since he began studying geology as an undergraduate 20 years ago. With his research group at the California Institute of Technology, Gurnis now develops computer programs that simulate the evolving motions of the mantle and reveal how those motions have shaped the planet over time. Gurnis's research highlights over the past three years have been deciphering the mysteries of the present-day African superswell and the bobbings of Australia during the Cretaceous period.

Article 7

# Nuclear power

## A renaissance that may not come

**This week, the Bush administration unveiled an energy policy that strongly supports nuclear power. This may revive a flagging industry, but the doubts remain as strong as ever**

THREE MILE ISLAND

THE gently rolling farmlands of central Pennsylvania do not prepare the casual visitor for what lies outside Middletown. Farmers tend cows and corn, diners serve simple food, and the occasional Amish buggy saunters by. But suddenly, there on the horizon, loom the cooling towers of the nuclear plant at Three Mile Island.

The words still send a shiver down the spine. It was here, early in the morning of March 28th 1979, that a reactor started to overheat. A combination of mechanical failure and human error sent the temperature in the reactor core soaring, threatening a blast that would have released huge quantities of lethal radiation. With the lives of perhaps half a million people at stake, politicians and scientists argued over what to do. In the end, disaster was averted; but the world did not forget.

For a while, it seemed that the accident at Three Mile Island (TMI) had killed off nuclear power. No new plants have been built in the United States since then. In Europe, too, people began to have second thoughts. TMI led directly to a referendum in Sweden in 1980 that demanded an end to nuclear power.

In 1986, an even worse accident, at Chernobyl in the Soviet Union, seemed to put the nail in the coffin of nuclear power in Europe. A number of countries, following Sweden's lead, campaigned for a ban. In Germany, the greens succeeded: the government has just agreed to end reprocessing of nuclear fuel by mid-2005. Moreover, Germany and Belgium have decided to ban new nuclear plants, although existing ones may serve out their useful lives. Even pro-nuclear France seemed to lose its enthusiasm for new plants.

For a while, Asia remained a bright spot for the nuclear industry. But the Asian financial crisis of 1998 cooled that enthusiasm. In recent months the new government in Taiwan, once a big fan of nuclear power, has tried to reverse course. In Japan, an accident soured public opinion: shoddy management practices at an experimental fuel-reprocessing plant in Tokaimura led in September 1999 to the deaths of two workers after they were exposed to radiation over 10,000 times the level considered safe. The Japanese government quietly scaled back its plans for 20 new plants.

The industry also hurt itself. In 1999, it emerged that British Nuclear Fuels (BNFL) had falsified records relating to shipments of nuclear fuel to Japan, sparking outrage in both countries. The firm had also understated the cost of nuclear clean-ups in Britain by some $13 billion. Clumsiness with deadly stuff, and now mendacity; in one way or another, nuclear power seemed to spell nothing but trouble.

### Votes of confidence

Yet some did not lose faith, and are even looking at nuclear power in a new way. South Africa's Eskom is working with BNFL and Exelon, America's biggest nuclear-energy firm, to build a reactor using new "pebble-bed" technology. In Finland, a power company is now requesting permission to build a €2.5 billion ($2.2 billion) nuclear plant.

Robin Jeffrey, the new chairman-designate of British Energy, Britain's largest nuclear operator, also sees a bright future for his industry. "The mood is buoyant," he says. "Utilities with a nuclear portfolio are seen to be attractive places to put your money." In fact, he has just completed a deal to take over the operation and maintenance of several reactors in Canada. Through AmerGen, a joint venture with Exelon, British Energy already manages a number of plants in the United States.

This week has produced the biggest boost of all for nuclear power: a strong endorsement from the Bush administration. On May 17th, a cabinet-level task-force unveiled a new energy policy that firmly supports the nuclear option. Vice-President Dick Cheney, the head of the taskforce, argues forcefully not only for giving existing nuclear plants a new lease of life, but also for building more: "We'd like to see an increase in the percentage of our electricity generated from nuclear power."

To understand why fans of nuclear power have become so optimistic, look back to Three Mile Island itself—now managed, as it happens, by AmerGen. The accident destroyed one of the plant's two reactors, but the surviving reactor has been back in service for some years now. In that time, TMI has become one of the most efficiently run and safest nuclear plants in America, as well as one of the most profitable. Corbin McNeill, Exelon's boss, is certain that financial success and operational safety are closely linked. TMI, he thinks, is a shining symbol of the future for nuclear power.

What explains this burst of enthusiasm? The short answer is the arrival, at long last, of market forces in the electricity business. After decades of being run as monopolies, either by the state or the private sector, the electricity industry is being deregulated the world over. As a result of this, and of the current high price of fossil fuels, existing nuclear plants look attractive, and are beginning to be run as proper businesses by serious managers.

## The charm of consolidation

This is best seen in America, which deregulated its wholesale markets for power in 1996. The result of deregulation was a painful squeeze on America's dozens of nuclear plants, many of which were run as one-off investments by local utilities. That is rapidly changing, however, thanks to the flurry of deals that have led to mega-mergers (like the one that created Exelon), joint ventures (like AmerGen), and other sorts of management coalitions. Nearly 30 gigawatts, about a quarter of the country's nuclear capacity, has already been affected by this consolidation. In the near future, today's 50 nuclear utilities will probably be reduced to a dozen.

The advantages of such consolidation are many. Plant managers can benefit from economies of scale and can apply best practices more widely. As a result, plants are running at higher capacity-utilisation rates and making better use of their fuel. Plant operators have also tried to expand capacity by upgrading their steam generators and turbines. Last winter, America's nuclear plants cranked out power at an operating cost of just 1.8 cents per kilowatt (kW)-hour; coal plants produced it for 2.1 cents per kW-hour, while those using natural gas (the price of which soared last winter) did no better than 3.5 cents.

Such improvements, argue nuclear fans, make a clear case for extending the licences of existing nuclear-power plants beyond their original limit of 40 years or so. In America, for example, a number of these permits will start expiring in 2006, and nearly all will have gone by 2030. The story is similar in Europe and elsewhere.

The fans are surely correct. Plants have been able to achieve such low operating costs because they are better managed and more efficient, and that, in turn, is linked to improved operational safety. When plants are safe, people do not mind living near them. Two plants have already received approvals from America's Nuclear Regulatory Commission (NRC) for another 20 years of operation. More will follow.

## Cheaper tomorrow...

Tomorrow, the advocates say, nuclear power will be even cheaper. They point to promising new designs (such as pebble-bed technology) and argue that power plants are on the way that are safer and more cost-effective than today's. The industry is also mature now, they say; both companies and regulators know how to avoid the costly bureaucratic quagmire that followed the TMI accident. In future, new plants will be "cheaper than coal".

Maybe; maybe not. The new designs for nuclear plants are undoubtedly improvements. Technical experts agree that they are probably inherently safer, as they use "passive" safety features that make a TMI-style meltdown virtually impossible. The NRC has already given its blessing to

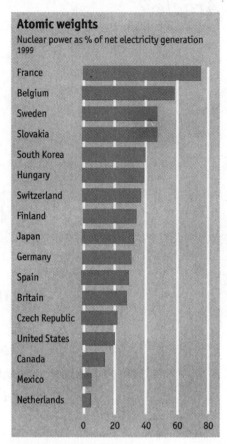

Source: Intrenational Energy Agency

three advanced designs. However, critics argue that some new designs "put all the safety eggs in the prevention basket", while short-changing systems that might limit an accident if one occurred.

Even if they do prove safer, the new designs may not necessarily be cheaper. By the reckoning of the International Energy Agency (IEA), which has just produced a new analysis of the economics of nuclear power*, the capital cost for today's nuclear designs runs at about $2,000 per kW, against about $1,200 per kW for coal and just $500 per kW for a combined-cycle gas plant. History also suggests that not everything goes as planned when turning clever paper designs into real-life nuclear plants. What is more, the debts of any new plants, unlike the debts of existing plants, will not be written off. In fact, the true cost of power from today's plants is at least double the apparent figure, argues Florentin Krause, an American economist, once debt write-offs, government subsidies and externalities are accounted for. More on that later.

Capital cost clearly remains a big hurdle for nuclear power. When considering the full life-cycle costs of a new project in today's money, some 60–75% of a nuclear plant's costs may be front-loaded; for a gas plant, about a quarter may be. That is before considering interest accrued during construction, which, over the many years it takes to build nuclear plants, can make or break a project. All these dismal sums explain why nuclear projects are exceptionally sensitive to the cost of capital.

## ...and cheap forever?

Never mind that, the nuclear industry argues; tomorrow's plants will prove cheaper to operate in several ways. For a start, some say, plants will be bigger, to take advantage of economies of scale. Second, there will be a whole series of plants, rather than the uneconomic one-off structures of the past, and these (in imitation of France's *dirigiste* approach) will be standardised replicas of new designs, rather than endless permutations.

Each of these bright ideas has problems. Building bigger plants introduces more complexity, which, experts say, means greater uncertainty and cost. The idea of building many plants is thwarted by the fact that the electricity market in the developed world is not growing fast enough to need them. In fact, the trend since the mid-1970s has been towards smaller plants. It is micropower, not megapower, that the market favours, thanks to the far smaller financial risk involved.

It is true that regulators are becoming lighter-handed. They no longer drag out the completion of plants for ten or 15 years. Even so, careful analysis of the delays after the TMI accident shows that technical hitches were largely to blame, rather than red tape. Many new plants happened to come on line at around that time, and a number had generic technical faults. Even the French programme, touted as a model, suffered from such problems as late as the 1990s: its latest N4 design developed cracks in its heat-removal systems.

## A dubious special case

The industry's advocates point to other benefits: security of supply, environmental benefits, and so on. In some countries and in some circumstances, such arguments might have merit. But, taken together, do they make nuclear energy a special case that justifies subsidies or other forms of government intervention?

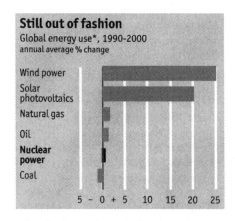

*Based on installed capacity for wind and nuclear power, shipments for solar PV and consumption for natural gas, oil and coal

Source: Worldwatch Institute

The energy-security arguments vary, but mostly involve reduced reliance on fossil fuels, less vulnerability to an OPEC embargo, or a smaller bill for imported fuel. Whatever the political soundness of such arguments, an analysis done in 1998 by several agencies affiliated with the OECD is quite clear about the relative costs and benefits: "For many countries, the additional energy security obtained from investing in non-fossil-fuelled generation options is likely to be worth less than the cost of obtaining that security."

On environmental grounds, too, nuclear power does not emerge a clear winner. It is true that nuclear energy does not produce carbon dioxide, the chief culprit behind man-made global warming. That, say fans like Mr Cheney, means that the world "ought to" build more such plants. But handing out public money to the nuclear-power industry (through production or investment tax credits, for example) is an inefficient way for governments to discourage global warming.

The better way would be through some sort of carbon tax, which would penalise fossil fuels but not any fuel free of carbon emissions, whether nuclear power or renewables. The IEA's boffins have analysed how much of a boost nuclear power could get from a carbon tax, and the answer is quite a lot—possibly enough to compete with coal. Assume that the carbon tax falls between $25 and $85 per tonne of carbon, the level many experts think may be needed if industrialised countries are serious about the emissions targets agreed on in 1997 at Kyoto (some would set it higher still). The IEA thinks that the highest of those taxes would boost the competitiveness of nuclear electricity against coal by two cents per kW-hour, and against natural gas by one cent. At the low end, nuclear gains an advantage of half a cent and a quarter-cent respectively.

But nuclear energy, even if boosted by a carbon tax, also carries grave environmental liabilities. Radiation is a threat to human health at every stage of the process, from uranium mining to plant operation (even in those new ultra-safe plants) to waste disposal. And waste disposal, despite decades of research and politicking, remains a farce. No country has yet built a "permanent" waste-disposal site. The United States hopes to have one finished at Yucca Mountain, in the Nevada desert, in a decade's time; the European countries are a decade behind that. Even if these geological storage sites are completed, they are not the final answer. Nuclear waste may remain deadly for 100,000 years. To bury it in a big hole in the ground, and pray that some future generation may discover how to make it safe, is simply passing the buck.

Lastly, when costing nuclear power, it is essential to remember the scope, scale and subtlety of the subsidies it has received. The IEA analysis of nuclear economics shows that various OECD governments subsidise the industry's fuel-supply services, waste disposal, fuel reprocessing and R&D. They also limit the liability of plants in case of accident, and help them clean up afterwards. Antony Froggatt, an industry expert who has advised Greenpeace, points to export loans and guarantees as another unfair boost.

How much does all this add up to? The IEA's otherwise comprehensive analysis falls strangely silent on this topic, doubtless for political reasons. Reliable, comprehensive and up-to-date global figures from neutral analysts are scarce. Estimates by nuclear opponents typically carry too many zeroes to fit on this page. They are about as rigorous and credible as those put forth by the industry, which usually maintains the outrageous fiction that it no longer receives any subsidies at all.

## Coddled to the hilt

Liability insurance is a good example of this. The American industry's official position is that there is no subsidy involved in the Price-Anderson Act, by which Congress limits the civilian nuclear industry's liability for nuclear catastrophes to less than $10 billion (a small fraction of what a Chernobyl-scale disaster would cost in America). Since there is no subsidy involved, why not let the act lapse when it comes up for renewal next year? Mr Cheney's response is revealing: "It needs to be renewed...[if not], nobody's going to invest in nuclear-power plants."

One concrete figure gives an idea of how enormous the overall subsidy pie might be. According to official figures, OECD governments poured $159 billion in today's money into nuclear research between 1974 and 1998. Some of that breathtaking sum is a sunk cost from the early days of the industry, but not all: governments still shell out about half their energy R&D budgets on this mature industry. Even so, says the industry, those have been tapering off over time. But so will all the other subsidies, as the liberalisation of markets advances. The pillar on which the strength of the nuclear industry has rested is crumbling away.

In the end, nuclear energy's future may be skewered by the same sword that is making it fashionable today: the deregulation of electricity markets. This liberalising movement has put a shine on old nuclear plants that are already paid for. TMI, for example, was bought for a pittance and so cranks out power for virtually nothing.

Yet liberalisation is also exposing the true economics of new plants, and is aiming a fierce spotlight at the hefty subsidies that nuclear power has long enjoyed. As these fade, the industry will once again be brought down to earth.

---

"Nuclear Power in the OECD". International Energy Agency, 2001.

---

From *The Economist*, May 19, 2001, pp. 49-50. © 2001 by The Economist, Ltd. Distributed by the New York Times Special Features. Reprinted by permission.

Article 8

# Filling the Void

Clearing the wreckage was just the beginning. For those rebuilding Ground Zero, the stakes are high, and the whole world is watching.

BY CATHLEEN MCGUIGAN

ONE YEAR AFTER THE TERRORIST ATTACKS, THE PLACE where the Twin Towers once loomed looks like any big-city construction site. Swept clean of the last pieces of twisted steel, the 16 acres now thrum with the noise of bulldozers and cranes as work goes on to rebuild underground subway lines. Near the southeast corner of Ground Zero is one unofficial monument: a simple cross forged by workmen from two pieces of those once mighty steel columns. Gone are the twin beams of the Tribute in Light, which soared into the night sky last spring. Now tourists—thousands a day—clog the perimeter of the site. Some are quiet and reverent, others smile for the family video camera or wolf down hot dogs. Two weeks ago crews began erecting a special mesh fence—rather than plywood—to give visitors a better view when the reconstruction actually begins.

But exactly what will be built there is still anyone's guess. One of the many remarkable stories of the last year is how public opinion, galvanized by the tragedy, has derailed the business-as-usual development plans that called for replacing the World Trade Center's 11 million square feet of office space. In mid-July, the city-state agency charged with planning, along with the Port Authority—which owns the land—rushed out a half-dozen half-baked schemes. In the look-alike models, Styrofoam boxes choke the site, interspersed with green patches that vary in shape from plan to plan ("Memorial Triangle," one was called, just to give you an idea). The reaction was swift. Newspapers labeled the plans "dreary," "leaden" and "retarded." But the most damning criticism came at a public "listening session" held on July 20 at New York's Javits Center, where the 4,500 attendees could punch their comments into computers that collated responses and flashed them on JumboTron screens. At the sight of one clunky scheme, somebody typed furiously: "Looks like Albany."

Just how deeply the wider world cares about rebuilding Ground Zero was driven home when the six plans were first unveiled. The Lower Manhattan Development Corporation had 50 million hits on its Web site that very day, and thousands of unsolicited schemes have poured into the agency. A private project called Imagine New York has collected more than 20,000 ad hoc proposals since last spring. This month the Venice Biennale in architecture will show WTC plans from dozens of the world's top architects; next weekend The New York Times Magazine will publish schemes for the site by such designers as Richard Meier and Zaha Hadid. And architects including Frederic Schwartz, who designed the Staten Island Ferry Terminal, have been working on their own. (Schwartz's plan: put a deck over the highway west of the site, build all 11 million square feet of office space there and leave Ground Zero for whatever—a memorial, museums, parks—or nothing at all.) "People aren't going to stand for something not positive happening on the site," Schwartz says. "It has to be the most magnificent thing for all time."

No one would disagree. If the LMDC and Port Authority planning officials were embarrassed by the nearly unanimous rejection of their first efforts, they didn't admit it. "These are concept plans, not schemes for the site," explains Alex Garvin, vice chairman of the LMDC and a professor of planning at Yale. OK, whatever. Garvin is one of several officials who turn a conversation quickly to aspects of the plans the public actually liked. Those include keeping the "footprints" of the Twin Towers empty and creating more public spaces—especially a tree-lined memorial promenade—billed as Manhattan's own Champs-Elysees—that could be built atop a submerged highway west of the site, linking the new WTC to Battery Park City. People also wanted culture, and LMDC chairman John Whitehead agrees. "Museums and various performing arts must be part of the future of lower Manhattan," he says.

But the biggest beef with the plans was that the dense assemblage of office towers would dominate any memorial and make a humane mix of offices, housing, retail and such cultural amenities impossible. "What the public said is, they just don't want to see this driven by a real-estate deal," says Robert Yaro, president of the Regional Plan Association. Larry Silverstein, who holds the redevelopment rights under the terms of his 99-year lease of the Twin Towers, is still battling insurance companies over their payout for the disaster. (A trial date is set for No-

# Brainstorming a New Downtown

Officials asked New Yorkers how to rebuild the World Trade Center site. The public gave them an earful. Now their ideas will help guide a new round of design proposals. Here are eight principles likely to be part of any future plan.

**1 STREET GRID:** The Trade Center turned a big chunk of the old downtown grid into a windy, empty plaza. New plans would restore some streets and bring back pedestrians—and maybe cars, too.
TEXT BY JOHN SPARKS

**2 CULTURAL SITES:** Redesigning downtown could give the arts capital a chance to add new museums and maybe an opera house to the neighborhood's mix.
ROBERT SPENCER—AP, TEXT BY JOHN SPARKS

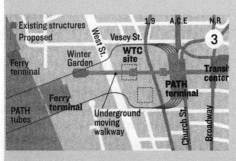

**3 GRAND CENTRAL:** A dramatic landmark station, on the scale of midtown's great terminal, could link PATH trains from New Jersey, subway lines, buses and the ferry boats that ply the harbor.
TEXT BY JOHN SPARKS

**4 PUBLIC OPEN SPACES:** Besides the main memorial on the Trade Center site, new plans for the area should include a mix of smaller parks and plazas, each with its own unique design.
SUZANNE PLUMKETT—AP, TEXT BY JOHN SPARKS

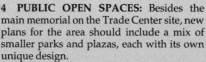

**5 TOWER FOOTPRINTS:** The square outlines that mark where the Twin Towers stood—one acre each—should be preserved and kept free of any new buildings.
PETER J. ECKEL—AP, TEXT BY JOHN SPARKS

**6 COMMERCIAL/RETAIL SPACE:** Negotiations are under way to reduce the 11 million square feet of office space formerly on the site. People in the neighborhood say they want new stores.
JJEFF ZELEVANSKY—AP, TEXT BY JOHN SPARKS

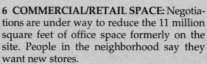

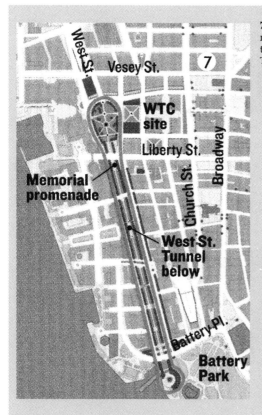

**7 GRAND PROMENADE:** West Street's traffic isolates the Trade Center site. By submerging the street and creating a landscaped promenade above it, the site could be linked to Battery Park and the waterfront.
TEXT BY JOHN SPARKS

**8 RENEWING THE DISTINCTIVE SKYLINE:** When the Twin Towers fell, they left a hole in the face the city shows the world. People want a new and dramatic presence to symbolize New York.
BEBETO MATTHEWS—AP, TEXT BY JOHN SPARKS

vember.) But his landlord, the Port Authority, in response to the public outcry, is considering options to reduce commercial space at the site. One idea came from Mayor Michael Bloomberg's office: a land swap with the PA, in which the city would get the 16-acre plot in exchange for land it owns under New York's airports, with their lucrative retail potential.

At the same time, the PA has also been talking with the Development Corporation about possible ways to cut down the office space by 2 million or 3 million square feet, according to Roland Betts, a member of the LMDC board. Even before the public trashed the six schemes, Betts was emerging as the leader in the planning process. Last month the agency announced a new competition to seek ideas from a wider range of designers; Betts will help choose five finalists at the end of September. "There's a little bit of feeling our way here," admits Betts. "It became clear that we're missing the creativity and experience of some of the best firms in the world, and we have to go get it." (The LMDC will also hold a separate competition for a design for the memorial, but not until next year.) It doesn't hurt that Betts has been a close friend of President George W. Bush's since they were at Yale almost 40 years ago. In August, Betts took a long ramble with the president through the canyons of the ranch at Crawford, Texas, and filled him in on the competition. "He's rooting for us," says Betts.

Meanwhile, despite the very public rebuke, the firm behind most of the rejected plans, Beyer Blinder Belle, continues to work as part of a consulting team under a $3 million contract. Along with giant engineering firm Parsons Brinckerhoff, they're envisioning a vast transit hub—a "grand central station," Garvin calls it—much of it underground, that would link subways, commuter rail and ferries. Last month $4.5 billion in federal funds was earmarked for such improvements. And already, the consultants are mapping out the grand promenade, even engaging landscape architect Laurie Olin to think about the trees.

There's never been a design project like this in history. The great urban places of the world are the products of kings (Place des Vosges in Paris) or tycoons (Rockefeller Center). And none has had to deal with the tragic symbolism of this spot. Yet here are ordinary citizens trying to bring democracy to the design of something great—and a transcendent new space may be the best way to honor those who died on September 11. But who in the end will decide what gets built? You still need a scorecard to sort out all the vested interests—including the governors of New York and New Jersey, and the mayor, who's readying his own wish list for the site. "The next step isn't going to be easy," says Holly Leicht of Imagine New York. No, it won't. And the public, from New York City and far beyond, will be watching the sky.

With JULIE SCELFO

From *Newsweek*, September 9, 2002, pp. 57-60. © 2002 by Newsweek, Inc. All rights reserved. Reprinted by permission.

# UNIT 2
# Human-Environment Relationships

## Unit Selections

9. **Global Warming: The Contrarian View**, William K. Stevens
10. **The Pollution Puzzle**, Tom Arrandale
11. **Human Modification of the Geomorphically Unstable Salt River in Metropolitan Phoenix**, Martin Roberge
12. **Texas and Water: Pay Up or Dry Up,** The Economist
13. **Beyond the Valley of the Dammed**, Bruce Barcott
14. **Environmental Enemy No. 1**, The Economist
15. **Carbon Sequestration: Fired Up With Ideas**, The Economist
16. **Past and Present Land Use and Land Cover in the USA**, William B. Meyer
17. **Operation Desert Sprawl**, William Fulton and Paul Shigley
18. **A Modest Proposal to Stop Global Warming**, Ross Gelbspan
19. **A Greener, or Browner, Mexico?** The Economist

## Key Points to Consider

- What are the long-range implications of atmospheric pollution? Explain the greenhouse effect.
- How can the problem of regional transfer of pollutants be solved?
- The manufacture of goods needed by humans produces pollutants that degrade the environment. How can this dilemma be solved?
- Where in the world are there serious problems of desertification and drought? Why are these areas increasing in size?
- What will be the major forms of energy in the twenty-first century?
- How are you as an individual related to the land? Does urban sprawl concern you? Explain.
- Can humankind do anything to ensure the protection of the environment? Describe.

 **Links: www.dushkin.com/online/**
These sites are annotated in the World Wide Web pages.

**Alliance for Global Sustainability (AGS)**
http://www.global-sustainability.org

**Environment News Service: Global Warming Could Make Water a Scarce Resource**
http://ens.lycos.com/ens/dec2000/2000L-12-15-06.html

**Human Geography**
http://www.geog.le.ac.uk/cti/hum.html

**The North-South Institute**
http://www.nsi-ins.ca/ensi/index.html

**United Nations Environment Programme (UNEP)**
http://www.unep.ch

**World Health Organization**
http://www.who.int

The home of humankind is Earth's surface and the thin layer of atmosphere enveloping it. Here the human populace has struggled over time to change the physical setting and to create the telltale signs of occupation. Humankind has greatly modified Earth's surface to suit its purposes. At the same time, we have been greatly influenced by the very environment that we have worked to change.

This basic relationship of humans and land is important in geography and, in unit 1, William Pattison identified it as one of the four traditions of geography. Geographers observe, study, and analyze the ways in which human occupants of Earth have interacted with the physical environment. This unit presents a number of articles that illustrate the theme of human-environment relationships. In some cases, the association of humans and the physical world has been mutually beneficial; in others, environmental degradation has been the result.

At the present time, the potential for major modifications of Earth's surface and atmosphere is greater than at any other time in history. It is crucial that the environmental consequences of these modifications be clearly understood before such efforts are undertaken.

The first article in this unit contends that greenhouse gas accumulations in the atmosphere are highly exaggerated. The next article reports on state initiatives to address pollution problems. The channeling of the Salt River in Phoenix is discussed next, followed by treatment of water shortages along the Rio Grande. The next article deals with recent demands to reduce the number of dams in the United States.

Two short articles follow with speculations on ways to reduce carbon dioxide accumulations in the atmosphere. William Meyer then addresses changing land use in the United States. "Operation Desert Sprawl" follows with an analysis of Las Vegas's reliance on transportation systems and the availability of fresh water. The next article provides a reasoned approach to global warming. Finally, the question of NAFTA's impact on economics and the environment is covered.

This unit provides a small sample of the many ways in which humans interact with the environment. The outcomes of these interactions may be positive or negative. They may enhance the position of humankind and protect the environment, or they may do just the opposite. We human beings are the guardians of the physical world. We have it in our power to protect, to neglect, or to destroy.

*Article 9*

# Global Warming: The Contrarian View

By WILLIAM K. STEVENS

Over the years, skeptics have tried to cast doubt on the idea of global warming by noting that measurements taken by earth satellites since 1979 have found little or no temperature rise in large parts of the upper atmosphere. The satellites' all-encompassing coverage yields more reliable results than temperature samplings showing a century-long warming trend at the earth's surface, they argued.

In January, a special study by the National Research Council, the research arm of the National Academy of Sciences, declared that the "apparent disparity" between the two sets of measurements over the 20-year history of the satellite measurements "in no way invalidates the conclusion that surface temperature has been rising." The surface warming "is undoubtedly real," the study panel said.

But the dissenters are a long way from conceding the debate, and they have seized on other aspects of the panel's report in an effort to bolster their case.

To be sure, according to interviews with some prominent skeptics, there is now wide agreement among them that the average surface temperature of the earth has indeed risen.

"I don't think we're arguing over whether there's any global warming," said Dr. William M. Gray, an atmospheric scientist at Colorado State University, known for his annual predictions of Atlantic hurricane activities as well as his staunch, longtime dissent on global climate change. "The question is, 'What is the cause of it?'"

On that issue, and on the remaining big question of how the climate might change in the future, skeptics continue to differ sharply with the dominant view among climate experts.

The dominant view is that the surface warming is at least partly attributable to emissions of heat-trapping waste industrial gases like carbon dioxide, a product of the burning of fossil fuels like coal, oil and natural gas. A United Nations scientific panel has predicted that unless

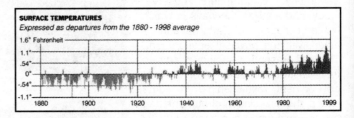

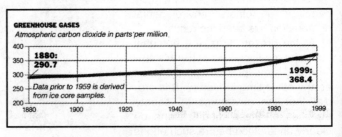

these greenhouse gas emissions are reduced, the earth's average surface temperature will rise by some 2 to 6 degrees Fahrenheit over the next century, with a best estimate of about 3.5 degrees, compared with a rise of 5 to 9 degrees since the depths of the last ice age 18,000 to 20,000 years ago. This warming, the panel said, would touch off widespread disruptions in climate and weather and cause the global sea level to rise and flood many places.

Dr. Gray and others challenge all of this. To them, the observed surface warming of about 1 degree over the last century—with an especially sharp rise in the last quarter century—is mostly or wholly natural, and there is no significant human influence on global climate. They also adhere firmly to their long-held opinion that any future warming will be inconsequential or modest at most, and that its effects will largely be beneficial.

In some ways, though, adversaries in the debate are not so far apart. For instance, some dissenters say that future warming caused by greenhouse gases will be near the low end of the range predicted by the United Nations

# Article 9. Global Warming: The Contrarian View

## Two sides, two data sets

*In support of their position, skeptics in the debate over climate change cite an apparent disparity between two sets of temperature data.*

❶ THERMOMETER READINGS TAKEN AT THE SURFACE

❷ TEMPERATURES TAKEN BY SATELLITES IN THE FREE ATMOSPHERE

*Sources: National Research Council; Carbon Dioxide Information Analysis Center*

scientific panel. And most adherents of the dominant view readily acknowledge that the size of the human contribution to global warming is not yet known.

The thrust-and-parry of the climate debate goes on nevertheless.

With the National Research Council panel's conclusion that the surface warming is real, "one of the key arguments of the contrarians has evaporated," said Dr. Michael Oppenheimer, an atmospheric scientist with Environmental Defense, formerly the Environmental Defense Fund.

But those on the other side of the argument see things differently.

To them, the most important finding of the panel is its validation of satellite readings showing less warming, and maybe none, in parts of the upper atmosphere, said Dr. S. Fred Singer, an independent atmospheric scientist who is an outspoken dissenter.

## The surface warming is real; humanity's role is at issue.

For him and other climate dissenters, this disparity is key, in that it does not show up in computer models scientists use to predict future trends. The fact that these models apparently missed the difference in warming between the surface and the upper air, the skeptics say, casts doubt on their reliability over all.

Experts on all sides of the debate acknowledge that the climate models are imperfect, and even proponents of their use say their results should be interpreted cautiously.

A further problem is raised by the divergent temperatures at the surface and the upper air, said Dr. Richard S. Lindzen, an atmospheric scientist at the Massachusetts Institute of Technology, who is a foremost skeptic. "Both are right." But, he said, increasing levels of greenhouse gases should warm the entire troposphere (the lower 6 to 10 miles of the atmosphere). That they have not, he said, suggests that "what's happening at the surface is not related to the greenhouse effect."

Skeptics also argue that the lower temperatures measured by the satellites are confirmed by instruments borne aloft by weather balloons. But over a longer period, going back 40 years, there is no discrepancy between surface readings and those obtained by the balloons, said Dr. John Michael Wallace, an atmospheric scientist at the University of Washington in Seattle, who was chairman of the research council study panel.

Although the climate debate has usually been portrayed as a polarized argument between believers and contrarians, there is actually a broad spectrum of views among scientists. And while the views of skeptics display some common themes, there are many degrees of dissent, many permutations and combinations of individual opinion. The views of some have changed materially over the years, while others have expressed essentially the same basic opinions all along.

One whose views have evolved is Dr. Wallace, who describes himself as "more skeptical than most people" but "fairly open to arguments on both sides" of the debate. He says the especially sharp surface warming trend of the

# 1. TEMPERATURE CHANGES AT OR NEAR THE EARTH'S SURFACE, 1979–1998

## CLIMATE DEBATE:

### THE DOMINANT VIEW

Surface warming is at least partly attributable to emissions of heat-trapping gases, like carbon dioxide, produced by the burning of fossil fuels.

### THE SKEPTICS

Although the surface does appear to be warming, the cause is wholly or mostly natural.

Thermometers at land stations and aboard ships show a general global warming trend. The data are subject to distortion, but a recent study by the National Research Council says, nevertheless, that the surface warming is "undoubtedly real."

The *New York Times*; Images by NASA (Earth); National Academy of Science (Temperature maps)

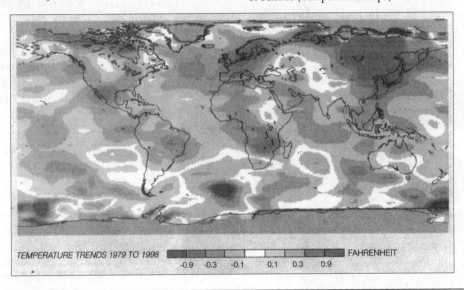

TEMPERATURE TRENDS 1979 TO 1998   −0.9  −0.3  −0.1  0.1  0.3  0.9   FAHRENHEIT

1990's has "pulled me in a mainstream direction." While he once believed the warming observed in recent decades was just natural variation in climate, he said, he is now perhaps 80 percent sure that it has been induced by human activity—"but that's still a long way from being willing to stake my reputation on it."

A decade ago, Dr. Wallace said, many skeptics questioned whether there even was a surface warming trend, in part because what now appears to be a century-long trend had been interrupted in the 1950's, 1960's and early 1970's, and it had not yet resumed all that markedly. But the surge in the 1980's and 1990's changed the picture substantially.

Today, Dr. Wallace said, few appear to doubt that the earth's surface has warmed. One prominent dissenter on the greenhouse question, Dr. Robert Balling, a climatologist at Arizona State University, says, "the surface temperatures appear to be rising, no doubt," and other skeptics agree.

There also appears to be general agreement that atmospheric concentrations of greenhouse gases are rising. At 360 parts per million, up from 315 parts in the late 1950's, the concentration of carbon dioxide is nearly 30 percent higher than before the Industrial Revolution, and the highest in the last 420,000 years.

Mainstream scientists, citing recent studies, suggest that the relatively rapid warming of the last 25 years cannot be explained without the greenhouse effect. Over that period, according to federal scientists, the average surface temperature rose at a rate equivalent to about 3.5 degrees per century—substantially more than the rise for the last century as a whole, and about what is predicted by computer models for the 21st century.

But many skeptics, including Dr. Gray and Dr. Singer, maintain that the warming of the past 25 years can be explained by natural causes, most likely changes in the circulation of heat-bearing ocean waters. In fact, Dr. Gray says he expects that over the next few decades, the warming will end and there will be a resumption of the cooling of the 1950's and 1960's.

At bottom, people on all sides of the debate agree, the question of the warming's cause has not yet been definitively answered. In December, a group of 11 experts on the question looked at the status of the continuing quest to detect the greenhouse signal amid the "noise" of the climate's natural variability.

## 2. TEMPERATURE CHANGES IN THE FREE ATMOSPHERE, UP TO 5 MILES, 1979–1998

### CLIMATE DEBATE:

#### THE SKEPTICS

Computer models on which projections of future warming are based failed to reflect the surface-satellite disparity, casting doubt on the projections' accuracy. They say future warming will be small to modest, and largely beneficial.

#### THE DOMINANT VIEW

Scientists adhere to the computer models which say the average global surface temperature will rise by 2 to 6 degrees Fahrenheit over the next century if greenhouse gas emissions are not reduced.

Instruments aboard satellites show little warming, and even some cooling, in the lower to middle part of the free atmosphere, mainly in the tropics. Although scientists cannot fully explain the disparity, some say that over a longer period it might disappear.

The *New York Times*; Images by NASA (Earth); National Academy of Science (Temperature maps)

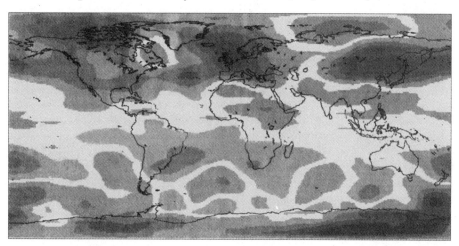

The lead author of the study was Dr. Tim P. Barnett, a climatologist at the Scripps Institution of Oceanography in La Jolla, Calif. Dr. Barnett, who has long worked on detecting the greenhouse signal, describes himself as a "hard-nosed" skeptic on that particular issue, even though he believes that global warming in the long run will be a serious problem.

The study by Dr. Barnett and others, published in The Bulletin of the American Meteorological Society, concluded that the "most probable cause" of the observed warming had been a combination of natural and human-made factors. But they said scientists had not yet been able to separate the greenhouse signal from the natural climate fluctuations. This state of affairs, they wrote, "is not satisfactory."

Two big questions complicate efforts to predict the course of the earth's climate over the next century: how sensitive is the climate system, inherently, to the warming effect of greenhouse gases? And how much will atmospheric levels of the gases rise over coming decades?

The mainstream view, based on computerized simulations of the climate system, is that a doubling of greenhouse gas concentrations would produce a warming of about 3 to 8 degrees. But Dr. Lindzen and Dr. Gray, pointing to what they consider the models' problems with the physics of the atmosphere, say they overestimate possible warming. It "will be extremely little," Dr. Gray said. How little? Dr. Lindzen pegs it at about half a degree to a bit less than 2 degrees, if atmospheric carbon dioxide doubles.

Other factors in the climate system modify the response to heat-trapping gases, and the United Nations panel's analysis included these to arrive at its projection of a 2- to 6-degree rise in the average global surface temperature by 2100. One factor is various possible levels of future carbon dioxide emissions. Dr. Singer, saying that improving energy efficiency will have a big impact on emissions, predicts a warming of less than 1 degree by 2100.

Dr. Balling projects a warming just shy of 1 degree for the next 50 years, not out of line with the United Nations panel's lower boundary. Another skeptic, Dr. Patrick J. Michaels, a climatologist at the University of Virginia, similarly forecasts a greenhouse warming rise of 2.3 degrees over the next century. As is now the case, he says, the warming would be most pronounced in the winter, at night, and in sub-Arctic regions like Siberia and Alaska. A warming of that magnitude, he and others insist, could not be very harmful, and would in fact confer benefits like longer growing seasons and faster plant growth.

# ANNUAL EDITIONS

"It should be pretty clear," he said, that the warming so far "didn't demonstrably dent health and welfare very much," and he said he saw no reason "to expect a sudden turnaround in the same over the next 50 years." After that, he said, it is impossible to predict the shape of the world's energy system and, therefore, greenhouse gas emissions.

A warming in the low end of the range predicted by the United Nations panel may well materialize, said Dr. Oppenheimer, the environmentalist. But, he said, the high end may also materialize, in which case, mainstream scientists say, there would be serious, even catastrophic, consequences for human society. "There is no compelling evidence to allow us to choose between the low end, or the high end, or the middle," Dr. Oppenheimer said.

If business continues as usual, the world is likely at some point to find out who is right.

---

From the *New York Times*, February 29, 2000, pp. F1, F6. © 2000 by The New York Times Company. Reprinted by permission.

# Article 10

# The Pollution Puzzle

The federal government isn't solving it. States are giving it a shot.

BY TOM ARRANDALE

Among all the states, New Hampshire hasn't usually ranked as a trailblazer on protecting the environment. But in the past few years, Granite State officials and residents have grown alarmed that global warming could kill off its colorful maple forests and wreak havoc with coastal resorts. This spring, New Hampshire responded with a groundbreaking state law that could pave the way for cost-effective national controls on climate-changing greenhouse gases.

As far as New Hampshire is concerned, "ignoring global warming is not the answer," says Kenneth A. Colburn, the state air-quality director who helped draft the program. But since ignoring it is precisely what the federal government has done, New Hampshire decided to craft its own "multi-pollutant" emission controls for the state's three aging coal-fired electric generating stations.

Colburn, now director of the air-quality association for eight Northeastern states, expects the East Coast's big industrial states to follow suit and perhaps extend multi-pollutant regulations to other industrial sectors. That, in turn, will put pressure on Midwestern states, such as Ohio and Michigan, that currently balk at carbon dioxide controls they fear will cost their economies dearly. In time, Northeastern states are counting on a domino effect that will prod the federal government to fashion a comprehensive response to global warming.

On the other side of the country, the biggest state domino of all is tackling another primary source of greenhouse gases. In July, the California legislature passed a bill to limit carbon dioxide from automobile exhaust. If it survives an expected challenge in the courts, the measure would force dramatic changes in the cars Americans drive in the future. The Golden State, which one auto industry official recently called "the center of the environmental regulatory universe," already has the country's toughest restrictions on ozone, soot and other air pollutants.

That's because California wrote many of its environmental regulations before federal laws went into effect, and it has gone its own way ever since with minimal interference from EPA. Increasingly, other states have exercised the option to adopt California standards. But they also are no longer content to merely follow the leader. Even the smallest states have begun flexing their muscles on a variety of fronts because the federal government hasn't been agile enough to cope with emerging environmental threats of national proportions.

At the same time, state regulators are trying to adapt conventional regulatory strategies to deal more effectively with serious hazards right in their own backyards. In those efforts, however, they've run into a major obstacle—the federal government's own hidebound system for by-the-book environmental regulation that's still stuck in the 1970s.

During the 1990s, for instance, New Hampshire's environmental services commissioner, Robert W. Varney, grew all too aware that raw sewage was seeping down coastal streams to contaminate shellfish beds and toxic mercury was building up in New Hampshire's lakes and rivers. But because the federal government hadn't deemed those to be the country's most imminent environmental risks, neither could New Hampshire.

That's the way the nation's pollution-control system has always operated. The U.S. Environmental Protection Agency for years compelled state governments to concentrate on carrying out federal cleanup rules, rather than on investigating what's actually harming the local environment and coming up with sensible solutions. In New Hampshire, Varney finally persuaded EPA to let the state shift funds from federal program grants to clean up shellfish beds and implement a mercury-control strategy. The state hasn't exactly turned into environmental paradise, but at least it has gotten a start on correcting previously overlooked problems "that are very high priorities for New Hampshire," Varney says.

A year ago, Varney left state government and started commuting to Boston as EPA's New England regional administrator. He saw the new job as a chance to build on New Hampshire's record and help the Northeast make even more impressive strides toward more effective environmental protection. But when Varney is asked how much authority he's got to rework EPA's rigid regulatory rules so that New England can focus on its most pressing problems, he replies, more than a bit ruefully, "not enough."

Varney's not the type to give up on trying new approaches. But doubt is growing in many quarters about the ability of government at all levels to tackle the nation's remaining pollution challenges. Critical policy decisions at the federal level have been mired in political deadlock for the past decade. And even as states have begun using some new and nimbler tactics to address environmental threats, many regulators feel they have made only marginal progress and are getting discouraged. It's becoming clear that the whole system needs to be overhauled before more substantial progress can be made.

In a 2000 report, a National Academy of Public Administration panel concluded that the nation's present pollution-control system "will not solve the most pressing of the nation's outstanding environmental problems." Mary A. Gade, Illinois' former environmental protection director and a member of the NAPA panel, says, "It's not that the current system hasn't done good things, but now it's just limping along."

In addition to atmospheric pollution, the nation's unresolved environmental issues include the consequences of traffic congestion, suburban sprawl and contaminants that flow from farmlands, lawns and public parks to choke rivers, lakes and coastal bays. EPA's prescriptive top-down procedures may work fine to curb industrial emissions, but they're not adaptable enough to manage these widely dispersed—but cumulatively severe—environmental hazards.

The NAPA panel recommended that whoever took office as EPA administrator in 2001 break new ground with innovative campaigns to curb greenhouse gases, reduce urban smog and keep fertilizers from polluting waterways. If anything, though, national environmental policy has become even more bogged down in partisan strife. Last year, Bush pulled back from a campaign pledge to implement a "four-pollutant program" to simultaneously curtail utility discharges of smog-forming nitrogen oxide, sulfur dioxide, toxic mercury and carbon dioxide.

And his EPA administrator, former New Jersey Governor Christine Todd Whitman, has moved painfully slowly despite being handed the best chance in years to transform EPA's approach and turn state governments loose to deal with new kinds of challenges. So to the extent that they are able, governors, legislators and state regulators have begun taking matters into their own hands, crafting laws and regulations on threats from hog farms to global warming.

Oregon imposed $CO_2$ limits on newly built power plants five years ago. Illinois, New York, and Connecticut have also begun fashioning multi-pollutant programs to cut $NO_x$ and $SO_2$ below what federal rules require. This June, North Carolina Governor Mike Easley signed a law that requires the state's 14 coal-fired powerplants to cut smog-forming $SO_2$ emissions 78 percent by 2009 and $NO_x$ emissions 74 percent by 2013. With federal action on nitrogen oxides stalled in legal maneuvering, "quite frankly, we felt we needed to do it ourselves," says Senator Stephen M. Metcalf, one of the North Carolina law's sponsors.

Through a variety of measures, state governments are forging ahead to inventory the greenhouse gases they create and design some ways to curb them. Minnesota and Montana provide incentives to plant trees; Nebraska is exploring farming methods to lock carbon into soil; North Carolina and Wisconsin are working on controlling methane from manure collecting in animal feedlots. During Whitman's tenure as governor, New Jersey incorporated greenhouse gases in industrial permits and negotiated an agreement with the Netherlands for trading $CO_2$ emission cutbacks. New Jersey is on track to meet its 2005 goal of cutting greenhouse gases 3.5 percent below 1990 levels.

On the other hand, 16 state legislatures went on record opposing U.S. ratification of the Kyoto Treaty on climate change that the Clinton administration negotiated. The Michigan legislature prohibited state agencies from drafting greenhouse-gas-reduction plans, and West Virginia adopted a law barring the state from working with EPA on voluntary climate-change programs. Vice President Dick Cheney is widely assumed to have taken charge of how environmental policy influences energy production, and environmental groups have concluded that "clearly, Whitman's an absentee landlord," says Frank O'Donnell, a lobbyist for the Clean Air Trust.

When the White House picked Varney to take over EPA's office in Boston and several other experienced state regulators moved into EPA regional posts, it seemed to signal a changing of the old guard. And Whitman and her aides say they want the new regional administrators to work hand-in-hand with state pollution commissioners to implement innovative regulation. But so far Whitman hasn't followed through by instructing EPA's career staff to give states the benefit of the doubt when they come up with new ideas for meeting pollution goals. "That would take an incredible commitment to change the bureaucracy, and I think the administrator's talking more about it than she's committed to it," says Russell J. Harding, Michigan's environmental quality director.

In fact, not much has changed in the way that EPA operates since President Nixon created the agency to administer the federal environmental standards imposed in the 1970s. Congress enacted the Clean Air Act, the Clean Water Act and several follow-up laws because it was clear that state health authorities weren't protecting the public against raw sewage and industrial discharges into the nation's water and air. In most cases, the statutes direct EPA to issue permits that specify what equipment a facility

# Cities Take the Sewer Plunge

When it comes to wastewater, the U.S. Environmental Protection Agency isn't hesitating to tell local governments what they have to do.

Municipal wastewater agencies have spent billions of dollars on new sewer lines and improved wastewater-treatment plants to comply with standards that EPA drafts and imposes. EPA regulates most municipal pollution sources by requiring local government agencies to get federal discharge permits—the same as factories, power plants and other industrial facilities. Now, federal engineers and lawyers are pushing local governments hard to fix leaky or overloaded systems that spill raw sewage from time to time and violate federal water pollution standards.

In April, for instance, Baltimore agreed to pay a $600,000 EPA fine and spend $940 million over the next 14 years to correct defective sewer pipes and connections that discharge untreated wastes into the Chesapeake Bay. Baltimore's chronic overflows have released more than 100 million gallons of raw sewage over the past six years, and addressing the problem will force governments to double the rates for 1.6 million users of the regional sewage system. "We're not the first city to be whacked by the federal government on this, and we won't be the last," Baltimore Mayor Martin O'Malley said when announcing the settlement with EPA and U.S. Department of Justice environmental lawyers.

Around the country, EPA is using its water-quality authority to force localities to raise billions more dollars to make sure that their treatment systems can handle surges of sewage in even the foulest weather. After spending $1.7 billion to upgrade its system to meet federal standards, Cleveland's regional sewer district expects it will have to commit another $1 billion or so to prevent overflows from antiquated combined sewers that collect both sewage and stormwater running into city streets. The mixed streams both funnel through treatment plants; rainstorms or sudden snowmelts can overwhelm the system, carrying raw sewage into the Cuyahoga River and its tributaries. EPA has begun using its enforcement powers to get tough with mayors and city councils that balk at the cost of remedying the problem.

More than 700 U.S. cities and towns still rely on combined sewer systems built as long as a century ago—before sewage was handled separately and treated before being discharged. In 1994, EPA approved new rules for combined sewer systems that will require local governments to correct the problem at a cost of at least $45 billion, and probably much more. It would be prohibitively expensive to tear up streets to install completely separate systems, so most cities are building huge underground tunnels to collect and hold stormwater surges and then release the water slowly when treatment plants can handle the load.

Atlanta, for instance, plans to bore three tunnels a combined 20 miles long, 100 feet beneath the city's streets. The tanks could store 300 million gallons of tainted water, then funnel it into two new treatment plants. Atlanta's average wastewater bill is expected to climb from $31 to about $65 a month to pay for the system. Threatened with $275 million in EPA fines, the Pittsburgh regional sewage system is working on a $3 billion project. EPA and the U.S. Department of Justice have also reached settlements that order Boston, New Orleans, San Diego, Honolulu, Miami, Cincinnati and Mobile, Alabama, to correct chronic sewage overflows.

What's more, EPA is following up rules for combined-sewer overflows with new regulations requiring local governments to curb overflows from aging sanitary sewers that aren't connected to stormwater systems but nonetheless are springing leaks or can be inundated by surges after rainstorms. That has become a problem for cities that grew rapidly after World War II and used brick or unreinforced concrete pipes that are now prone to corrode or collapse, creating blockages that back up into homes or spill onto streets. To correct its overflows, Oklahoma City has begun replacing 1 percent of its 2,200 miles of sewer lines every year with more durable PVC plastic pipe. Fairfax County, Virginia, has reduced overflows by two-thirds through stepped up monitoring, maintenance and repairs to its sewage lines.

In addition, EPA requires municipalities with 100,000 or more residents to get federal permits for the runoff they discharge from streets into separate stormwater-collection systems. They must devise plans to control the contaminants that run off the surface into the system. A pending set of regulations for smaller communities will set forth best management practices they'll be required to adopt to limit polluted runoff.

Mayors contend that EPA ignores what meeting all those requirements will cost the ratepayers whose monthly bills must pay off construction loans and cover operating expenses of upgraded sewage systems. In the 1970s, the federal government picked up 80 percent of the cost of the first round of sewage-treatment improvements through direct grants, but now it offers low-interest loans that sewage systems must pay back. That is reviving local officials' complaints from a few years ago about unfunded environmental mandates. As Baltimore's O'Malley contends, "Unfortunately, our federal government is a lot better at sending lawyers to cities than they are in sending dollars."

In some cases, EPA has agreed to stretch out deadlines for controlling sewage overflows. Three years ago, EPA's Boston regional office negotiated a compromise two-step program for correcting combined-sewer overflows into the Merrimack River from Manchester, New Hampshire. The city agreed to spend $52.4 million over 10 years on projects that will capture 93 percent of wet-weather overflows. EPA and the city will monitor the improvements during that time before settling on whatever additional steps will be necessary. As part of the deal, Manchester is also investing $5.6 million on preserving rare and sensitive swamps, controlling streambank erosion and other related projects.

EPA has made tentative moves toward linking sewer controls to its efforts to deal with upstream contamination from septic tanks, animal feedlots and other "non-point" sources. On the Charles River in Boston and Cambridge, Massachusetts, for instance, the New England regional office launched a coordinated Clean Charles 2005 program that has brought citizens groups into a project to make the river suitable for swimming. The program used traditional permit enforcement to target combined sewer overflows and industrial discharges. But intensive monitoring by volunteers also found that part of the river's pollution was coming from sewage pipes clogged with grease and from illegal hookups that piped raw sewage straight into stormwater drains.

—*Tom Arrandale*

must install to meet federal standards. EPA or state regulators conduct periodic inspections to make sure factory managers are keeping the controls operating, and government lawyers enforce those limits by levying fines for violations and sometimes pressing legal charges in court.

EPA's one-size-fits-all technology standards brought major sources of contamination under control. But uniform rules give businesses no incentive to develop cheaper controls that cut emissions more than the federal government requires. In addition, it's simply not practical for government regulators to issue permits and then inspect for millions of small, widely dispersed sources that cumulatively cause serious problems.

This statutory scheme has other troublesome consequences. Power to write and enforce standards remains concentrated in semi-independent air, water, drinking water and waste offices inside EPA headquarters. Parallel offices in EPA's 10 regions distribute federal grants to state agencies, then keep an eye on how closely they follow permitting and enforcement procedures.

Although it made sense administratively when separate federal pollution laws first went onto the books, EPA's media-by-media specialization inhibits its ability to respond holistically to today's bigger environmental picture. The stovepipe structure fragments decision-making authority and keeps headquarters, regional and state program staffs focused on following by-the-book procedures to enforce "end-of-pipe" standards for what regulated facilities discharge. EPA's organization "deals poorly with complex, multilayered environmental and economic problems," the NAPA panel's report found. "As constituted, the agency cannot possibly be the protector of the nation's environment, despite the expansive responsibility implied by its name."

States now have established full-fledged regulatory agencies that assumed most day-to-day responsibility for making sure national standards are met. State agencies write 90 percent of pollution permits, handle three-quarters of enforcement actions and cover nearly 75 percent of the cost of pollution regulation. But EPA hasn't adjusted to the states' increasing competence. Nearly half of EPA's 18,000 employees are still based in the regional offices. State agencies complain that EPA headquarters and the regional staffs meddle too often, second-guess too much and balk when states come up with new approaches they think will produce better results.

"EPA looks at an issue as a question of air or water or toxic waste, but states look at it as a question of what it does to community X or river valley Y," says University of Wisconsin political scientist Donald F. Kettl, a NAPA panel member. "The difference has been extremely difficult for EPA to address. It's one of the hardest problems we've got in government."

As part of the Clinton administration's reinventing government initiatives, EPA Administrator Carol M. Browner approved experimental programs that granted leeway for state agencies and some industries to try "cleaner, cheaper, smarter" compliance mechanisms if they pledged to achieve better—not lesser—pollution-control performance. EPA regional offices and the Environmental Council of the States, whose members are state pollution-control chiefs, also negotiated "performance partnership agreements" designed to give state agencies more flexibility to identify neglected problems and figure out ways to correct them.

In the 1990s, EPA's New England region demonstrated the potential for innovative thinking. John DeVillars, Varney's predecessor at the Boston regional office, broke up the region's separate air, water and waste offices and assigned nearly half the staff to new ecosystem protection and environmental stewardship units responsible for dealing with all forms of pollutants. The reorganization also set up state units within the EPA region to work directly with the six state environmental agencies.

The Massachusetts Department of Environmental Protection, meanwhile, persuaded EPA to accept a groundbreaking strategy that has brought emissions from 2,300 dry cleaners, printers and photo-processing shops under government supervision for the first time. Working with small business trade associations, the Massachusetts Environmental Results Program drafted workbooks that explain potential problems, including some in Korean aimed at the state's large number of Korean-owned dry cleaners. It holds owners personally responsible for auditing their discharges and certifying they're complying, and the DEP staff double-checks compliance by inspecting a percentage of the businesses. "We've had great success when we're able to show EPA that we can achieve equivalent or superior results with less process," says Lauren Liss, the current Massachusetts environmental protection commissioner.

Other states have begun taking advantage of self-regulating audits by private firms that are adopting environmental management systems to account for wasteful pollution discharges. New Jersey replaced separate air, water and waste permits with facility-wide permits for companies that demonstrate superior pollution-control results. Wisconsin and Oregon are working on "performance track" compacts with conscientious companies. The firms commit to seeking superior results, and state regulators stand back and let plant managers devise the most efficient ways they can get there.

It's taken so much effort to put new approaches in place, however, that some state officials have all but given up. In retrospect, "we all thought we could innovate more than we have, that it would be an easier thing to do," says Robert E. Roberts, a former state environmental secretary and ECOS director who recently took over as EPA's Rocky Mountain regional administrator. Before launching successful experiments, state officials have to brave a gauntlet of EPA program officials and enforcement lawyers who resist procedural changes that they fear will weaken federal enforcement authority. Says Michigan's Russell Harding, "anything you do is going to be second-guessed, and the lawyers are never going to be happy."

As a result, Harding adds, "some innovations may be more cumbersome than the program we had before." Since 1995, 35 state pollution-control agencies have signed performance partnership agreements with EPA regions that state leaders hoped would clear the way for more experimentation. But Delaware canceled its agreement because EPA's staff fought so tenaciously to keep channeling funds through their own air, water and waste programs. "It was almost like the United Nations Security Council—each individual program could veto something," says Robert Zimmerman, Delaware's deputy natural resources director. Instead of combining regulatory efforts in sensible ways, the partnership "almost became an add-on to what we were already doing."

Whitman's staff wants to reinvigorate the partnership system and is working to integrate state agencies into EPA's nationwide planning process. The agency also is drawing on state expertise to draft a nationwide state of the environment report to set benchmarks to judge future progress. And Whitman is reconsidering the Clinton administration rule to tighten EPA supervision of new state programs for controlling polluted runoff to rivers and lakes.

Nevertheless, more than a few state officials say EPA's entrenched staff goes out of its way to sabotage state initiatives they think stray too far from tried-and-true procedures. Of course, rank-and-file EPA regulators think they're simply enforcing federal pollution control laws the way Congress wants them to. Harding, Michigan's DEQ director, thinks it would take congressional legislation endorsing regulatory innovation to change the agency's antagonism to state inventiveness. Gade, a former EPA attorney, suggests going further—abolishing the regional offices and using resources the agency now spends overseeing the states to perfect scientific research and data-processing technology to accurately target priority environmental problems.

"The states in the past 10 years have come into their own and are capable of doing virtually everything," Gade says. "It's time to say, 'What new tools do we have available?' You could see a federal-state partnership that's equipped to deal with the challenges of this century in a collaborative, creative and flexible way that really could accomplish environmental goals."

From *Governing*, August 2002, pp. 22-26. © 2002 by Governing. Reprinted by permission.

Article 11

# Human Modification of the Geomorphically Unstable Salt River in Metropolitan Phoenix

**Martin Roberge**
*Towson University*

In-stream gravel mining, massive bridge piers, and channelization have all contributed to the geomorphic instability of the Lower Salt River channel in Arizona. Dam closure, changing dam operating rules, and the frequent modification of the channel bed have decreased our ability to predict the Salt River hydrology. Engineering practice has adapted to this situation and to a public that is increasingly intolerant of service disruptions by constructing larger bridges and extending levees. Building these larger structures may be counterproductive; future construction should not constrict the channel and should re-establish a braided river to decrease the energy available to the system. **Key Words: Arizona, fluvial geomorphology, human impacts, Salt River, urbanization.**

## Introduction

The city of Phoenix and the Lower Salt River, both located in central Arizona (Figure 1), compete with one another for space and materials. Phoenix alters the Salt River with its growing demand for water, land, river crossings, and construction materials. In turn, the river has altered the city, changing its course, scouring its bed, and repeatedly destroying bridges and structures along its banks. Bridge design, engineering practice, and management of the Lower Salt River have evolved against this background of rapid population growth and startling changes to the hydrology of the Salt River.

This investigation uses historical methods to review how planners and engineers around the Salt River have been forced to adapt to both changes in hydrology and changes in public opinion. Two related questions guide this research. First, how has the Salt River responded to increasing human modification of the river? Second, how has engineering practice adapted to these changes?

One pattern to emerge from this historical analysis is that human modifications to the Salt River have made the channel less stable and the hydrology more difficult to model or predict. Local governments have attempted to reduce the resulting uncertainty by completely reengineering the Salt River through channelization and by dramatically increasing the size of bridge designs.

## Study Site

### Description of Physical System

The Salt River basin drains the southern rim of the Colorado Plateau and covers an area of 50,436 km$^2$ (19,473 mi$^2$). The Lower Salt River, which starts at the Granite Reef Dam, east of Phoenix, divides the metropolitan area in half before joining with the Gila and Colorado Rivers. Since construction of a series of six dams upstream from Phoenix, the Salt River below Granite Reef Dam is a dry channel that receives water only during exceptional floods or from local storm runoff (Central Arizona Water Survey 1983; Figure 2). Infrequent high flows are typically contained within a 400-meter wide braided channel, while the occasional winter or spring low flows are conveyed in a meandering channel set into the larger braided flood channel. This compound channel form is typical of arid region and dammed rivers (Graf 1983, 1988a).

Floods along the Lower Salt River have had a history of being destructive, necessitating an array of flood-control struc-

# Article 11. Human Modification of the Geomorphically Unstable Salt River in Metropolitan Phoenix

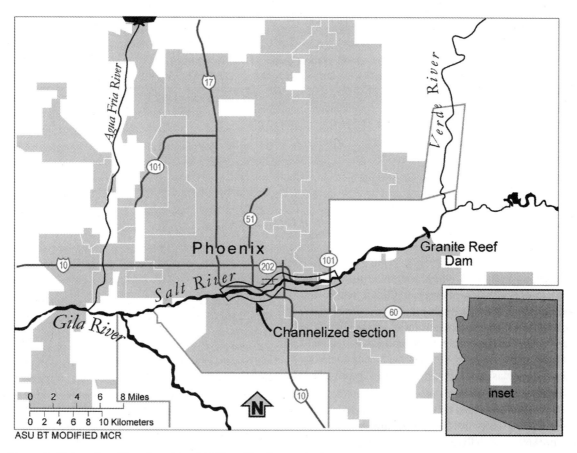

**Figure 1** *Metropolitan Phoenix and the Salt River. The Granite Reef Dam separates the Upper and Lower Salt River watershed. The Lower Salt River is dry in most years.*

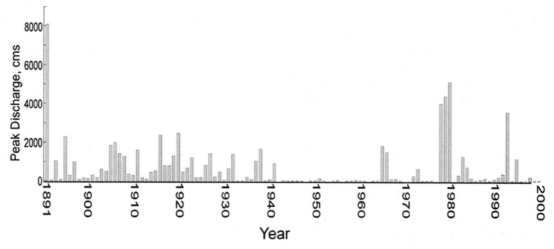

**Figure 2** *Annual peak flow at Granite Reef Dam, 1891–2000. Discharges were measured at Priest Drive Bridge after 1994.*

tures that can be astonishing to visitors familiar only with the dry riverbed. In the past forty years, at least two structures have failed in each of the last five major floods over 1,900 cubic meters per second (cms; equivalent to 67,000 cubic feet per second, or cfs; see Table 1). Except for the most recent flood in 1993, there have been major transportation disruptions in each of the these five flood events, requiring bridge closure and replacement at all but three of the approximately fourteen bridge crossing sites in the Phoenix area. Figure 3 presents the lifespan of various crossings in diagrammatic form. Lines terminating in an "X" represent bridges destroyed by a flood.

## River Development during the Hohokam and Territorial Periods

The Hohokam (850–1450 CE) made the earliest attempts to control the Salt River. Small diversion dams supplied several large towns and as much as 5500 ha of irrigated fields with water (Masse 1981). Although these dams supplied the largest

*Table 1* Damage in Phoenix Due to Salt River Flooding

| Date of Flood | Peak Discharge | Total Damage [a] | Number of Open Bridges | Number of Closed Bridges | Cost of Travel Delays reported [a] | Cost of Travel Delays calculated [c] |
|---|---|---|---|---|---|---|
| April 1965 | 1,897 cms | 31.8 | 3 | not reported (17)[b] | 1.7 | — |
| March 1978 | 3,450 cms | 63.6 | 2 | 9 | 1.1 | 11.36 |
| December 1978 | 3,964 cms | 103.0 | 2 | 10 | 31.7 | 20.73 |
| February 1980 | 4,813 cms | 103.7 | 2 | 10 | 17.6 | 17.85 |
| January 1993 | 3,500 cms | not reported[d] | 18 | 3 | — | — |

Source: U.S. Army Corps of Engineers (1966, 1979a, 1979b, 1981); Waters (1993).
[a] In millions of 2000 dollars, using annualized Consumer Price Index values to adjust reported values.
[b] Seventeen crossings were reported closed, including an unknown number of bridges.
[c] Calculated from lost work hours times $8.50 per hour (in millions of dollars).
[d] Not reported, but likely under 5 million dollars.

prehistoric irrigation system in North America, they are unlikely to have disturbed the flow of the river. For this reason, the natural flood regime of the Salt River may have resembled the discharges during the Hohokam and territorial eras. Under these nearly natural conditions, flow through the arid Salt River could occur year-round, but was highly variable, with occasional dry reaches and sudden, massive, turbulent floods that prevented navigation (Littlefield 1996).

Settlement of the Phoenix area by U.S. citizens occurred after Arizona was made a territory in 1863. At the time of Arizona's initial survey in 1868, the Salt River was a broad, braided channel lined with cottonwoods, willows, and occasional marshy land (Pierce and Engalls 1868). Soon after the survey, Anglo-Americans rebuilt ancient Hohokam dams and canals (Graf, Haschenburger, and Lecce 1988, 81; Zarbin 1997).

The highly variable discharge of the Lower Salt River is capable of mobilizing the riverbed under flows of only 350 cms (12,300 cfs) (Parker 1992) to 700 cms (25,000 cfs) (Graf 1983), potentially undermining any channel structures. This constant threat forced the Hohokam and Anglo settlers to rebuild their dams and canals repeatedly (Huckleberry 1997), and was responsible for at least three early railway bridge collapses at the most secure bridge site (Figure 4), in 1888, 1891, and 1905 (Lykes 1993). In 1891, just before the end of this period of relatively unaltered hydrology, the largest discharge on record, a flood of 8,500 cms (300,000 cfs), ripped through the Salt River channel at Phoenix (Figure 5).

## Dam Closure and Encroachment onto the Channel

The Theodore Roosevelt Dam was the first and largest of a series of irrigation and water supply dams built upstream from Phoenix around the time of Arizona's statehood, in 1911 (Table 2). Alone, the dam is capable of storing two years of the Upper Salt River's mean discharge (calculated from Central Arizona Water Survey 1983 and Federal Emergency Management Agency 1996); collectively, the storage dams upstream from Phoenix can store 2.8 billion cubic meters (2.3 million acre-feet) of water for irrigation. As more water is needed, it is released down the Salt River to the Granite Reef Dam, which diverts water into the Arizona and the Southern canal systems. On those rare occasions when the upstream dams release water too rapidly for the canals to accept, the excess pours over Granite Reef Dam and into the channel below. Although it is assumed that these dams will provide some relief from flooding, that is not their primary purpose.

During a twenty-four-year period of no flows through the Lower Salt River from 1941 to 1965, the Phoenix metropolitan

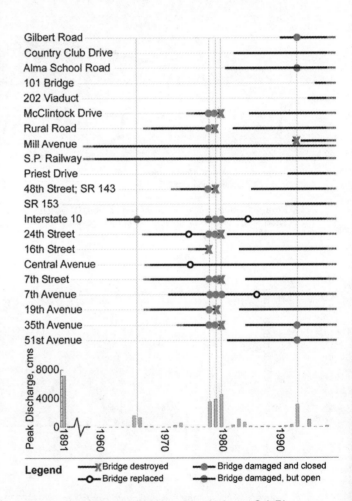

Figure 3 The lifespan of bridges along the Lower Salt River.

### Article 11. Human Modification of the Geomorphically Unstable Salt River in Metropolitan Phoenix

**Figure 4** *The Santa Fe Rail Road at Tempe, Arizona, c. 1888. This bridge site is one of the most secure because it is one of the only sites in the valley where it is possible for the piers to reach bedrock. Courtesy Luhrs Family Collection, Arizona Collection, Arizona State University Libraries. CP LFPC 425.*

**Figure 5** *The Santa Fe Rail Road at Tempe, Arizona, 1891. The largest flow on record once again destroyed all bridges over the Lower Salt River and flooded much of the valley. Courtesy Herb and Dorothy McLaughlin Collection, Arizona Collection, Arizona State University Libraries. MCL 97222.PHX39*

area grew tremendously in population and in infrastructure (Sargent 1988). Perhaps inevitably, some of this growth took place on the once dangerous but now dry Salt River. At the time of the 1965 floods (1900 cms; 67,000 cfs), the city of Phoenix only had one bridge designed for flows greater than 600 cms (20,000 cfs), and the south runway of Phoenix's major airport, Sky Harbor, extended 850 meters (2600 feet) into the river channel (U.S. Army Corps of Engineers 1966). In Tempe, a sewage treatment plant and other structures occupied 50 percent of the width of the Salt River, and in 1969, dikes pinched the channel at one point down to a clearly inadequate 13-meter (40-foot) opening (Ruff 1971).

### The 1978–1980 Floods

In March 1978, rainfall onto the snow pack in the upper Salt watershed produced a torrent of water that was unprecedented in the postdam period. Authorities closed all of the automobile bridges over the Salt River except for the Mill Avenue, Central Avenue, and Interstate 10 bridges (see Figure 6 for bridge locations). The flood caused 3.2 million dollars of damage to the south runway extension at Sky Harbor Airport, and damaged a parking lot at Arizona State University. The 16th Street Bridge was undermined by scour and then destroyed by the river. At the time, the 3,450-cms (122,000- cfs) flood was estimated to have a recurrence interval of slightly more than 40 years (U.S. Army Corps of Engineers 1979a).

In December 1978, winter rains again brought water to the dry urban reaches of the Salt River. In addition to the bridges that were still damaged from the earlier flood, two more bridges at 48th Street and 19th Avenue were damaged and later replaced with grade-level crossings, and I-10 was closed to traffic due to an undermined pier (U.S. Army Corps of Engineers 1979b). This round of closings effectively split the city in two, as thousands of workers were separated from their jobs by long commutes over the remaining bridges (Table 1).

The 4,000-cms (140,000-cfs) flow in December 1978 was followed a year later by a larger flood of 4,800 cms (170,000 cfs) in February 1980. This third flood crippled the city by wiping out twelve of the fifteen bridges over the Salt River, requiring major repairs or replacement in every case. Only three bridges remained open in 1980: the two-lane Mill Avenue Bridge, the Central Avenue Bridge, and the Southern Pacific Railroad Bridge. A fortunate turn of events allowed the I-10 bridge to reopen after a closure of only thirteen days, when a large scour hole was largely refilled during the waning stages of the flood. For the two weeks that the interstate was closed, Phoenix enjoyed a never-to-return vision of what a commuter rail system might look like, as the rail bridge was used to ferry commuters back and forth across the river for their jobs. The Central Avenue Bridge, built in 1975, probably survived after repairs because it was designed to withstand floods of up to 5,700 cms (200,000 cfs; U.S. Army Corps of Engineers 1981).

The bridge design process for the Lower Salt River transformed radically within weeks of the 1980 flood. A Governor's Special Task Force made recommendations that the Salt River be channelized from McClintock Drive in Tempe to 40th Street in Phoenix, and that Roosevelt Dam be raised to provide flood protection (Flood Control District of Maricopa County 1994). Within a month of the 1980 flood, the Army Corps of Engineers revised the standard flood frequency estimates upward using a new method that included all of the floods on record, modified by a process model of how the flow would have been affected had the existing dams been in place. These methods transformed the 1980 flood from a 100-year flood to a 65-year flood (Figure 7; Leach 1980a).

Since 1980, construction in the Lower Salt River has proceeded at a rapid pace; however, most structures are now designed to withstand floods as large as the 1980 flood. The added costs associated with the larger structures have been covered by

## ANNUAL EDITIONS

**Table 2** History of Dam Closure along the Salt River and Tributaries

| Dam Name | Other Name | Year | Length[a] | Height[a] | Storage[b] |
|---|---|---|---|---|---|
| Granite Reef Dam | | 1908 | 344 | 9 | 0.9 |
| Roosevelt Dam | | 1911 | 220 | 87 | 1,917.5 |
| Roosevelt Extension | | 1996 | | +21 | |
| Mormon Flat Dam | Canyon Lake | 1925 | 116 | 71 | 83.9 |
| Horse Mesa Dam | Apache Lake | 1927 | 201 | 94 | 322.2 |
| Stewart Mountain | Saguaro Lake | 1930 | 384 | 65 | 87.6 |
| Bartlett Dam[c] | | 1939 | 244 | 87 | 219.7 |
| Horseshoe Dam[c] | | 1945 | 457 | 44 | 162.0 |
| Orme Dam[d] (proposed for flood control) | | ~~1983~~ | 1,737 | 59 | 2,034.5 |

*Source:* Federal Emergency Management Agency (1996).
[a] *In meters.*
[b] *In millions of cubic meters.*
[c] *On the Verde River, a major tributary of the Salt River.*
[d] *Bureau of Reclamation (1976).*

the 1978–1980 disaster assistance (Sowers 1980), and by a greater reliance on state and federal money.

## Discussion

### Aggressive versus Conservative Approaches to Development

Since the completion of Roosevelt Dam in 1911, there have been two opposing approaches to developing the once braided channel of the Lower Salt River. The more conservative of the two approaches has been to leave the former channel area undeveloped, or to at least design structures in the anticipation that not every flood will be contained upstream. This viewpoint was difficult to maintain in the face of the 1941–1965 no-flow period. The alternative approach has been to develop the bed of the Lower Salt River. This approach has been strengthened by the usefulness of the Salt River as a corridor through Phoenix and as a source of cheap, centrally located land. Power lines, highways, and airplane approach corridors all follow the riverbed. Other uses include gravel mines, sewage treatment plants, land-fills, parking lots, government office buildings (including the Maricopa County Flood Control offices), and now the Tempe Town Lake, created in the riverbed using inflatable dams and a clay liner.

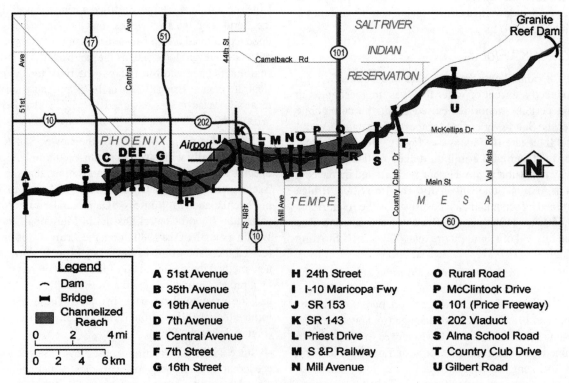

**Figure 6** *Structures along the Lower Salt River. Channelized sections are indicated with a box. Refer to the legend to find the name of each bridge or crossing.*

# Article 11. Human Modification of the Geomorphically Unstable Salt River in Metropolitan Phoenix

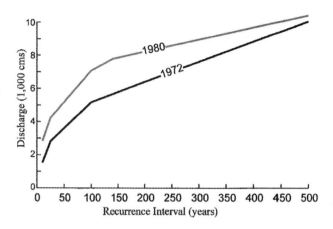

**Figure 7** Two flood-frequency diagrams for the Lower Salt River. The black line indicates a flood-frequency diagram established in 1972; the gray line is the official flood-frequency curve released in 1980, after a major flood. Source: Central Arizona Water Survey (1983).

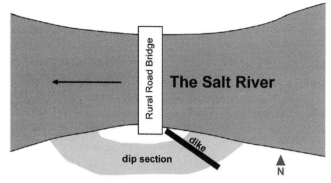

**Figure 8** A schematic drawing of the Scottsdale Road Bridge shortly before its destruction in 1980. Note the dip section on the south end of the bridge. The dike was added later to prevent water from flowing through this section, and instead funneled the water under the span. Unfortunately, the span was designed for a discharge of only 1133 cms (40,000 cfs). Based on aerial photography taken 15 December 1979.

The Scottsdale Road Bridge, built in 1970, is a good example of the interplay between the conservative and aggressive development forces. This bridge typified the bridges designed at the time. It was designed to handle a flood of only 1133 cms (40,000 cfs) and had a bridge span of only 213 meters (650 feet). Like all of the bridges across the Salt River except for those near Mill Avenue, its footings were set into alluvial fill, since bedrock is too deep to reach at this site. To provide for large floods, a dip was built into the road approaching the bridge, allowing water to flow over the approach. However, by early 1978, Maricopa County had built low dikes across this dip in order to keep the bridge open as much as possible during high flows (Figure 8). This strategy disabled the bridge in 1978, when the dikes rerouted the two floods of 1978 (3,450 cms and 3,964 cms) through the span, thereby scouring a large hole to within a foot of the pier footings (Bridge Scour Committee 1979). In response, it was recommended that Maricopa County remove the dikes and double the length of the span. Before this could happen, another larger flood destroyed what remained of the bridge in 1980.

## Bridge Design and Public Outrage

As in the Scottsdale Road example, the majority of bridges and river crossings built between 1945 and 1980 were intentionally designed to be small. During most years, culverts were sufficient to protect a road from any local runoff. Planners also reasoned that it made no sense to build a bridge that could withstand a 100-year flood when it would have to be replaced in 50 years to accommodate more traffic. A study commissioned by Phoenix after the first flood in 1978 balanced the potential costs of various sized bridges against the possibility that they might be damaged in a flood (Advance Transportation Planning Team 1978; Figure 9). Despite the closure of all but three river crossings in the floods of 1965 and 1978 (U.S. Army Corps of Engineers 1966, 1979a) it was recommended that major river crossings still be designed to withstand floods with only a thirty-year recurrence interval (Advance Transportation Planning Team 1978).

The public disagreed with this strategy. Massive traffic delays at the two remaining auto crossings in December 1978 and February 1980 separated thousands of workers from their jobs and cost an estimated $20.4 million dollars in lost time ($49.3 million in 2000 dollars; U.S. Army Corps of Engineers 1979b, 1981; Table 1). The two-week traffic slowdown was not acceptable to the public; it was no consolation that the chance that it would happen again was less than 2 percent in any given year. Instead, once the public and elected officials understood the consequences of bridge failure, it seemed preferable to spend the extra money and to build monolithic bridges rather than face the traffic delays again. Robert C. Esterbrooks, the Maricopa County Engineer, recommended that longer bridge spans be used at river crossings, because he said he was tired of receiving criticism about damage to bridges that had deliberately been designed to withstand only smaller floods (Leach 1980b). In the immediate aftermath of the 1980 flood, several bridges in the planning phases were quickly redesigned to handle much larger floods of 5,700 cms (200,000 cfs) (Table 3). Although a flood of this size is roughly equivalent to the 1972 estimates of the 100-year flood, the new 1980 estimates placed the 100-year flood much higher (Central Arizona Water Survey 1983; Figure 7).

The 202 Viaduct, built in 1995, is an extreme example of the new design standards being applied to construction. This massive bridge has a span of 1.6 km, and is supported by 182 three-meter wide piles that extend 40 m below the surface. Construction of the 202 Viaduct cost $38 million dollars (Walsh, Schock, and Jimenez 1996).

## The Lower Salt River Still Poses a Threat

The remaining destructive power of the Lower Salt River is surprising, considering the increasing number of control structures and the added ability to regulate the discharge of the Salt River. Uncertainty still confronts the design process along the Lower Salt River, notwithstanding the fact that we now have over a hundred years of experience in working with the river. Despite a history of attempts to control it, the Lower Salt River remains an uncontrolled river.

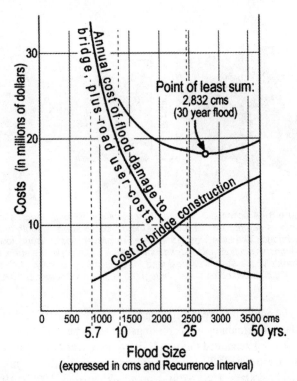

**Figure 9** A study on the optimal size for a bridge over the Salt River. Adapted from Advance Transportation Planning Team (1978).

Why has our increased knowledge and ability to regulate the flow of the Salt River been unable to prevent the regular destruction of bridges and levees? In the following two sections, this article describes two possible explanations for this fact. First, the gravel mines, protective structures, and levees built in the Lower Salt River have had individual impacts on the flow of the river that have collectively destabilized the channel, allowing the river to move laterally and to scour its channel more deeply. Secondly, the proliferation of control structures along the Salt River has made the hydrology of the total system less predictable. The most common methods used to predict the future behavior of the river are based on previous behavior. These empirical methods are inappropriate in a system where constant modifications to the channel and human participation in the behavior of the system render the concept of an "equilibrium" condition meaningless and the use of statistical averages misleading.

## Argument One: Development Has Decreased Channel Stability

In-stream gravel mining, scour of the channel bed, and channelization all threaten the stability of the Lower Salt River. The gravel mines, channel structures, and levees in the Lower Salt River channel each affect the hydraulics of the river in ways that can increase erosion downstream. As development in the channel bed intensifies, structures are built closer to one another and start to affect each other. The resulting physical environment is more mobile and less stable than it was before. This instability makes the Lower Salt River a more difficult place to build a structure.

### In-Stream Gravel Mining

In-stream gravel mining is a major threat to structures in the Salt (Li et al. 1989) and other rivers (Bull and Scott 1974; Kondolf and Swanson 1993; Kondolf 1994). River channels provide cheap, clean aggregate that is especially easy to remove in a dry channel such as the Salt River. Unfortunately, they also cause lateral channel migration (Mossa and McLean 1997). A report for the Arizona Department of Transportation (Bruesch 1980) linked the majority of damage to the I-10 bridge in 1978 and 1980 to an upstream mining pit owned by Tanner Industries, which had redirected the channel thalwag (the line of fastest water flow), causing it to flow out of alignment with the wall piers of the I-10 bridge. In-stream gravel mining can also cause areas upstream from the mining pit to degrade through headward erosion (Lee, Fu, and Song 1993). Figure 10 displays an example of headward migration caused in 1993 by a mining pit near the Alma School crossing.

Channel mines are regulated through a permitting system, but this system is not responsive enough to the concerns of bridge engineers. In the Tanner case, the Arizona Department of Transportation could do little about the mining occurring just upstream from the bridge right-of-way except to buy the mining rights to the property or to threaten Tanner Industries with a lawsuit (Bruesch 1980). Recently, the Maricopa County Department of Transportation was forced to buy the mining rights for the 1.1-kilometer (.7-mile) stretch of the river shown in Figure 10, which threatened the Alma School Crossing (Andrzej Wojakiewicz, Maricopa County Bridge Engineer, personal communication, 21 May 1999). Channel mines still pose a threat in the downtown reaches of the Salt River, where five pits are located along the channelized river sections between the I-10 and 16th Street crossings, with more mines located west of 19th Avenue. One gravel mine, located near 20th Street, had a pit in 1999 that appeared from the levee to be deeper than the river channel. The levee at this site is susceptible to collapse due to piping, because the materials of the area are highly permeable, and the pit is separated from the channel by less than twenty meters.

### Channel Scour

A second major threat to the stability of the Lower Salt River channel comes from increased scour. Scour is the process whereby sediment is removed from the channel and transported downstream, resulting in net erosion at a site. Only after the 1987 failure of the Schoharie Creek/New York State Thruway bridge did scour start receiving more national media and research attention, despite the fact that it had long been acknowledged to be the leading cause of bridge failure (Harrison 1991). Recently, scour at bridges has been highlighted by a major engineering society as one of the most important research agendas (American Society of Civil Engineers and the Committee on Hydraulic Engineering Research Advocacy 1996). Despite this attention, scour is still poorly understood (Hoffmans and Ver-

### Article 11. Human Modification of the Geomorphically Unstable Salt River in Metropolitan Phoenix

**Table 3** Salt River Bridge Design Revisions Following the 1980 Flood

| Name of Bridge | Original[a] Price | Revised Price | Original Design Flood | Revised Design Flood |
|---|---|---|---|---|
| Country Club[b] | — | $5 million | 3,700 cms | — |
| 16th Street[b] | — | $6.6 million | 3,700 cms | 5,700 cms |
| 24th Street[c] | $2 million | $3 million | — | — |
| 19th Avenue[b] | — | $5.2 million | 3,700 cms | 5,700 cms |

[a] "Original" refers to design specifications from before the 1980 flood.
[b] Source: Sowers (1980).
[c] Source: Staff (1980)

heij 1997) and often receives only limited treatment in bridge design textbooks (e.g., Xanthakos 1994, 1995; Melaragno 1998).

Scour is difficult to study; the rising stages of a flood tend to excavate sediment, while the waning stages of a flood will often fill these holes in, minimizing the apparent extent of the scour. In-situ devices that measure scour in the Lower Salt River must survive bombardment by cobble-sized particles and water velocities estimated to reach up to 6 meters (18 feet) per second in narrow reaches during the 100-year flood (Michael Baker Consulting, Inc. 1997).

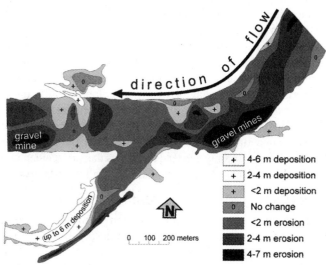

**Figure 10** A map of erosion and deposition in the Salt River Channel, 1992–97, between Alma School Road and Country Club Drive, Mesa. Erosion is shown in dark tones, while deposition is shown in light tones. In-stream mining pits may have been responsible for limiting flow into the south channel, deepening the upstream segment of the low-flow channel, and extending the upstream extent of the gravel pits. Modified from drawing by Mike Henze in Graf (1999).

Scour can be classified into two categories, general and local (Figure 11). General scour is the removal of sediment over a broad area, while turbulent forces around a structure cause local scour (Hoffmans and Verhij 1997). General scour may result from faster flows associated with channelization, clear water downstream from dams that has more potential energy available to move sediment, and by a sediment shortage that causes water to pick up more local sources of sediment. Local scour may result at locations where water is forced to speed up as it squeezes through a narrow space, such as the water rerouted under the Scottsdale Road Bridge in 1978. Scour may also be increased when water is not aligned with a solid wall-type pier, increasing turbulence as the water flows around the leading edge of the pier (Figure 12). This process was responsible for much of the damage to the I-10 Bridge in 1978 and 1980 (Bruesch 1980).

A number of structural solutions exist for the protection of bridges from flood erosion. On the Salt River, these typically add an additional 50 percent to the cost of a bridge (Andrzej Wojakiewicz, Maricopa County Bridge Engineer, personal communication, 21 May 1999). Revetments or rip-rap are massive, loose objects that are too large for the river to mobilize and that prevent scour of piers and channel walls. Drop structures are low steps placed in the channel bed that prevent the lowering of the channel bed upstream from the structure. Unfortunately, once the water passes over the drop structure, it "drops" to a lower level, picking up energy and increasing turbulence. Two doctors were drowned in the 1993 flood when they attempted to canoe over a low drop structure and were caught in a roll vortex (Waters 1993). This same turbulence can also contribute to downstream scour.

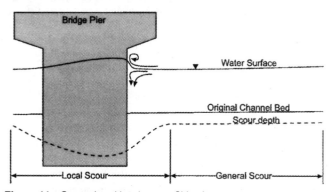

**Figure 11** General and local scour. Side view.

### The Downstream Effects of Channelization

A final protective measure against scour is channelization, in which a smooth, straight channel is built using levees on each side of the channel. There are a couple of reasons for channelizing a river. First, the channel is less likely to move laterally due to the levees. Second, the smoother, straighter channel provides less resistance to flow, reducing scour from turbulence.

However, channelization also has some well documented risks (Brookes 1988; Brookes and Gregory 1988; Goudie 2000, 215). The reduced friction in a channelized river results in faster flows. This is demonstrated in a model of the Lower Salt River (Figure 13), in which the average velocity of the flow tends to be higher in the channelized reach of the river, and the fastest flows occur under and downstream from bridges. Faster flows mean that the water can move more sediment, increasing general scour. The Rillito channel in Tucson has deepened significantly since its channelization (Graf 1984; Chang 1988, 379), and channelization of the Santa Rosa Wash south of Phoenix has led to widening and deepening of the wash (Rhoads 1990). Increased scour in the Lower Salt River would help to undermine bridge piers and could lead to bridge failure.

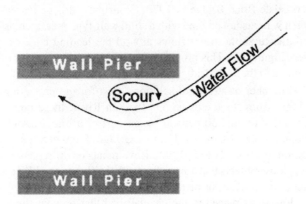

**Figure 12** *Local scour caused by nonalignment of flow. When the stream flow is shifted out of alignment with a wall pier, local scour can develop from turbulent eddies. Plan view.*

In the short term, faster flows within a channelized reach can lower the water depth for a given discharge. The city of Tempe used this principle to justify placing hotels and other buildings associated with the Rio Salado/Town Lake project adjacent to the newly channelized Salt River (CRSS Civil Engineers 1992). This occurred despite earlier studies based on scaled-down physical models of the river that indicated the levees might not survive the 100-year flood (Chen, Fiuzat, and Roberts 1985). Tempe's reliance on levees to lower the river's stage may be further called into question by recent research that suggests that flood stages along extensively engineered humid-region rivers may have risen two to four meters for a given discharge over the past century (Criss and Shock 2001).

A number of difficulties exist with channelization. First, the Tempe example illustrates how a levee may encourage development by fostering a new sense of security. Levees are still susceptible to scour and failure. The levees protecting the Sky Harbor Airport failed during the floods of 1965, twice in 1978, and in 1980. In 1993, this same section of levee was undermined by a flow of only 3500 cms (12,400 cfs), causing $2.1 million dollars in damage. The second concern over channelization is that the decision to construct such a channel structure is self-reinforcing. Downstream from the levee system, the rapidly moving water will be transferred into a rougher, unprotected channel. This could produce increased erosion or lateral movement of the channel. One solution to either of these problems is to extend the levees further. In 1980 after the floods, the old I-10 bridge was threatened by the newly built levee system protecting Sky Harbor Airport—so the levee was extended (Roberts 1980). Similar arguments may soon be used to protect the area between the 101/202 interchange upstream of the Alma School Road crossing. Gravel mines threaten this area, which has a new highway built along the channel banks. Extending the levees this far would create a 24-km channelized reach. Los Angeles has already gone through this process of levee extension and has re-engineered the Los Angeles River into a simple concrete trough (Cooke 1984; Gumprecht 1999). Before extending its levees, Phoenix must decide if it too wants to transform its river into a storm drain.

## Argument Two: Development Has Decreased Our Ability to Predict the System

Bridge design along any river depends upon construction of the "flood-frequency curve," which estimates the size (discharge) and frequency of floods for a given section of a river. This will determine the size of the "design flood," or the largest flood that a bridge should be expected to withstand. Once the size of the design flood is estimated, it is necessary to model the behavior of this imaginary flood. This will determine the depth and speed of the currents at the proposed bridge site, so that measures may be taken to protect the structure from scour.

The Lower Salt River is a difficult system to predict for two reasons. The first difficulty is that the discharge is controlled in part by the decisions of the dam operators and in part by natural processes. Second, behavior of a given flood is difficult to model due to constant modifications of the channel. The following paragraphs explain how these two issues translate into increased uncertainty in the minds of engineers as they design a bridge. This uncertainty can lead to designs that are accidentally too small, or to massive designs that negate all uncertainty.

### Establishing the Flood-Frequency Relationship

Uncertainty confronts the bridge design team at the outset, when the team must determine the size of the design flood. Typically, the design flood is set to the same size as the 100-year flood, so that there is only a 1 percent chance that the design flood will be exceeded in any given year. However, the concept of a 100-year flood requires rethinking in a system where variations in dam operating rules or mistakes in dam operation can dramatically alter the size of a flood. Once the Salt River was dammed, lower discharges were expected through the urban reaches of the river, but the exact size of this effect could not be predicted using an empirical relationship until a new, postdam record of flooding could be collected. After the disastrous floods of 1978 and 1980, the Army Corps of Engineers revamped these empirical flood frequency relationships by conducting a "what if" scenario. Using numeric process models, they estimated how the Salt River dams would have affected earlier floods and established a new flood-frequency relationship using the adjusted discharges. Unfortunately, it is difficult to assess the accuracy of the new model, since flows are so infrequent. In any case, both the old and the new flood frequency

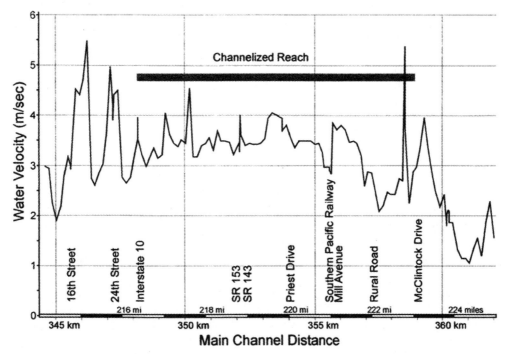

**Figure 13** *A plot of water velocity as calculated by HEC-RAS for the Lower Salt River. Source: Michael Baker Consulting, Inc. (1997).*

estimates (Figure 7) would predict more than the only three flows of over 1000 cms (35,000 cfs) that have occurred since 1980.

The predictive powers of flood-frequency diagrams and other empirical relationships suffer further when applied to a system that has the ability to "learn." While the data for a new, postdam record are being collected, dam operating procedures change as experience accrues and hydrologic models improve. During the 1980 floods, the Salt River Project (SRP), which operates the dams upstream from Phoenix, used a hydrologic model, HEC-5, to conduct 35 dam-release simulations (Salt River Project 1983), allowing them to select the optimal release schedule for that particular storm event. A statistical model cannot accurately portray such decision-making processes, because SRP presumably becomes better at optimizing the release schedule as the hydrologic models improve and as dam operators gain experience. Changes in dam operating procedures have meant that attempts to calculate flood recurrence intervals for the Salt River have not met with much success (Chin, Aldridge, and Longfield 1991, 71).

### Process Models and the Lower Salt River

Process models are also faced with significant challenges in an urban/human-modified river system. HEC-RAS is a program commonly used to model the height of the 100-year flood during bridge design. However, frequent construction in the channel of the Lower Salt River means that the expensive topographic surveys used to create a model in HEC-RAS are out of date soon after they are completed. In fact, the use of HEC-RAS in the highly mobile Lower Salt River may be problematic at the onset of a flood, since this model assumes that the channel floor does not change during a flood event (Graf 1988b). The Lower Salt River is fraught with uncertainty because the initial boundary conditions for process models are outdated before they can be tested against an actual (and rare) channel discharge (Roberge 1999, 53).

The infrequency of flows in the Lower Salt River has made the uncertainty in this system apparent, and yet most other urban river systems experience human-induced changes that are too rapid and exceed the ability of the river to "equilibrate." Human changes to rivers are cumulative, and do not allow a river to vary randomly around an equilibrium or average state. Despite these issues, concepts of equilibrium and "steady state" abound in models used throughout the United States.

## Conclusions

### The Competing Needs of Society

Societal and geomorphic forces interact with one another as they shape the Salt River. Demands for less-expensive bridges with shorter spans compete with the need to build larger structures to deal with increased uncertainty. It was not considered economical to build large bridges to prevent a rare two-week service disruption, so bridges were built to withstand only up to a 30-year flood (Figure 9 illustrates the reasoning behind this decision). In 1980, after thousands of workers were separated from their work for two weeks, planners began to value reliability more highly than economy. Trade-offs occur again when levees are built to meet the need for riverside land. Channelizing the river this way has increased the river's power and ability to threaten downstream structures. The ability to regulate larger flows using upstream dams has weakened the ability to predict the size and frequency of future downstream floods. In part, this may be linked to the competing demands placed on dams to provide both water supplies and flood control. Instream gravel pits are needed for new construction, but threaten the channel's stability near older structures. In the end, all of these competing demands originate in society—the river plays the role of mediator. In order to balance these competing demands ourselves, we must better understand how society and hydrologic systems interact with one another.

### Implications for Modeling an Urban Environment

To support the growth of Phoenix, the Salt River has been modified upstream by dams, and downstream by the placement of structures and mining pits within the channel. In response to these changes, the river has become more geomorphically active and more difficult to predict. The hydrology of the Lower Salt River is controlled by natural variations and by human behavior, resulting in changes over time that cannot be modeled accurately using static empirical relationships. The constant addition of drop structures, rip-rap, or channel mines means that there are more changes to the channel bed than there are opportunities for the channel to respond. Process models made for this environment must make predictions before they can be tested, and once tested will only remain valid until the next channel modification. Finally, turbulence initiated by new structures will produce scour around older structures that were not designed to anticipate these changes. These unintentional effects will become more common as the development of the riverbed becomes more intensive.

### Stability through River Engineering, or through River Planning?

Engineering practice has evolved over the years to incorporate the changing conditions of the Lower Salt River. The mobile channel and unpredictable conditions have led to massive bridge designs padded with a large margin of error to withstand the increased uncertainty. Bridges have become monolithic, designed to withstand scour from repeated large floods. Attempts have also been made to reduce uncertainty by confining the Lower Salt River within levees. This practice is unlikely to succeed, because it encourages construction near the river, and because it will lead to increased erosion downstream from the levees. As the levees are extended to deal with each new threat, the river is transformed into a simple storm drain.

Pressure to develop the Salt River bed will not go away. Instead, it must be directed and planned in such a way that it is still possible to maintain control over the flow of the river. The natural variability of a desert river does not pose a problem for society until structures have been placed in harm's way. In Phoenix, the cheapest strategy to prevent flood damage may be through a phased-in removal of structures placed too close to the channel.

## Literature Cited

Advance Transportation Planning Team. 1978. *Salt River Bridge concept study*. Phoenix: City of Phoenix.
American Society of Civil Engineers and the Committee on Hydraulic Engineering Research Advocacy. 1996. Environmental hydraulics: New research directions for the 21st century. *Journal of Hydraulic Engineering* 122 (4): 180–83.
Bridge Scour Committee. 1979. *A study of waterway bridges in Arizona with potential scour-related foundation problems*. Phoenix: Arizona Department of Transportation and the Federal Highway Agency.
Brookes, Andrew. 1988. *Channelized rivers: Perspectives for environmental management*. Chichester: John Wiley.
Brookes, Andrew, and Ken Gregory. 1988. Channelization, river engineering, and geomorphology. In *Geomorphology in environmental planning*, ed. J. M. Hooke, 145–167. New York: John Wiley & Sons Ltd.
Bruesch, W. R. 1980. *An evaluation of effects of excavations in the vicinity of the I–10 Salt River Bridge on the flow regime and local scour at the bridge*. Phoenix: Arizona Department of Transportation Highways Division Structures Section.
Bull, William B., and Kevin M. Scott. 1974. Impact of mining gravel from urban stream beds in the southwestern United States. *Geology* 2:171–74.
Bureau of Reclamation. 1976. *Draft environmental statement: Orme Dam and Reservoir, Central Arizona Project*, Arizona-New Mexico. Phoenix: Bureau of Reclamation.
Central Arizona Water Survey. 1983. *Central Arizona water control survey*. Phoenix: Army Corps of Engineers.
Chang, Howard H. 1988. *Fluvial processes in river engineering*. New York: Wiley.
Chen, Yung Hai, Abbas A. Fiuzat, and Benjamin R. Roberts. 1985. Salt River channelization project: Model study. *Journal of Hydraulic Engineering* 111 (2): 267–83.
Chin, E. H., B. N. Aldridge, and R. J. Longfield. 1991. *Floods of February 1980 in southern California and central Arizona*. Washington, DC: United States Geological Survey.
Cooke, R. U. 1984. *Geomorphological hazards in Los Angeles*. London: Allen and Unwin.
Criss, Robert E., and Everett L. Shock. 2001. Flood enhancement through flood control. *Geology* 29 (10): 875–78.
CRSS Civil Engineers, Inc. 1992. *Salt River channelization floodplain delineation study*. Phoenix: Rio Salado Task Force, City of Tempe.
Federal Emergency Management Agency. 1996. *National inventory of dams*. Washington, DC: U.S. Army Corps of Engineers.
Flood Control District of Maricopa County. 1994. *Annual report*. Phoenix: FCDMC.
Goudie, Andrew. 2000. *The human impact on the natural environment*. 5th ed. Oxford: Blackwell.
Graf, W. L. 1983. Flood-related channel change in an arid-region river. *Earth Surface Processes and Landforms* 8:125–39.
—. 1984. A probabilistic approach to the spatial assessment of river channel instability. *Water Resources Research* 20 (7): 953–62.
—. 1988a. Definition of flood plains along arid-region rivers. In *Flood geomorphology*, ed. V. R. Baker, R. C. Kochekl, and P. C. Patton, 231–42. New York: Wiley.
—. 1988b. Science, engineering, and the law on western Sunbelt rivers. *Journal of Soil and Water Conservation* 43 (3): 221–25.
Graf, W. L., Judith K. Haschenburger, and Scott A. Lecce, eds. 1988. *The Salt and Gila Rivers in central Arizona*. Department of Geography Publication no. 3. Tempe: Arizona State University.
Gumprecht, Blake. 1999. *The Los Angeles River: Its life, death, and possible rebirth*. Baltimore: Johns Hopkins University Press.
Harrison, Lawrence J. 1991. Federal Highway Administration bridge scour practice. *Transportation Research Record* 1290:212–17.
Hoffmans, G. J. C. M., and H. J. Verheij. 1997. *Scour manual*. Rotterdam: A. A. Balkema.
Huckleberry, Gary A. 1997. Abstract: Paleoflood impacts to prehistoric agriculturists in the Sonoran Desert. *Geological Society of America*, 1997 Annual Meeting, Salt Lake City 29 (6): 242.
Kondolf, G. Mathias. 1994. Geomorphic and environmental effects of instream gravel mining. *Landscape and Urban Planning* 28:225–43.

Kondolf, G. M., and M. L. Swanson. 1993. Channel adjustments to reservoir construction and gravel extraction along Stony Creek, California. *Environmental Geology* 21:256–69.

Leach, John. 1980a. Corps revises standards for measuring future Salt floods. *Arizona Republic* 9 March:A1.

—. 1980b. County engineer urges longer spans over Salt. *Arizona Republic* 4 March:A2.

Lee, Hong-Yuan, Deng-Tsuang Fu, and Ming- Huang Song. 1993. Migration of rectangular mining pit composed of uniform sediments. *Journal of Hydraulic Engineering* 119 (1): 64–80.

Li, Ruh-Ming, George K. Cotton, Michael E. Zeller, Daryl B. Simons, and Patricia O. Deschamps. 1989. *Effects of in-stream mining on channel stability.* Phoenix: U.S. Department of Transportation, Federal Highway Administration.

Littlefield, Douglas R. 1996. *Assessment of the Salt River's navigability prior to and on the date of Arizona's statehood,* February 14, 1912. Phoenix: Salt River Project.

Lykes, Aimee DePotter. 1993. A hundred years of Phoenix history. In *Phoenix in the twentieth century,* ed. G. W. Johnson, Jr., 222. Norman: University of Oklahoma Press.

Masse, W. Bruce. 1981. Prehistoric irrigation systems in the Salt River Valley, Arizona. *Science* 214:408–15.

Melaragno, Michele. 1998. *Preliminary design of bridges for architects and engineers.* New York: Marcel Dekker.

Michael Baker Consulting, Inc. 1997. *Flood insurance study.* Phoenix: Flood Control District of Maricopa County.

Mossa, Joann, and Mark McLean. 1997. Channel planform and land cover changes on a mined river floodplain. *Applied Geography* 17 (1): 43–54.

Parker, J. T. C. 1992. Channel change in desert rivers from moderate flows: Initial results of a monitoring program, Maricopa County, Arizona. *Eos* 73 (43): 226.

Pierce, W. M., and W. Engalls. 1868. *Plat map field* notes. Phoenix: Bureau of Land Management.

Rhoads, Bruce L. 1990. The impact of stream channelization on the geomorphic stability of an arid-region river. *National Geographic Research* 6 (2): 157–77.

Roberge, Martin Craig. 1999. *Physical interactions between Phoenix and the Salt River, Arizona.* Ph.D. diss., Department of Geography, Arizona State University, Tempe.

Roberts, Benjamin. 1980. *A physical model of the Salt River for Phoenix, with additional study for I–10 Bridge.* Phoenix: Arizona Department of Transportation.

Ruff, Paul F. 1971. *A history of the Salt River channel in the vicinity of Tempe, Arizona 1868–1969.* Tempe: Arizona State University.

Salt River Project. 1983. *Emergency preparedness plan for reservoir operations.* Phoenix: Salt River Project.

Sargent, Charles. 1988. *Metro Arizona.* Scottsdale, AZ: Biffington Books.

Sowers, Carol. 1980. Designers may alter 3 bridges. *Arizona Republic* 29 February:A1.

Staff. 1980. Construction of fortified bridge at 24th Street to begin in April. *Arizona Republic* 5 March:A1.

U.S. Army Corps of Engineers. 1966. *Flood damage report on flood of December 1956–January 1966.* Phoenix: Corps of Engineers.

—. 1979a. *Flood damage report: 28 February–6 March 1978 on the storm and floods in Maricopa County, Arizona.* Los Angeles: United States Army Corps of Engineers.

—. 1979b. *Flood damage report: Phoenix metropolitan area, December 1978 Flood.* Los Angeles: United States Army Corps of Engineers.

—. 1981. *Flood damage survey, Phoenix, Arizona, February 1980.* Los Angeles: United States Army Corps of Engineers.

Walsh, Kenneth D., Robert E. Schock, and Steven A. Jimenez. 1996. Riddle of the riverbed. *Civil Engineering* 66:64–67.

Waters, Stephen D. 1993. *After-action report: The floods of January 1993.* Phoenix: Flood Control District of Maricopa County.

Xanthakos, Petros P. 1994. *Theory and design of bridges.* New York: J. Wiley and Sons.

—. 1995. *Bridge substructure and foundation design.* Upper Saddle River, NJ: Prentice-Hall.

Zarbin, Earl. 1997. *Two sides of the river: Salt River Valley canals, 1867–1902.* Phoenix: Salt River Project.

---

MARTIN C. ROBERGE is an Assistant Professor of Geography at Towson University, Towson, MD, 21252–0001. E-mail: mroberge@towson.edu. His research interests include fluvial geomorphology, GIS, and the human impact on hydrologic systems.

# Article 12

## Texas and Water

# Pay up or dry up

Another state trying not to go thirsty

Austin

OIL built Texas, but water will shape its future. With five droughts in the past four years, water is running out fast. The Rio Grande, one of the state's main sources, has failed to reach the Gulf of Mexico for the first time in about 50 years. And demand is soaring, with the number of Texans expected to double by 2050. If this goes on, El Paso will run out of water in 20 years and other cities not much later.

Can the supply be increased? Texas has been squabbling with its neighbours about the division of the Rio Grande's water. There is talk of new reservoirs, but they may not be ready before El Paso and other cities have gone dry. Since surface water is so scarce, many people want to pump more water from underground aquifers. This would help—55% of the state's population already depends on such water for drinking and agriculture—but reliance on well water, if not handled carefully, could create vast environmental problems and leave the state with no water reserves.

So Texas is again wondering how to cut demand. Nobody can stop the state's population growing. But there are surely ways to limit the amount of water each Texan uses. Americans consume more water per head than most people, yet the price (a third of a cent per gallon) is lower than in most other rich countries. Water takes up less than 0.8% of the average American household budget. In theory, a price increase could achieve two things. It could raise money to help build better pipelines. It could also tame consumption, perhaps by as much as 20% in some areas.

Meanwhile, however, T. Boone Pickens, an erstwhile corporate raider, has been working on the supply side. He has been busy buying water rights in western Texas, setting up a consortium of landowners who have agreed to sell him their water at a given price. Mr Pickens hopes to pipe 150,000-200,000 acre-feet of water from the vast Ogallala aquifer in the Texas Panhandle to the state's parched cities. (An acre-foot, enough water to cover an acre a foot deep, is 326,000 gallons, or some 1.2m litres.)

The cities would pay on a sliding scale, depending on the distance Mr Pickens has to pump the stuff. Far-away El Paso would pay $1,400 per acre-foot, Dallas around $800. Much of the projected revenue of $200m would go into building pipelines to carry the stuff. But these are still stiff prices: the state of California recently paid $260 an acre-foot in a similar deal.

Environmentalists have reason to worry, too. The Ogallala replenishes itself at a rate of less than one acre-

## Article 12. Texas and Water: Pay up or dry up

foot a year. In these circumstances water extraction is little different from mining. Indeed, Texas law treats water much as it does oil or gold: anybody who has access to the stuff can extract as much as he wants.

In theory, such water is protected by Texas's Groundwater Conservation Districts. But the state's strong tradition of property rights limits the GCDs' ability to control men like Mr Pickens. C.E. Williams, general manager of the Panhandle's Groundwater Conservation District No. 3, which deals with Mr Pickens, thinks the current crisis justifies pumping up to 50% of the Ogallala. But Mr Pickens's methods, he argues, would empty it within 25 years.

Mr Williams is pinning his hopes on a bill that is currently being discussed in the state legislature. This would allow more GCDs to be established, and empower them to levy a fee of at least two-and-a-half cents per 1,000 gallons to study the effects of pumping and set up replenishment projects. The only trouble is, Mr Pickens may have already got himself exempted from the fee.

From *The Economist*, May 26, 2001, p. 33. © 2001 by The Economist, Ltd. Distributed by the New York Times Special Features. Reprinted by permission.

# Beyond the Valley of
# THE DAMMED

*A strange alliance of fish lovers, tree huggers, and bureaucrats
say what went up must come down*

**By God but we built some dams!** We backed up the Kennebec in Maine and the Neuse in North Carolina and a hundred creeks and streams that once ran free. We stopped the Colorado with the Hoover, high as 35 houses, and because it pleased us we kept damming and diverting the river until it no longer reached the sea. We dammed our way out of the Great Depression with the Columbia's Grand Coulee; a dam so immense you had to borrow another fellow's mind because yours alone wasn't big enough to wrap around it. Then we cleaved the Missouri with a bigger one still, the Fort Peck Dam, a jaw dropper so outsized they put it on the cover of the first issue of *Life*. We turned the Tennessee, the Columbia, and the Snake from continental arteries into still bathtubs. We dammed the Clearwater, the Boise, the Santiam, the Deschutes, the Skagit, the Willamette, and the McKenzie. We dammed Crystal River and Muddy Creek, the Little River and the Rio-Grande. We dammed the Minnewawa and the Minnesota and we dammed the Kalamazoo. We dammed the Swift and we dammed the Dead.

One day we looked up and saw 75,000 dams impounding more than half a million miles of river. We looked down and saw rivers scrubbed free of salmon and sturgeon and shad. Cold rivers ran warm, warm rivers ran cold, and fertile muddy banks turned barren.

And that's when we stopped talking about dams as instruments of holy progress and started talking about blowing them out of the water.

## BY BRUCE BARCOTT

Surrounded by a small crowd, Secretary of the Interior Bruce Babbitt stood atop McPherrin Dam, on Butte Creek, not far from Chico, California, in the hundred-degree heat of the Sacramento Valley. The constituencies represented—farmers, wildlife conservationists, state fish and game officials, irrigation managers—had been wrangling over every drop of water in this naturally arid basin for most of a century. On this day, however, amity reigned.

With CNN cameras rolling, Babbitt hoisted a sledgehammer above his head and—with "evident glee," as one reporter later noted—brought this tool of destruction down upon the dam. Golf claps all around.

The secretary's hammer strike in July 1998 marked the beginning of the end for that ugly concrete plug and three other Butte Creek irrigation dams. All were coming out to encourage the return of spring-run chinook salmon, blocked from their natural spawning grounds for more than 75 years. Babbitt then flew to Medford, Oregon, and took a swing at 30-year-old Jackson Street Dam on Bear Creek. Last year alone, Babbitt cracked the concrete at four dams on Wisconsin's Menominee River and two dams on Elwha River in Washington state; at Quaker Neck Dam on North Carolina's Neuse River; and at 160-year-old Edwards Dam on the Kennebec in Maine.

## Article 13. Beyond the Valley of THE DAMMED

By any reckoning, this was a weird inversion of that natural order. Interior secretaries are supposed to christen dams, not smash them. Sixty years ago, President Franklin D. Roosevelt and his interior secretary, Harold Ickes, toured the West to dedicate four of the largest dams in the history of civilization. Since 1994, Babbitt, who knows his history, has been following in their footsteps, but this secretary is preaching the gospel of dam-going-away. "America overshot the mark in our dam-building frenzy," he told the Ecological Society of America. "The public is now learning that we have paid a steadily accumulating price for these projects.... We did not build them for religious purposes and they do not consecrate our values. Dams do, in fact, outlive their function. When they do, some should go."

Many dams continue, of course, to be invaluable pollution-free power plants. Hydroelectric dams provide 10 percent of the nation's electricity (and half of our renewable energy). In the Northwest, dams account for 75 percent of the region's power and bestow the lowest electrical rates in the nation. In the past the public was encouraged to believe that hydropower was almost free; but as Babbitt has been pointing out, the real costs can be enormous.

> What we know now that we didn't know then is that **a river isn't a water pipe.**

What we know now that we didn't know in 1938 is that a river isn't a water pipe. Dam a river and it will drop most of the sediment it carries into a still reservoir, trapping ecologically valuable debris such as branches, wood particles, and gravel. The sediment may be mixed with more and more pollutants—toxic chemicals leaching from abandoned mines, for example, or naturally occurring but dangerous heavy metals. Once the water passes through the dam it continues to scour, but it can't replace what it removes with material from upstream. A dammed river is sometimes called a "hungry" river, one that eats its bed and banks. Riverbeds and banks may turn into cobblestone streets, large stones cemented in by the ultrafine silt that passes through the dams. Biologists call this "armoring."

Naturally cold rivers may run warm after the sun heats water trapped in the reservoir; naturally warm rivers may run cold if their downstream flow is drawn from the bottom of deep reservoirs. Fish adapted to cold water won't survive in warm water, and vice versa.

As the toll on wild rivers became more glaringly evident in recent decades, opposition to dams started to go mainstream. By the 1990s, conservation groups, fishing organizations, and other river lovers began to call for actions that had once been supported only by environmental extremists and radical groups like Earth First! Driven by changing economics, environmental law, and most of all the specter of vanishing fish, government policy makers began echoing the conservationists. And then Bruce Babbitt, perhaps sensing the inevitable tide of history, began to support decommissioning as well.

So far, only small dams have been removed. Babbitt may chip away at all the little dams he wants, but when it comes to ripping major federal hydropower projects out of Western rivers, that's when the politics get national and nasty. Twenty-two years ago, when President Jimmy Carter suggested pulling the plug on several grand dam projects, Western senators and representatives politically crucified him. Although dam opponents have much stronger scientific and economic arguments on their side in 1999, the coming dam battles are apt to be just as nasty.

Consider the Snake River, where a major confrontation looms over four federal hydropower dams near the Washington-Idaho state line. When I asked Babbitt about the Snake last fall, he almost seemed to be itching for his hammer. "The escalating debate over dams is going to focus in the coming months on the Snake River," he declared. "We're now face to face with this question: Do the people of this country place more value on Snake River salmon or on those four dams? The scientific studies are making it clear that you can't have both."

Brave talk—but only a couple of weeks later, after a bruising budget skirmish with congressional dam proponents who accused him of planning to tear down dams across the Northwest, Babbitt sounded like a man who had just learned a sobering lesson in the treacherous politics of dams. The chastened interior secretary assured the public that "I have never advocated, and do not advocate, the removal of dams on the main stem of the Columbia-Snake river system."

## Showdown on the Snake

Lewiston, Idaho, sits at the confluence of the Snake and Clearwater Rivers. It's a quiet place of 33,000 solid citizens, laid out like a lot of towns these days: One main road leads into the dying downtown core, the other to a thriving strip of Wal-Marts, gas stations, and fast-food greaseries. When Lewis (hence the name) and Clark floated through here in 1805, they complained about the river rapids—"Several of them verry bad," the spelling-challenged Clark scrawled in his journal. Further downriver, where the Snake meets the Columbia, the explorers were amazed to see the local Indians catching and drying incredible numbers of coho salmon headed upriver to spawn.

The river still flows, though it's been dammed into a lake for nearly 150 miles. Between 1962 and 1975, four federal hydroelectric projects were built on the river by the Army Corps of Engineers: Ice Harbor Dam, Lower Monumental Dam, Little Goose Dam, and Lower Granite Dam. The dams added to the regional power supply, but more crucially, they turned the Snake from a whitewater roller-coaster into a navigable waterway. The surrounding wheat farmers could now ship their grain on barges to Portland, Oregon, at half the cost of overland transport, and other industries also grew to depend on this cheap highway to the sea.

Like all dams, however, they were hell on the river and its fish—the chinook, coho, sockeye, and steelhead. True, some salmon species still run up the river to spawn, but by the early 1990s the fish count had dwindled from 5 million to less than 20,000. The Snake River coho have completely disappeared, and the sockeye are nearing extinction.

In and around Lewiston, the two conflicting interests—livelihoods that depend on the dams on the one side, the fate of the fish on the other—mean that just about everyone is either a friend of the dams or a breacher. The Snake is the dam-breaching movement's first major test case, but it is also the place where dam defenders plan to make their stand. Most important, depending in part on the results of a study due later this year, the lower Snake could become the place where the government orders the first decommissioning of several big dams.

In the forefront of those who hope this happens is Charlie Ray, an oxymoron of a good ol' boy environmentalist whose booming Tennessee-bred baritone and sandy hair lend him the aspect of Nashville Network host. Ray makes his living as head of salmon and steelhead programs for Idaho Rivers United, a conservationist group that has been raising a fuss about free-flowing rivers since 1991. At heart he's not a tree hugger, but a steelhead junkie: "You hook a steelhead, man, you got 10,000 years of survival instinct on the end of that line."

Despite Ray's bluff good cheer, it's not easy being a breacher in Lewiston. Wheat farming still drives a big part of the local economy, and the pro-dam forces predict that breaching would lead to financial ruin. Lining up behind the dam defenders are Lewiston's twin pillars of industry: the Potlatch Corporation and the Port of Lewiston. Potlatch, one of the country's largest paper producers, operates its flagship pulp and paper mill in Lewiston, employing 2,300 people. Potlatch executives will tell you the company wants the dams mainly to protect the town's economy, but local environmentalists say the mill would find it more difficult to discharge warm effluent into a free-flowing, shallow river.

Potlatch provides Charlie Ray with a worthy foil in company spokesman Frank Carroll, who was hired after spending 17 years working the media for the U.S. Forest Service. Frankie and Charlie have been known to scrap. At an anti-breaching rally in Lewiston last September, Carroll stood off-camera watching Ray being interviewed by a local TV reporter. Fed up with hearing Ray's spin, Carroll started shouting "Bullshit, Charlie, that's bullshit!" while the video rolled. Ray's nothing more than a "paid operative," Carroll says. Ray's reaction: "Yeah, like Frankie's not."

"A lot of people are trying to trivialize the social and economic issues," Carroll says, "trying to tell us the lives people have here don't count, that we'll open up a big bait shop and put everyone to work hooking worms. We resent that. Right now, there's a blanket of prosperity that lies across this whole region, and that prosperity is due to the river in its current state—to its transportation."

Ever since the dams started going up along the Snake River, biologists and engineers have been trying to revive the rapidly declining salmon runs. Their schemes include fish ladders, hatcheries, and a bizarre program in which young smolts are captured and shipped downriver to the sea in barges. By the late 1980s, it was clear that nothing was working; the fish runs continued to plummet. In 1990, the Shoshone-Bannock Indians, who traditionally fished the Snake's sockeye run, successfully petitioned the National Marine Fisheries Service to list the fish as endangered. Every salmon species in the Snake River is now officially threatened or endangered, which means the agencies that control the river must deal with all kinds of costly regulations.

In 1995, under pressure from the federal courts, the National Marine Fisheries Service and the Army Corps of Engineers (which continues to operate the dams) agreed to launch a four-year study of the four lower Snake River dams. In tandem with the Fisheries Service, the Corps made a bombshell announcement. The study would consider three options: maintain the status quo, turbocharge the fish-barging operation, or initiate a "permanent natural river drawdown"—breaching. The study's final report is due in December, but whatever its conclusions, that initial statement marked a dramatic shift. Suddenly, an action that had always seemed unthinkable was an officially sanctioned possibility.

Two separate scientific studies concluded that breaching presented the best hope for saving the river. In 1997 the *Idaho Statesman*, the state's largest newspaper, published a three-part series arguing that breaching the four dams would net local taxpayers and the region's economy $183 million a year. The dams, the paper concluded, "are holding Idaho's economy hostage."

"That series was seismic," says Reed Burkholder, a Boise-based breaching advocate. Charlie Ray agrees. "We've won the scientific argument," he says. "And we've won the economic argument. We're spending more to drive the fish to extinction than it would cost to revive them."

In fact, the economic argument is far from won. The *Statesman*'s numbers are not unimpeachable. The key to their prediction, a projected $248 million annual boost in recreation and fishing, assumes that the salmon runs will return to pre-1960s levels. Fisheries experts say that could take up to 24 years, if it happens at all. The $34 million lost at the Port of Lewiston each year, however, would be certain and immediate.

The Northwest can do without the power of the four lower Snake River dams: They account for only about 4 percent of the region's electricity supply. The dams aren't built for flood control, and contrary to a widely held belief, they provide only a small amount of irrigation water to the region's farmers. What the issue comes down to, then, is the Port of Lewiston. You take the dams out, says port manager Dave Doeringsfeld, "and transportation costs go up 200 to 300 percent."

## To breach or to blow?

The pro-dam lobbyists know they possess a powerful, not-so-secret weapon: Senator Slade Gorton, the Washington Republican who holds the commanding post of chairman of the Subcommittee on Interior Appropriations. Gorton has built his political base by advertising himself as the foe of liberal Seattle environmentalists, and with his hands on Interior's purse strings, he can back up the role with real clout. As determined as Bruce Babbitt is to bring down a big dam, Slade Gorton may be more determined to stop him.

During last October's federal budget negotiations, Gorton offered to allocate $22 million for removing two modest dams in the Elwha River on the Olympic Peninsula, a salmon-restoration project dear to the hearts of dam-breaching advocates. But Gorton agreed to fund the Elwha breaching if—and only if—the budget included language forbidding federal officials from unilaterally ordering the dismantling of any dam, including those in the Columbia River Basin. Babbitt and others balked at Gorton's proposal. As a result, the 1999 budget includes zero dollars for removal of the Elwah dams.

Gorton's Elwha maneuver may have been hardball politics for its own sake, but it was also a clear warning: If the Army Corps and the National Marine Fisheries Service recommend breaching on the Snake in their study later this year, there will be hell to pay.

Meanwhile, here's a hypothetical question: If you're going to breach, how do you actually do it? How do you take those behemoths out? It depends on the dam, of course, but the answer on the Snake is shockingly simple.

"You leave the dam there," Charlie Ray says. We're standing downstream from Lower Granite Dam, 35 million pounds of steel encased in concrete. Lower Granite isn't a classic ghastly curtain like Hoover Dam; it resembles nothing so much as an enormous half-sunk harmonica. Ray points to a berm of granite boulders butting up against the concrete structure's northern end. "Take out the earthen portion and let the river flow around the dam. This is not high-tech stuff. This is front-end loaders and dump trucks."

It turns out that Charlie is only a few adjectives short of the truth. All you do need are loaders and dump trucks—really, really big ones, says Steve Tatro of the Army Corps of Engineers. Tatro has the touchy job of devising the best way to breach his agency's own dams. First, he says, you'd draw down the reservoir, using the spillways and the lower turbine passages as drains. Then you'd bypass the concrete and steel entirely and excavate the dam's earthen portion. Depending on the dam, that could mean excavating as much as 8 million cubic yards of material.

Tatro's just-the-facts manner can't disguise the reality that there is something deeply cathartic about the act he's describing. Most environmental restoration happens at the speed of nature. Which is to say, damnably slow. Breaching a dam—or better yet, blowing a dam—offers a rare moment of immediate gratification.

## The Glen Canyon story

From the Mesopotamian canals to Hoover Dam, it took the human mind about 10,000 years to figure out how to stop a river. It has taken only 60 years to accomplish the all-too-obvious environmental destruction.

Until the 1930s, most dam projects were matters of trial and (often) error, but beginning with Hoover Dam in 1931, dam builders began erecting titanic riverstoppers that approached an absolute degree of reliability and safety. In *Cadillac Desert*, a 1986 book on Western water issues, author Marc Reisner notes that from 1928 to 1956, "the most fateful transformation that has ever been visited on any landscape, anywhere, was wrought." Thanks to the U.S. Bureau of Reclamation, the Tennessee Valley Authority, and the Army Corps, dams lit a million houses, turned deserts into wheat fields, and later powered the factories that built the planes and ships that beat Hitler and the Japanese. Dams became monuments to democracy and enlightenment during times of bad luck and hunger and war.

Thirty years later, author Edward Abbey became the first dissenting voice to be widely heard. In *Desert Solitaire* and *The Monkey Wrench Gang*, Abbey envisioned a counterforce of wilderness freaks wiring bombs to the Colorado River's Glen Canyon Dam, which he saw as the ultimate symbol of humanity's destruction of the American West. Kaboom! Wildness returns to the Colorado.

Among environmentalists, the Glen Canyon Dam has become an almost mythic symbol of riparian destruction. All the symptoms of dam kill are there. The natural heavy metals that the Colorado River used to disperse into the Gulf of California now collect behind the dam in Lake Powell. And the lake is filling up: Sediment has reduced the volume of the lake from its original 27 million acre-feet to 23 million. One million acre-feet of water are lost to evaporation every year—enough, environmentalists note, to revive the dying upper reaches of the Gulf of California. The natural river ran warm and muddy, and flushed its channel with floods; the dammed version runs cool, clear, and even. Trout thrive in the Colorado. This is like giraffes thriving on tundra.

> In 1963, the most beautiful of all the canyons of the Colorado **began disappearing beneath Lake Powell.**

Another reason for the dam's symbolic power can be traced to its history. For decades ago, David Brower, then executive director of the Sierra Club, agreed to a compromise that haunts him to this day: Conservationists would not oppose Glen Canyon and 11 other projects if plans for the proposed Echo Park and Split Mountain dams, in Utah and Colorado, were abandoned. In 1963, the place Wallace Stegner once called "the most serenely beautiful of all the canyons of the Colorado" began disappearing beneath Lake Powell. Brower led the successful fight to block other dams in the Grand Canyon area, but he remained bitter about the compromise. "Glen Canyon died in 1963," he later wrote, "and I was partly responsible for its needless death."

In 1981 Earth First! inaugurated its prankster career by unfurling an enormous black plastic "crack" down the face of Glen Canyon Dam. In 1996 the Sierra Club rekindled the issue by calling for the draining of Lake Powell. With the support of Earth Island Institute (which Brower now chairs) and other environmental groups, the proposal got a hearing before a subcommittee of the House Committee on Resources in September 1997. Congress has taken no further action, but a growing number of responsible voices now echo the monkey-wrenchers' arguments. Even longtime Bureau of Reclamation supporter Barry Goldwater admitted, before his death last year, that he considered Glen Canyon Dam a mistake.

Defenders of the dam ask what we would really gain from a breach. The dam-based ecosystem has attracted peregrine falcons, bald eagles, carp, and catfish. Lake Powell brings in $400 million a year from tourists enjoying houseboats, powerboats, and personal watercraft—a local economy that couldn't be replaced by the thinner wallets of rafters and hikers.

"It would be completely foolhardy and ridiculous to deactivate that dam," says Floyd Dominy during a phone conversation from his home in Boyce, Virginia. Dominy, now 89 years old and retired since 1969, was the legendary Bureau of Reclamation commissioner who oversaw construction of the dam in the early 1960s. "You want to lose all that pollution-free energy? You want to destroy a world-renowned tourist attraction—Lake Powell—that draws more than 3 million people a year?"

It goes against the American grain: the notion that knocking something down and returning it to nature might be progress just as surely as replacing wildness with asphalt and steel. But 30 years of environmental law, punctuated by the crash of the salmon industry, has shifted power from the dam builders to the conservationists.

The most fateful change may be a little-noticed 1986 revision in a federal law. Since the 1930s, the Federal Energy Regulatory Commission has issued 30- to 50-year operating licenses to the nation's 2,600 or so privately owned hydroelectric dams. According to the revised law, however, FERC must consider not only power generation, but also fish and wildlife, energy conservation, and recreational uses before issuing license renewals. In November 1997, for the first time in its history, FERC refused a license against the will of a dam owner, ordering the Edwards Manufacturing Company to rip the 160-year-old Edwards Dam out of Maine's Kennebec River. More than 220 FERC hydropower licenses will expire over the next 10 years.

If there is one moment that captures the turning momentum in the dam wars, it might be the dinner Richard Ingebretsen shared with the builder of Glen Canyon Dam, Floyd Dominy himself. During the last go-go dam years, from 1959 to 1969, this dam-building bureaucrat was more powerful than any Western senator or governor. Ingebretsen is a Salt Lake City physician, a Mormon Republican, and a self-described radical environmentalist. Four years ago, he founded the Glen Canyon Institute to lobby for the restoration of Glen Canyon. Ingebretsen first met Dominy when the former commissioner came to Salt Lake City in 1995 to debate David Brower over the issue of breaching Glen Canyon Dam. To his surprise, Ingebretsen found that he liked the man. "I really respect him for his views," he says.

Their dinner took place in Washington, D.C., in early 1997. At one point Dominy asked Ingebretsen how serious the movement to drain Lake Powell really was. Very serious, Ingebretsen replied. "Of course I'm opposed to putting the dam in mothballs," Dominy said. "But I heard what Brower wants to do." (Brower had suggested that Glen Canyon could be breached by coring out some old water bypass tunnels that had been filled in years ago.) "Look," Dominy continued, "those tunnels are jammed with 300 feet of reinforced concrete. You'll never drill that out."

With that, Dominy pulled out a napkin and started sketching a breach. "You want to drain Lake Powell?" he asked. "What you need to do is drill new bypass tunnels. Go through the soft sandstone around and beneath the dam and line the tunnels with waterproof plates. It would be an expensive, difficult engineering feat. Nothing like this has ever been done before, but I've done a lot of thinking about it, and it will work. You can drain it."

The astonished Ingebretsen asked Dominy to sign and date the napkin. "Nobody will believe this," he said. Dominy signed.

Of course, it will take more than a souvenir napkin to return the nation's great rivers to their full wildness and health. Too much of our economic infrastructure depends on those 75,000 dams for anyone to believe that large numbers of river blockers, no matter how obsolete, will succumb to the blow of Bruce Babbitt's hammer anytime soon. For one thing, Babbitt himself is hardly in a position to be the savior of the rivers. Swept up in the troubles of a lame-duck administration and his own nagging legal problems (last spring Attorney General Janet Reno appointed an independent counsel to look into his role in an alleged Indian casino-campaign finance imbroglio), this interior secretary is not likely to fulfill his dream of bringing down a really big dam. But a like-minded successor just might. It will take a president committed and powerful enough to sway both Congress and the public, but it could come to pass.

Maybe Glen Canyon Dam and the four Snake River dams won't come out in my lifetime, but others will. And as more rivers return to life, we'll take a new census of emancipated streams: We freed the Neuse, the Kennebec, the Allier, the Rogue, the Elwah, and even the Tuolumne. We freed the White Salmon and the Souradabscook, the Ocklawaha and the Genesee. They will be untidy and unpredictable, they will flood and recede, they will do what they were meant to do: run wild to the sea.

---

*Bruce Barcott is the author of* The Measure of a Mountain: Beauty and Terror on Mount Rainier *(Sasquatch, 1997).*

---

From *Utne Reader*, May/June 1999, pp. 50–57. Originally appeared in *Outside*, February 1999. © 1999 by Bruce Barcott. Reprinted by permission.

# Environmental enemy No. 1

Cleaning up the burning of coal would be the best way to make growth greener

IS GROWTH bad for the environment? It is certainly fashionable in some quarters to argue that trade and capitalism are choking the planet to death. Yet it is also nonsense. As our survey of the environment this week explains, there is little evidence to back up such alarmism. On the contrary, there is reason to believe not only that growth can be compatible with greenery, but that it often bolsters it.

This is not, however, to say that there are no environmental problems to worry about. In particular, the needlessly dirty, unhealthy and inefficient way in which we use energy is the biggest source of environmental fouling. That is why it makes sense to start a slow shift away from today's filthy use of fossil fuels towards a cleaner, low-carbon future.

There are three reasons for calling for such an energy revolution. First, a switch to cleaner energy would make tackling other green concerns a lot easier. That is because dealing with many of these—treating chemical waste, recycling aluminium or incinerating municipal rubbish, for instance—is in itself an energy-intensive task. The second reason is climate change. The most sensible way for governments to tackle this genuine (but long-term) problem is to send a powerful signal that the world must move towards a low-carbon future. That will spur all sorts of innovations in clean energy.

The third reason is the most pressing of all: human health. In poor countries, where inefficient power stations, sooty coal boilers and bad ventilation are the norm, air pollution is one of the leading preventable causes of death. It affects some of the rich world too. From Athens to Beijing, the impact of fine particles released by the combustion of fossil fuels, and especially coal, is among today's biggest public-health concerns.

## Dethroning King Coal

The dream of cleaner energy will never be realised as long as the balance is tilted toward dirty technologies. For a start, governments must scrap perverse subsidies that actually encourage the consumption of fossil fuels. Some of these, such as cash given by Spain and Germany to the coal industry, are blatantly wrong-headed. Others are less obvious, but no less damaging. A clause in America's Clean Air Act exempts old coal plants from complying with current emissions rules, so much of America's electricity is now produced by coal plants that are over 30 years old. Rather than closing this loophole, the Bush administration has announced measures that will give those dirty old clunkers a new lease on life. Nor are poor countries blameless: many subsidise electricity heavily in the name of helping poor people, but rich farmers and urban elites then get to guzzle cheap (mostly coal-fired) power.

That points to a second prescription: the rich world could usefully help poorer countries to switch to cleaner energy. A forthcoming study by the International Energy Agency estimates that there are 1.6 billion people in the world who are unable to use modern energy. They often walk many miles to fetch wood, or collect cow dung, to use as fuel. As the poor world grows richer in coming decades, and builds thousands of power plants, many more such unfortunates will get electricity. That good news will come with a snag. Unless the rich world intervenes, many of these plants will burn coal in a dirty way. The resultant surge in carbon emissions will cast a grim shadow over the coming decades. Ending subsidies for exporters of fossil-fuel power plants might help. But stronger action is probably needed, meaning that the rich world must be ready to pay for the poor to switch to low-carbon energy. This should not be regarded as mere charity, but rather as a form of insurance against global warming.

The final and most crucial step is to start pricing energy properly. At the moment, the harm done to human health and the environment from burning fossil fuels is not reflected in the price of those fuels, especially coal, in most countries. There is no perfect way to do this, but one good idea is for governments to impose a tax based on carbon emissions. Such a tax could be introduced gradually, with the revenues raised returned as reductions in, say, labour taxes. That would make absolutely clear that the time has come to stop burning dirty fuels such as coal, using today's technologies.

## The dawning of the age of hydrogen

None of these changes need kill off coal altogether. Rather, they would provide a much-needed boost to the development of low-carbon technologies. Naturally, renewables such as solar and wind will get a boost. But so too would "sequestration", an innovative way of using fossil fuels without releasing carbon into the air.

This matters for two reasons. For a start, there is so much cheap coal, distributed all over the world, that poor countries are bound to burn it. The second reason is that sequestration offers a fine stepping-stone to squeaky clean hydrogen energy. Once the energy trapped in coal is unleashed and its carbon sequestered, energy-laden hydrogen can be used directly in fuel cells. These nifty inventions can power a laptop, car or home without any harmful emissions at all.

It will take time to get to this hydrogen age, but there are promising harbingers. Within a few years, nearly every big car maker plans to have fuel-cell cars on the road. Power plants using this technology are already trickling on to the market. Most big oil companies have active hydrogen and carbon-sequestration efforts under way. Even some green groups opposed to all things fossil say they are willing to accept sequestration as a bridge to a renewables-based hydrogen future.

Best of all, this approach offers even defenders of coal a realistic long-term plan for tackling climate change. Since he rejected the UN's Kyoto treaty on climate change, George Bush has been portrayed as a stooge for the energy industry. This week, California's legislature forged ahead by passing restrictions on emissions of greenhouse gases; a Senate committee has acted similarly. Mr Bush, who has made surprisingly positive comments about carbon sequestration and fuel cells, could silence the critics by following suit. By cracking down on carbon and embracing hydrogen, he could even lead.

From *The Economist*, July 6, 2002, p. 1. © 2002 by the Economist, Ltd. Distributed by the New York Times Special Features. Reprinted by permission.

# Article 15

## Carbon sequestration
# Fired up with ideas

Capturing and storing carbon dioxide could slow down climate change and also allow fossil fuels to be a bridge to a clean hydrogen-based future

IF THE world is to tackle the problem of climate change in earnest, "clean coal" has to become more than just an amusing oxymoron. All fossil fuels contain carbon, but coal is by far the most carbon-intensive. This is troubling, since global warming seems to be driven by an increase in the level of atmospheric greenhouse gases, of which carbon dioxide ($CO_2$) is the most worrisome. Coal is also the most abundant fossil fuel (see chart). If all known conventional oil and gas reserves (those in underground formations, obtained by drilling) were burned, the level of $CO_2$ in the atmosphere would still be less than twice what it was before the beginning of the industrial revolution. Climate change associated with that level of $CO_2$ might be tolerable. Burn all the coal, however, and it would be more than four times that starting-point—with larger, less predictable and quite likely more unpleasant climatic consequences.

Much of that coal will, nevertheless, be burned. In particular, poor countries such as China, India and South Africa have large reserves that are almost certain to be used to fuel economic growth. So it makes sense to consider possible technical fixes to the problem. These will never be the whole answer; unless the correct incentives are applied, burning coal without such fixes is likely to remain cheaper than burning it with them. But combined with the right incentives, in the form of such things as carbon taxes, they could help to keep the rise in atmospheric $CO_2$ to manageable proportions. They may also, surprisingly, help to usher in the green nirvana of a "hydrogen economy", in which the fuel of choice is that non-poisonous, non-greenhouse gas.

## Catch me if you can
Technological solutions to rising atmospheric $CO_2$ (as opposed to, say, planting forests to soak the stuff up through photosynthesis) come in two parts. The first is capture: extracting the gas from the machine that is burning it. The second is sequestration: putting it somewhere it cannot easily escape from.

Capture is the more expensive of the two, especially when it needs to be designed into a plant from the start. One method that can be retrofitted on to existing machinery (although it is probably worth doing so only for large emitters) is to "scrub" $CO_2$ from an exhaust by passing it through a chemical, such as mono-ethanolamine, which has a particular affinity for the gas.

Smaller $CO_2$ sources (such as car engines and homes), which account for about half the gas generated by burning fossil fuels, are unlikely to be susceptible to retrofitted scrubbing. And even in large plants, scrubbing is easier and cheaper when the exhaust has a high concentration of $CO_2$—which is not the case for fuels that have been burned in air. Some have suggested using pure oxygen rather than air, but that is hardly an economically practical solution.

In some ways, though, a truly practical solution is even more radical: to change the way that energy is extracted from fossil fuels by separating the $CO_2$ formation from the process of heat generation. That can be done by adapting a well-established chemical process known as steam reformation. This involves reacting a carbon-based fuel with oxygen and steam to produce a so-called "synthesis gas" that is composed of carbon monoxide and hydrogen (much of the latter comes from the water, rather than the fuel). That mixture can be separated quite easily, and the hydrogen burned in, for example, a gas turbine. Then, mixing the carbon monoxide with more steam in the presence of a suitable catalyst yields $CO_2$ and still more hydrogen; again a mixture that can be separated quite easily.

It is this idea, known as the integrated gasifier combined cycle (IGCC) approach, that has environmentalists excited. It would require big changes to the design of energy-generating equipment, and so could not be introduced quickly. But it is not pie in the sky. And true visionaries will notice that, since it produces hydrogen, it permits the generation of electricity by fuel cells (chemical reactors that create electrical current from the reaction between hydrogen and oxygen, without any harmful emissions) as well as conventional gas turbines. Fuel cells are a critical component of most projections of what a hydrogen economy might look like.

Although the whole package seems idealistic, most of the technologies involved are in fact already in common use, in such processes as ammonia production. Indeed, there are several IGCC plants already operating in Europe and America (although they do not bother to remove $CO_2$). Some firms are talking of building one in China. Robert Williams, head of energy-systems analysis at Princeton University, reckons that, even in countries such as China, where tackling global warming is not exactly a priority, such gasification could lead to "zero emissions from coal", helped on its way by the extra revenues that may come from so-called "polygeneration" of clean synthetic fuels, hydrogen and electricity together.

## Down under
Scrubbing and gasification can thus deliver $CO_2$ in a form that can be disposed of. Actually doing so, though, remains a challenge. Leaks from $CO_2$ repositories would hardly be as disastrous as leaks from a nuclear-waste dump. Nevertheless, to be effective, those repositories would have to stay gas-tight for centuries.

One way of disposing of the gas might be to **sink it beneath the waves**. The oceans already store a lot of $CO_2$, so they might be in-

## Carbon in, Carbon out

| Energy resources* | Resource base, GtC† | Potential $CO_2$ storage reservoirs | |
|---|---|---|---|
| | | Sequestration option | Worldwide capacity‡ GtC† |
| Oil (conventional) | 241 | | |
| Oil (unconventional, eg tar sands) | 407 | Ocean | 1,000s |
| Natural gas (conventional) | 253 | Deep saline formations | 100s-1,000s |
| Natural gas (unconventional, eg coal-bed methane) | 509 | Depleted oil and gas reservoirs | 100s |
| Coal | 5,151 | Coal seams | 10s-100s |

*Proven reserves plus likely future resources  †1 GtC = 1 billion tonnes of carbon equivalent  ‡Orders of magnitude estimates
Sources: UNDP World Energy Assessment, 2000; Howard Herzog, Massachusetts Institute of Technology

duced to accept a little more. But it would have to be buried in water that is not likely to come to the surface any time soon. Some scientists worry, though, that dissolving vast quantities of $CO_2$ in the bottom of the ocean could result in ecological damage; others fear that the gas will be regurgitated wherever it is put. An international research consortium planned to test such worries by releasing 60 tonnes of $CO_2$ on the seabed near Hawaii—but noisy protests forced it to cancel the plan last month. Howard Herzog of the Massachusetts Institute of Technology, a member of the consortium, says that the team now hopes to shift to the North Sea.

A less speculative option would be to use **depleted oil and gas reservoirs**. These are layers of porous rocks topped by a cap of impermeable rock, usually in the shape of a dome. After decades of oil exploitation, there are plenty of ageing fields around. The advantages are many: the geology and technology involved are well understood; the exploration costs are small; and the reservoirs in question have already proved they can hold liquids and gases for aeons, since that is how the extracted hydrocarbons built up in them. What is more, injecting $CO_2$ into such wells could produce a saleable by-product; a similar technique is already used by oilmen to squeeze extra output from declining sources. For example, EnCana, a Canadian oil company, pays the Dakota Gasification Company to pump $CO_2$ produced at Dakota's coal gasification plant by pipeline to some of its wells.

In what may prove to be a straw in the wind, an American emissions brokerage called CO2e.com announced this week the largest-ever public trade in the emerging greenhouse-gas market. Ontario Power Generation, a Canadian company, bought the right to the emissions-reduction "credits" associated with 9m tonnes of $CO_2$. That gas, produced as a by-product of natural-gas processing, would normally have been vented into the atmosphere, but it will now be injected instead into old oilfields in Wyoming, Texas and Mississippi. The power company has volunteered to cut its emissions of greenhouse gases, and sees these credits as a way to offset its fossil-fuel emissions.

A similar idea to burying $CO_2$ in old oil wells is to inject it into **coal seams** that are too deep and uneconomic to mine. This has two attractions. First, the injected gas will be absorbed on to the surface of the coal, and so locked up more or less permanently. Second, the incoming $CO_2$ often displaces methane that would otherwise not have seen the light of day. Capturing and selling that methane could turn this approach into a nice little earner. A scheme in New Mexico already uses $CO_2$ in this manner.

Also promising are **saline aquifers** located deep below the earth's surface. $CO_2$ pumped into such places would dissolve, at least in part, in the salt water. In some such formations, it would also react with local silicate minerals to form carbonates and bicarbonates that could stay put for millions of years. Statoil, a Norwegian oil firm, has been pumping $CO_2$ into a deep saline aquifer under the North Sea since 1996—the first time "geological" sequestration of this sort has been motivated by a fear of climate change, in the form of a Norwegian tax on carbon emissions. That tax created the incentive for Statoil to bury the stuff rather than continue releasing it into the atmosphere.

Geological sequestration, then, is not merely a speculative idea. In some cases it may even pay part or all of its own way, by releasing otherwise inaccessible fuel deposits. Recognising this, eight big energy companies, led by BP, have recently formed the $CO_2$ Capture Project to promote research. The Natural Resources Defence Council, a big American green group, says it is keeping an open mind about this sort of sequestration. And there is one other important endorsement: "We all believe technology offers great promise to significantly reduce emissions—especially carbon capture, storage and sequestration technologies." Thus spake George Bush, the man who said No to the Kyoto treaty on climate change.

From *The Economist*, July 6, 2002, pp. 78-79. © 2002 by The Economist, Ltd. Distributed by the New York Times Special Features. Reprinted by permission.

# Past and Present Land Use and Land Cover in the USA

WILLIAM B. MEYER

"Land of many uses," runs a motto used to describe the National Forests, and it describes the United States as a whole just as well. "Land of many covers" would be an equally apt, but distinct, description. *Land use* is the way in which, and the purposes for which, human beings employ the land and its resources: for example, farming, mining, or lumbering. *Land cover* describes the physical state of the land surface: as in cropland, mountains, or forests. The term land cover originally referred to the kind and state of vegetation (such as forest or grass cover), but it has broadened in subsequent usage to include human structures such as buildings or pavement and other aspects of the natural environment, such as soil type, biodiversity, and surface and groundwater. A vast array of physical characteristics—climate, physiography, soil, biota—and the varieties of past and present human utilization combine to make every parcel of land on the nation's surface unique in the cover it possesses and the opportunities for use that it offers. For most practical purposes, land units must be aggregated into quite broad categories, but the frequent use of such simplified classes should not be allowed to dull one's sense of the variation that is contained in any one of them.

Land cover is affected by natural events, including climate variation, flooding, vegetation succession, and fire, all of which can sometimes be affected in character and magnitude by human activities. Both globally and in the United States, though, land cover today is altered principally by direct human use: by agriculture and livestock raising, forest harvesting and management, and construction. There are also incidental impacts from other human activities such as forests damaged by acid rain from fossil fuel combustion and crops near cities damaged by tropospheric ozone resulting from automobile exhaust.

Changes in land cover by land use do not necessarily imply a degradation of the land. Indeed, it might be presumed that any change produced by human use is an improvement, until demonstrated otherwise, because someone has gone to the trouble of making it. And indeed, this has been the dominant attitude around the world through time. There are, of course, many reasons why it might be otherwise. Damage may be done with the best of intentions when the harm inflicted is too subtle to be perceived by the land user. It may also be done when losses produced by a change in land use spill over the boundaries of the parcel involved, while the gains accrue largely to the land user. Economists refer to harmful effects of this sort as *negative externalities*, to mean secondary or unexpected consequences that may reduce the net value of production of an activity and displace some of its costs upon other parties. Land use changes can be undertaken because they return a net profit to the land user, while the impacts of negative externalities such as air and water pollution, biodiversity loss, and increased flooding are borne by others. Conversely, activities that result in secondary benefits (or *positive externalities*) may not be undertaken by landowners if direct benefits to them would not reward the costs.

Over the years, concerns regarding land degradation have taken several overlapping (and occasionally conflicting) forms. *Conservationism* emphasized the need for careful and efficient management to guarantee a sustained supply of productive land resources for future generations. *Preservationism* has sought to protect scenery and ecosystems in a state as little human-altered as possible. Modern *environmentalism* subsumes many of these goals and adds new concerns that cover the varied secondary effects of land use both on land cover and on other related aspects of the global environment. By and large, American attitudes in the past century have shifted from a tendency to interpret human use as improving the condition of the land towards a tendency to see human impact as primarily destructive. The term "land reclamation" long denoted the conversion of land from its natural cover; today it is more often used to describe the restoration and repair of land damaged by human use. It would be easy, though, to exaggerate the shift in attitudes. In truth, calculating the balance of costs and benefits from many land use and land cover changes is enormously difficult. The full extent and consequences of proposed changes are often less than certain, as is their possible irreversibility and thus their lasting significance for future generations.

## WHERE ARE WE?

The United States, exclusive of Alaska and Hawaii, assumed its present size and shape around the middle of the 19th century. Hawaii is relatively small, ecologically distinctive, and profoundly affected by a long and distinctive history of human use; Alaska is huge and little affected to date by direct land use. In this review assessment we therefore survey land use and land cover change, focusing on the past century and a half, only in the conterminous or lower 48 states. Those states cover an area of almost 1900 million acres, or about 3 million square miles.

*The adjustments that are made in land use and land cover in coming years will in some way alter the life of nearly every living thing on Earth.*

How land is *used*, and thus how *land cover* is altered, depends on who owns or controls the land and on the pressures and incentives shaping the behavior of the owner. Some 400 million acres in the conterminous 48 states—about 21% of the total—are federally owned. The two largest chunks are the 170 million acres of western rangeland controlled by the Bureau of Land Management and the approximately equal area of the National Forest System. Federal land represents 45% of the area of the twelve western states, but is not a large share of any other regional total. There are also significant land holdings by state governments throughout the country.

Most of the land in the United States is privately owned, but under federal, state, and local restrictions on its use that have increased over time. The difference between public and private land is important in explaining and forecasting land use and land coverage change, but the division is not absolute, and each sector is influenced by the other. Private land use is heavily influenced by public policies, not only by regulation of certain uses but through incentives that encourage others. Public lands are used for many private activities; grazing on federal rangelands and timber extraction from the national forests by private operators are the most important and have become the most controversial. The large government role in land use on both government and private land means that policy, as well as economic forces, must be considered in explaining and projecting changes in the land. Economic forces are of course significant determinants of policy—perhaps the most significant—but policy remains to some degree an independent variable.

There is no standard, universally accepted set of categories for classifying land by either use or cover, and the most commonly used, moreover, are hybrids of land cover and land use. Those employed here, which are by and large those of the U.S. National Resources Inventory conducted every five years by relevant federal agencies, are cropland, forest, grassland (pasture and rangeland), wetlands, and developed land.

- *Cropland* is land in farms that is devoted to crop production; it is not to be confused with total farmland, a broad land use or land ownership category that can incorporate many forms of land cover.
- *Forest land* is characterized by a predominance of tree cover and is further divided by the U.S. Census into timberland and non-timberland. By definition, the former must be capable of producing 20 cubic feet of industrial wood per acre per year and remain legally open to timber production.
- *Grassland* as a category of land cover embraces two contrasting Census categories of use: pasture (enclosed and what is called improved grassland, often closely tied to cropland and used for intensive livestock raising), and range (often unenclosed or unimproved grazing land with sparser grass cover and utilized for more extensive production).
- *Wetlands* are not a separate Census or National Resources Inventory category and are included within other categories: swamp, for example, is wetland forest. They are defined by federal agencies as lands covered all or part of the year with water, but not so deeply or permanently as to be classified as water surface *per se*.
- The U.S. government classifies as *developed* land urban and built-up parcels that exceed certain size thresholds. "Developed" or "urban" land is clearly a use rather than a cover category. Cities and suburbs as they are politically defined have rarely more than half of their area, and often much less, taken up by distinctively "urban" land cover such as buildings and pavement. Trees and grass cover substantial areas of the metropolitan United States; indeed, tree cover is greater in some settlements than in the rural areas surrounding them.

By the 1987 U.S. National Resources Inventory, nonfederal lands were divided by major land use and land cover classes as follows: cropland, about 420 million acres (22% of the entire area of the 48 states); rangeland, about 400 million (21%); forest, 390 million (21%); pasture, 130 million (7%); and developed land, 80 million (4%). Minor covers and uses, including surface water, make up an-

### Land Use and Cover in the Conterminous U.S.

| Land Class | Area in Million Acres | Fraction of Total Area |
|---|---|---|
| **Privately Owned** (shown in diagram below) | | |
| Cropland | 422 | 22.4% |
| Rangeland | 401 | 21.3 |
| Forest | 391 | 20.8 |
| Pasture | 129 | 6.9 |
| Developed | 77 | 4.1 |
| Other Catagories[1] | 60 | 3.2 |
| **Federally Owned**[2] | 404 | 21.4 |
| TOTAL[3] | 1884 | 100% |

[1] Other minor covers and surface water
[2] Federal land is approximately half forest and half rangeland
[3] Included in various catagories is about 100 million acres of wetland, covering about 5% of the national area

**Table 1** Source: U.S. 1987 National Resources Inventory, published in 1989. U.S. Government Printing Office.

other 60 million acres (Table 1). The 401 million acres of federal land are about half forest and half range. Wetlands, which fall within these other Census classes, represent approximately 100 million acres or about five percent of the national area; 95 percent of them are freshwater and five percent are coastal.

These figures, for even a single period, represent not a static but a dynamic total, with constant exchanges among uses. Changes in the area and the location of cropland, for example, are the result of the *addition* of new cropland from conversion of grassland, forest, and wetland and its *subtraction* either by abandonment of cropping and reversion to one of these less intensive use/cover forms or by conversion to developed land. The main causes of forest *loss* are clearing for agriculture, logging, and clearing for development; the main cause of forest *gain* is abandonment of cropland followed by either passive or active reforestation. Grassland is converted by the creation of pasture from forest, the interchange of pasture and cropland, and the conversion of rangeland to cropland, often through irrigation.

Change in wetland is predominantly loss through drainage for agriculture and construction. It also includes natural gain and loss, and the growing possibilities for wetland creation and restoration are implicit in the Environmental Protection Agency's "no *net* loss" policy (emphasis added). Change in developed land runs in only one direction: it expands and is not, to any significant extent, converted to any other category.

Comparison of the American figures with those for some other countries sets them in useful perspective. The United States has a greater relative share of forest and a smaller relative share of cropland than does Europe as a whole and the United Kingdom in particular.

Though Japan is comparable in population density and level of development to Western Europe, fully two-thirds of its area is classified as forest and woodland, as opposed to ten percent in the United Kingdom; it preserves its largely mountainous forest area by maintaining a vast surplus of timber imports over exports, largely from the Americas and Southeast Asia.

Regional patterns within the U.S. (using the four standard government regions of Northeast, Midwest, South, and West) display further variety. The Northeast, though the most densely populated region, is the most heavily wooded, with three-fifths of its area in forest cover. It is also the only region of the four in which "developed" land, by the Census definition, amounts to more than a minuscule share of the total; it covers about eight percent of the Northeast and more than a quarter of the state of New Jersey. Cropland, not surprisingly, is by far the dominant use/cover in the Midwest, accounting for just under half of its expanse. The South as a whole presents the most balanced mix of land types: about 40 percent forest, 20 percent each of cropland and rangeland, and a little more than ten percent pasture. Western land is predominantly rangeland, with forest following and cropland a distant third. Wetlands are concentrated along the Atlantic seaboard, in the Southeast, and in the upper Midwest. Within each region, of course, there is further variety at and below the state level.

# WHERE HAVE WE BEEN?

The public domain, which in 1850 included almost two-thirds of the area of the present conterminous states, has gone through two overlapping phases of management goals. During the first, dominant in 1850 and long thereafter, the principal goal of management was to transfer public land into private hands, both to raise revenue and to encourage settlement and land improvements. The government often attached conditions (which were sometimes complied with) to fulfill other national goals, such as swamp drainage, timber planting, and railroad construction in support of economic development.

The second phase, that of federal retention and management of land, began with the creation of the world's first national park, Yellowstone, shortly after the Civil War. It did not begin to be a significant force, however, until the 1890s, when 40 million acres in the West were designated as federal forest reserves, the beginning of a system that subsequently expanded into other regions of the country as well. Several statutory vestiges of the first, disposal era remain (as in mining laws, for example), but the federal domain is unlikely to shrink noticeably in coming decades, in spite of repeated challenges to the government retention of public land and its regulation of private land. In recent years, such challenges have included the "Sagebrush Rebellion" in the rangelands of the West in the 1970s and 1980s calling for the withdrawal of federal control, and legal efforts to have many land use regulations classified as "takings," or as exercises of the power of eminent domain. This classification, where it is granted, requires the government to compensate owners for the value of development rights lost as a result of the regulation.

## Cropland

Total cropland rose steadily at the expense of other land covers throughout most of American history. It reached a peak during the 1940s and has subsequently fluctuated in the neighborhood of 400 million acres, though the precise figure depends on the definition of cropland used. Long-term regional patterns have displayed more variety. Cropland abandonment in some areas of New England began to be significant in some areas by the middle of the nineteenth century. Although total farmland peaked in the region as late as 1880 (at 50%) and did not decline sharply until the turn of the century, a steady decline in the subcategory of cropland and an increase in other farmland covers such as woodland and unimproved pasture was already strongly apparent. The Middle Atlantic followed a similar trajectory, as, more recently, has the South. Competition from other, more fertile sections of the country in agricultural production and within the East from other demands on land and labor have been factors; a long-term rise in agricultural productivity caused by technological advances has also exerted a

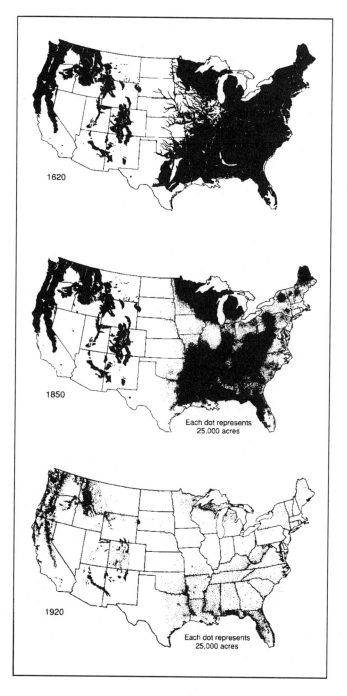

**Figure 1** Area of virgin forest: top to bottom 1620, 1850, and 1920 as published by William B. Greeley, "The Relation of Geography to Timber Supply," *Economic Geography*, vol. 1, pp. 1-11 (1925). The depiction of U. S. forests in the later maps may be misleading in that they show only old-growth forest and not total tree cover.

steady downward pressure on total crop acreage even though population, income, and demand have all risen.

Irrigated cropland on a significant scale in the United States extends back only to the 1890s and the early activities in the West of the Bureau of Reclamation. Growing rapidly through about 1920, the amount of irrigated land remained relatively constant between the wars, but rose again rapidly after 1945 with institutional and technological developments such as the use of center-pivot irrigation drawing on the Ogallala Aquifer on the High Plains. It reached 25 million acres by 1950 and doubled to include about an eighth of all cropland by about 1980. Since then the amount of irrigated land has experienced a modest decline, in part through the decline of aquifers such as the Ogallala and through competition from cities for water in dry areas.

*Forests*

At the time of European settlement, forest covered about half of the present 48 states. The greater part lay in the eastern part of the country, and most of it had already been significantly altered by Native American land use practices that left a mosaic of different covers, including substantial areas of open land.

Forest area began a continuous decline with the onset of European settlement that would not be halted until the early twentieth century. Clearance for farmland and harvesting for fuel, timber, and other wood products represented the principal sources of pressure. From an estimated 900 million acres in 1850, the wooded area of the entire U.S. reached a low point of 600 million acres around 1920 (Fig. 1).

It then rose slowly through the postwar decades, largely through abandonment of cropland and regrowth on cutover areas, but around 1960 began again a modest decline, the result of settlement expansion and of higher rates of timber extraction through mechanization. The agricultural censuses recorded a drop of 17 million acres in U.S. forest cover between 1970 and 1987 (though data uncertainties and the small size of the changes relative to the total forest area make a precise dating of the reversals difficult). At the same time, if the U.S. forests have been shrinking in area they have been growing in density and volume. The trend in forest biomass has been consistently upward; timber stock measured in the agricultural censuses from 1952 to 1987 grew by about 30%.

National totals of forested area again represent the aggregation of varied regional experiences. Farm abandonment in much of the East has translated directly into forest recovery, beginning in the mid- to late-nineteenth century (Fig. 2). Historically, lumbering followed a regular pattern of harvesting one region's resources and moving on to the next; the once extensive old-growth forest of the Great Lakes, the South, and the Pacific Northwest represented successive and overlapping frontiers. After about 1930, frontier-type exploitation gave way to a greater emphasis on permanence and management of stands by timber companies. Wood itself has declined in importance as a natural resource, but forests have been increasingly valued and protected for a range of other services, including wildlife habitat, recreation, and streamflow regulation.

*Grassland*

The most significant changes in grassland have involved impacts of grazing on the western range. Though data for many periods are scanty or suspect, it is clear that rangelands have often been seriously overgrazed, with deleterious consequences including soil erosion and compaction, increased streamflow variability, and floral and faunal biodiversity loss as well as reduced value for production. The net value of grazing use on the western range is nationally small, though significant locally, and pressures for tighter management have increasingly been guided by ecological and preservationist as well as production concerns.

*Wetland*

According to the most recent estimates, 53% of American wetlands were lost between the 1780s and the 1980s, principally to drainage for agriculture. Most of the conversion presumably took place during the twentieth century; between the 1950s and the 1970s alone, about 11 million acres were lost. Unassisted private action was long thought to drain too little; since mid-century, it has become apparent that the opposite is true, that unfettered private action tends to drain too much, i.e., at the expense of now-valued wetland. The positive externalities once expected from drainage—improved public health and beautification of an unappealing natural landscape—carry less weight today than the negative ones that it produces. These include the decline of wildlife, greater extremes of streamflow, and loss of a natural landscape that is now seen as more attractive than a human-modified one. The rate of wetland loss has now been cut significantly by regulation and by the removal of incentives for drainage once offered by many government programs.

*Developed land*

As the American population has grown and become more urbanized, the land devoted to settlement has increased in at least the same degree. Like the rest of the developed world, the United States now has an overwhelmingly non-farm population residing in cities, suburbs, and towns and villages. Surrounding urban areas is a classical frontier of rapid and sometimes chaotic land use and land cover change. Urban impacts go beyond the mere subtraction of land from other land uses and land covers for settlement and infrastructure; they also involve the mining of building materials, the disposal of wastes, the creation of parks and water supply reservoirs, and the introduction of pollutants in air, water, and soil. Long-term data on urban use and cover trends are unfortunately not available. But the trend in American cities has undeniably been one of residential dispersal and lessened settlement densities as transportation technologies have improved; settlement has thus required higher amounts of land per person over time.

## WHERE ARE WE GOING?

The most credible projections of changes in land use and land cover in the United States over the next fifty years have come from recent assessments produced under the federal laws that now mandate regular national inventories of resource stocks and prospects. The most recent inquiry into land resources, completed by the Department of Agriculture in 1989 (and cited at the end of this article), sought to project their likely extent and condition a half-century into the future, to the year 2040. The results indicated that only slow changes were expected nationally in the major categories of land use and land cover: a loss in forest area of some 5% (a slower rate of loss than was experienced in the same period before); a similarly modest decline in cropland; and an increase in rangeland of about 5% through 2040. Projections are not certainties, however: they may either incorrectly identify the consequences of the factors they consider or fail to consider important factors that could alter the picture. Because of the significant impacts of policy, its role—notoriously difficult to forecast and assess— demands increased attention, in both its deliberate and its inadvertent effects.

Trends in the United States stand in some contrast to those in other parts of the developed world. While America's forest area continues to decline somewhat, that of many comparable countries has increased in modest degree, while the developing world has seen significant clearance in the postwar era. There has been substantial stability, with slow but fluctuating decline, in cropland area in the United States. In contrast, cropland and pasture have declined modestly in the past several decades in Western Europe and are likely to decline sharply there in the future as longstanding national and European Community agricultural policies subsidizing production are revised; as a result, the European countryside faces the prospect of radical change in land use and cover and considerable dislocation of rural life.

## WHY DOES IT MATTER?

Land use and land cover changes, besides affecting the current and future supply of land resources, are important sources of many other forms of environmental change. They are also linked to them through synergistic connections that can amplify their overall effect.

Loss of plant and animal biodiversity is principally traceable to land transformation, primarily through the fragmentation of natural habitat. Worldwide trends in land use and land cover change are an important source of the so-called greenhouse gases, whose accumulation in the atmosphere may bring about global climate change. As much as 35% of the increase in atmospheric $CO_2$ in the last 100 years can be attributed to land use change, principally through deforestation. The major known sources of increased methane—rice paddies, landfills, biomass burning, and cattle—are all related to land use. Much of the increase in nitrous oxide is now thought due to a collection of sources that also depend upon the use of the land, including biomass burning, livestock raising, fertilizer application and contaminated aquifers.

*In most of the world, both fossil fuel combustion and land transformation result in a net release of carbon dioxide to the atmosphere.*

Land use practices at the local and regional levels can dramatically affect soil condition as well as water quality and water supply. And finally, vulnerability or sensitivity to existing climate hazards and possible climate change is very much affected by changes in land use and cover. Several of these connections are illustrated below by examples.

### Carbon emissions

In most of the world, both fossil fuel combustion and land transformation result in a net release of carbon dioxide to the atmosphere. In the United States, by contrast, present land use and land cover changes are thought to absorb rather than release $CO_2$ through such processes as the rapid growth of relatively youthful forests. In balance, however, these land-use-related changes reduce U.S. contributions from fossil fuel combustion by only about 10%. The use of carbon-absorbing tree plantations to help diminish global climate forcing has been widely discussed, although many studies have cast doubt on the feasibility of the scheme. Not only is it a temporary fix (the trees sequester carbon only until the wood is consumed, decays, or ceases to accumulate) and requires vast areas to make much of a difference, but strategies for using the land and its products to offset some of the costs of the project might have large and damaging economic impacts on other land use sectors of the economy.

### Effects on arable land

The loss of cropland to development aroused considerable concern during the 1970s and early 1980s in connection with the 1981 National Agricultural Lands Study, which estimated high and sharply rising rates of conversion. Lower figures published in the 1982 National Resource Inventory, and a number of associated studies, have led most experts to regard the conversion of cropland to other land use categories as representing something short of a genuine crisis, likely moreover to continue at slower rather than accelerating rates into the future. The land taken from food and fiber production and converted to developed land has been readily made up for by conversion of land from grassland and forest.

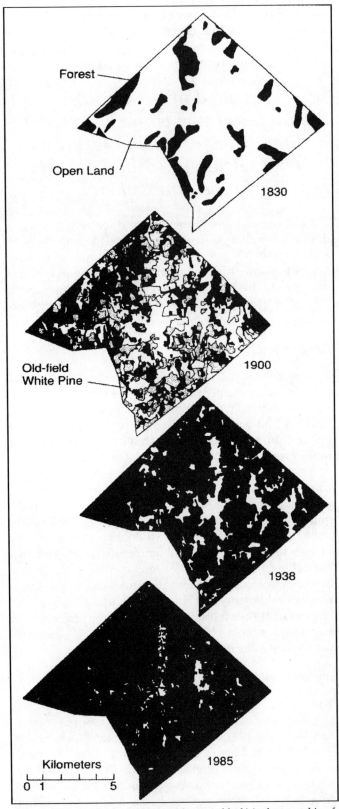

**Figure 2** Modern spread of forest (shown in black) in the township of Petersham, Massachusetts, 1830 through 1985. White area is that considered suitable for agriculture; shaded portions in 1900 map indicate agricultural land abandoned between 1870 and 1900 that had developed forest of white pine in this period. From "Land-use History and Forest Transformations," by David R. Foster, in *Humans as Components of Ecosystems,* edited by M. J. McDonnell and S.T.A. Pickett, Springer-Verlag, New York, pp. 91-110, 1993.

The new lands are not necessarily of the same quality as those lost, however, and some measures for the protection of prime farmland are widely considered justified on grounds of economics as well as sociology and amenities preservation.

*Vulnerability to climate change*

Finally, patterns and trends in land use and land cover significantly affect the degree to which countries and regions are vulnerable to climate change—or to some degree, can profit from it. The sectors of the economy to which land use and land cover are most critical—agriculture, livestock, and forest products—are, along with fisheries, among those most sensitive to climate variation and change. How vulnerable countries and regions are to climate impacts is thus in part a function of the importance of these activities in their economies, although differences in ability to cope and adapt must also be taken into account.

## Shifting patterns of land use in the U.S. and throughout the world are a proximate cause of many of today's environmental concerns.

These three climate-sensitive activities have steadily declined in importance in recent times in the U.S. economy. In the decade following the Civil War, agriculture still accounted for more than a third of the U.S. gross domestic product, or GDP. In 1929, the agriculture-forest-fisheries sector represented just under ten percent of national income. By 1950, it had fallen to seven percent of GDP, and it currently represents only about two percent. Wood in 1850 accounted for 90 percent of America's total energy consumption; today it represents but a few percent. These trends suggest a lessened macroeconomic vulnerability in the U.S. to climate change, though they may also represent a lessened ability to profit from it to the extent that change proves beneficial. They say nothing, however, about primary or secondary impacts of climate change on other sectors, about ecological, health, and amenity losses, or about vulnerability in absolute rather than relative terms, and particularly the potentially serious national and global consequences of a decline in U.S. food production.

The same trend of lessening vulnerability to climate changes is apparent even in regions projected to be the most exposed to the more harmful of them, such as reduced rainfall. A recent study examined agro-economic impacts on the Missouri-Iowa-Nebraska-Kansas area of the Great Plains, were the "Dust Bowl" drought and heat of the 1930s to recur today or under projected conditions of the year 2030. It found that although agricultural production would be substantially reduced, the consequences would not be severe for the regional economy overall: partly because of technological and institutional adaptation and partly because of the declining importance of the affected sectors, as noted above. The 1930s drought itself had less severe and dramatic effects on the population and economy of the Plains than did earlier droughts in the 1890s and 1910s because of land use, technological, and institutional changes that had taken place in the intervening period.

Shifting patterns in human settlement are another form of land use and land cover change that can alter a region's vulnerability to changing climate. As is the case in most other countries of the world, a disproportionate number of Americans live within a few miles of the sea. In the postwar period, the coastal states and counties have consistently grown faster than the country as a whole in population and in property development. The consequence is an increased exposure to hazards of hurricanes and other coastal storms, which are expected by some to increase in number and severity with global warming, and to the probable sea-level rise that would also accompany an increase in global surface temperature. It is unclear to what extent the increased exposure to such hazards might be balanced by improvements in the ability to cope, through better forecasts, better construction, and insurance and relief programs. Hurricane fatalities have tended to decline, but property losses per hurricane have steadily increased in the U.S., and the consensus of experts is that they will continue to do so for the foreseeable future.

## CONCLUSIONS

How much need we be concerned about changes in land use and land cover in their own right? How much in the context of other anticipated environmental changes?

As noted above, shifting patterns of land use in the U.S. and throughout the world are a proximate cause of many of today's environmental concerns. How land is used is also among the human activities most likely to feel the effects of possible climate change. Thus if we are to understand and respond to the challenges of global environmental change we need to understand the dynamics of land transformation. Yet those dynamics are notoriously difficult to predict, shaped as they are by patterns of individual decisions and collective human behavior, by history and geography, and by tangled economic and political considerations. We should have a more exact science of how these forces operate and how to balance them for the greatest good, and a more detailed and coherent picture of how land in the U.S. and the rest of the world is used.

The adjustments that are made in land use and land cover in coming years, driven by worldwide changes in population, income, and technology, will in some way alter the life of nearly every living thing on Earth. We need to understand them and to do all that we can to ensure that policy decisions that affect the use of land are made in the light of a much clearer picture of their ultimate effects.

## FOR FURTHER READING

*Americans and Their Forests: A Historical Geography,* by Michael Williams. Cambridge University Press, 599 pp, 1989.

An Analysis of the Land Situation in the United States: 1989–2040. USDA Forest Service General Technical Report RM-181. U.S. Government Printing Office, Washington, D.C., 1989.

*Changes in Land Use and Land Cover: A Global Perspective.* W. B. Meyer and B. L. Turner II, editors. Cambridge University Press, 537 pp, 1994.

"Forests in the Long Sweep of American History," by Marion Clawson. *Science,* vol. 204, pp 1168–1174, 1979.

---

**Dr. William B. Meyer** *is a geographer currently employed on the research faculty of the George Perkins Marsh Institute at Clark University in Worcester, Massachusetts. His principal interests lie in the areas of global environmental change with particular emphasis on land use and land cover change, in land use conflict, and in American environmental history.*

---

From *Consequences,* Spring 1995, pp. 25–33. © 1995 by Saginaw Valley State University. Reprinted by permission.

Article 17

INFRASTRUCTURE

# Operation Desert Sprawl

### The biggest issue in booming Las Vegas isn't growth. It's finding somebody to pay the staggering costs of growth.

BY WILLIAM FULTON AND PAUL SHIGLEY

On a late spring afternoon, the counters at the Las Vegas Development Services Center are only slightly less crowded than those at the nearby McDonald's. Here, in a nondescript office building some eight blocks from City Hall, a small army of planners occupies counters and cubicles, standing ready to process the daily avalanche of building projects. Several times a minute, people with blueprints tucked under their arms hurry in or out the door.

Upstairs, Tim Chow, the city's planning and development director, shakes his head and smiles. He says he has "the toughest planning job in the country," and he may be right. Two hundred new residents arrive in Las Vegas every day; a house is built every 15 minutes. Last year alone, the city issued 7,700 residential building permits, plus permits for $200 million worth of commercial construction—enough to build a good-sized Midwestern county seat from scratch.

Before he came to Nevada in April, Chow held a similar position in a smaller county in California. There, he points out, the planning process grinds slowly. State law—and local politics—require extensive environmental studies and public hearings before planners approve subdivisions and retail centers.

But this isn't California. It's Las Vegas, the nation's fastest-growing community, and the policy is to build first and ask questions later. "We just don't have time to do the kinds of rigorous analysis done in other places with regard to compatibility, impacts, infrastructure, coordination," Chow says. "Sometimes you make mistakes, and the impacts of those mistakes are felt many years later."

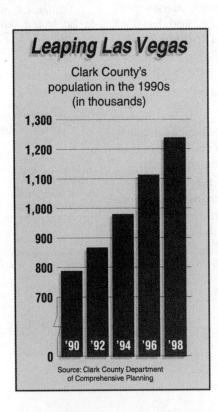

Tim Chow's planning counter isn't the only one in this area that looks like a fast-food restaurant. Fifteen miles to the southeast, his counterpart in Henderson, Mary Kay Peck, is presiding over the rapid creation of the second-largest city in Nevada. Henderson officials claim to have "the highest development standards in the Las Vegas Valley," but that hasn't stopped the city from growing seven-fold in the past two decades, to 170,000 people. The city recently annexed 2,500 acres to accommodate a new Del Webb project, and is lining up a federal land exchange that will allow the addition of 8,000 more acres near McCarran International Airport.

Henderson is in a running argument with well-established Reno over which city is second most populous in the state after Las Vegas. But that argument won't last much longer. "We have room," Peck says, "to grow and grow and grow." Henderson is expected to add another 100,000 people in the next decade—pushing it far past Reno. By 2010, Henderson will be as big as Las Vegas was in 1990.

The new residential subdivisions built all over the Las Vegas Valley in the past few years have created an enormous unsatisfied demand for parks, transportation, and water delivery systems. "Traffic is probably 100 times worse than it was 10 years ago," says Bobby Shelton, spokesman for the Clark County Public Works Department.

Local governments are trying to cope with the onslaught. A rail system connecting downtown, the casino-lined Strip and the airport is on the drawing board. Voters recently approved two tax increases—one for transportation projects, one for water projects—that together will produce some $100 million a year. But even with these projects moving forward Vegas-style—

ready, fire, aim—the problem is getting worse, not better. Even the local building industry acknowledges that growth has gotten far ahead of the infrastructure that is needed to support it.

"Someone back in the '70s and '80s should have said, 'Hey, we are going to need parks and schools and roads,'" says Joanne Jensen, of the Southern Nevada Home Builders, who moved to Las Vegas from Chicago in 1960. "We're playing catch-up." After two months in town, Planning Director Chow uses the same words to describe the situation.

Take the infrastructure needs faced by any fast-growing American community, multiply by a factor of about 20, and you get a rough idea of what is going on in Las Vegas. It is similar to other places in that it has come to realize that residential development does not pay for itself. The difference is in the magnitude of its problem. No other city of comparable size is taking on people at anything remotely close to the Las Vegas rate. No other city is being challenged to build so much so quickly.

The infrastructure will be built. That is not really the issue. Las Vegas and Henderson will continue to grow. The question is who will pay the bill.

As in other American cities, property owners have paid for community infrastructure during most of Las Vegas' history. But they have grown weary of the expense, and the one obvious way to grant them relief is to make developers cover more of the cost. Local homebuilders estimate that "impact fees"—fees paid by developers to cover the cost of community infrastructure—already account for a quarter of the cost of a new house in Las Vegas.

It was no surprise this year that a budding politician decided to make a name for himself by proposing that Las Vegas solve its growth problems by soaking its developers. The surprise was who that politician turned out to be, and how potent his message proved.

Nine months ago, Oscar Goodman was known around town as the classic Vegas mob lawyer—a veteran criminal defense attorney, given to wearing dark, double-breasted suits and a short-cropped silver beard. Today, on the strength of his soak-the-developer campaign rhetoric, Oscar Goodman is mayor of Las Vegas.

Ever since the mid 1980s, civic leaders here have worked hard to bury the Bugsy Siegel image—downplaying the gangster past, building suburban-style homes and office parks at a furious pace, and generally trying to re-position Vegas as an affordable, high-energy Sun Belt city attractive to everyone from retirees to young families. A few years ago, the *New York Times Magazine* heralded this new era in Las Vegas by reporting the city's transformation "from vice to nice."

It's true that the gambling industry remains the local economy's bedrock foundation. Las Vegas boasts 110,000 hotel rooms—one for every 11 residents of the region—and attracts 30 million visitors a year. But a fresh-scrubbed image was necessary to catapult Vegas past its sleazy resort-town reputation and support a new wave of mainstream urban growth. And so even as it tried to remain affordable, Las Vegas began to go upscale as well.

Perhaps the most highly publicized success story is Summerlin, the Howard Hughes Corp.'s 35-square-mile development on the valley's west side. On a tract of land that was nothing but empty desert when Hughes bought it, 35,000 people now live in the earth-tone houses and apartments lining cul-de-sacs. The company predicts the population could reach 180,000 by buildout. And lower-end Summerlin knock-offs line the roads leading from the big development toward downtown Las Vegas and the Strip.

Summerlin's sales literature boasts of a school and recreation system that starts with T-ball fields and continues on up to college scholarships for the local youth. "Summerlin," one of the brochures says, "offers a surprising number of public schools and private academies, offering a full range of close-to-home choices for preschool/kindergarten through high school education." It is clever promotion like this that has enabled Las Vegas to become the nation's fastest-growing Sun Belt city without sacrificing its economic base of tourism and gambling. But the cost of all of that growth has been high nevertheless. There was no way to finesse the need to build public infrastructure at an exponential rate. And for most of the past decade, the local governments swallowed hard and arranged for that infrastructure in the old-fashioned way: by raising taxes.

Three years ago, for example, traffic congestion had become so bad that it was decided to speed up completion of the 53-mile Las Vegas Beltway—by 17 years—so it could open in 2003. In Henderson, giant belly scrapers building the road rumble back and forth just beyond the walled-in backyards of brand-new houses.

Amazingly, the $1.5 billion beltway is being finished without federal funds. A 1 percent motor vehicle privilege tax and a "new home fee" of about $500 per house are generating $50 million a year, which goes toward bonds issued to raise capital for the construction.

Seven years ago, Las Vegas had no bus system at all. Today, the Citizen Area Transit system carries 128,000 passengers a day—the same volume as the busiest part of the beltway. Passengers pay 50 percent of the system's costs at the farebox—a higher percentage than in almost any other American city. The Regional Transportation System hopes to begin construction of a 5.2-mile rail line around downtown in 2001—mostly with federal funds and sales-tax revenues—and tie it to a privately funded monorail along the Strip.

Meanwhile, the Southern Nevada Water Authority, which serves as a wholesaler to cities and water districts in the Las Vegas area, is building a $2 billion water delivery project. The project involves a "second straw" from Lake Mead, which Hoover Dam creates about 30 miles southeast of Las Vegas, and 87 miles of large water mains in the Las Vegas Valley. The water agency's goal is to provide enough water for an additional 2 million people.

There is one catch so far as water supply is concerned: Las Vegas still lacks the legal right to draw additional water from Lake Mead. Changing this arrangement will require a massive political deal to overturn the 75-year-old agreement among seven Western states along the Colorado River. But that didn't stop Clark County voters from approving a quarter-cent sales-tax increase last year—expected to generate close to $50 million a year—to pay for the new straw and the other improvements.

So it can't be said that Las Vegas area residents have been unwilling to open their wallets and spend money for growth—they have spent heavily for it. But they realize all too well how much of the bill remains to be paid. This is the issue that is driving Las Vegas politics, and producing its unexpected results.

Even before this year's mayoral campaign, critics of the growth machine began speaking more loudly in the Las Vegas Valley, and their demand that developers pay impact fees had begun to resonate with the voters.

The ringleaders of this new movement have been Jan Laverty Jones, who was Goodman's predecessor as mayor, and

Dina Titus, a political science professor at the University of Nevada–Las Vegas, who is Democratic leader in the Nevada Senate. Because Las Vegas has a city manager form of government, the mayor's job was never viewed as important. But Jones, a former Vegas businesswoman who used to appear as Little Bo Peep in television car ads, adopted a high profile—especially on growth issues—at exactly the moment when growth was creating massive frustration.

Jones pushed to require developers to bear more of the cost of parks, roads and other community improvements. Meanwhile, Titus took an aggressive approach in the legislature. In 1997, she floated a "Ring Around the Valley" urban growth boundary concept. That bill failed, but last year she managed to push through a bill requiring local governments to conduct an impact analysis for large commercial and residential projects. Although the bill doesn't require mitigation of impacts, Titus hopes it will encourage local officials in the Las Vegas area to impose mitigation requirements more often and more rationally. "Right now, it's very capricious," she says. "Somebody has to put in street lights, another developer doesn't have to put in anything. It all depends on the political situation at hand."

Into this combustible situation stepped two mayoral candidates with exactly the wrong credentials to deal with it. Arnie Adamsen, a three-term city council member, was a title company executive; Mark Fine was one of the creators of Summerlin. Their ties to the real estate industry could not have been more conspicuous. Both Fine and Adamsen acknowledged that growth was a problem, but refused to recommend that developers pay more of its cost. That opened the door for Oscar Goodman, an unlikely political leader even in a town as unusual as Las Vegas.

At age 59, Goodman had lived in Las Vegas far longer than most of the voters—more than 30 years. A native of Philadelphia, he once clerked for Arlen Specter, then a prosecutor, now a U.S. senator. In the early '60s, however, Goodman headed for Las Vegas, looking for a place where he could make a reputation on his own.

Before long, he gained a reputation as a colorful—and effective—defender of accused mobsters. He once persuaded a judge to drop mob financier Meyer Lansky as a defendant in a case, apparently because of Lansky's failing health. In his most celebrated victory, he kept Tony "The Ant" Spilotro out of jail in the face of multiple murder and racketeering charges. Spilotro became the model for the character played by Joe Pesci in the movie "Casino"; Goodman played himself in the same movie.

Of course, Goodman always insisted that he had a wide-ranging criminal law practice, and only 5 percent of his clients were alleged mobsters. Furthermore, Goodman and his wife were regarded around town as good citizens who often contributed to, or even spearheaded, civic causes. Even so, when Goodman first announced his candidacy for mayor, it was surprising that he even wanted the job.

> **It wasn't ties to organized crime that concerned voters in the Las Vegas mayor's race; it was ties to real estate.**

Given the structure of Las Vegas' government, most mayors have been part-timers who served as glorified presiding officers for the city council. On the other hand, the most recent occupant of the seat had shown that it could be more than a ceremonial position. "Jones made it a more important job," says Eugene Moehring, author of *Resort City in the Sunbelt*, a history of Las Vegas. "She showed that you could use it as a platform. I think Oscar saw that and decided to go for it."

Goodman announced that, if elected mayor, he would give up his law practice. Even so, he wasn't taken seriously at first by the Las Vegas political establishment. Then he started talking about sprawl, real estate developers and impact fees. "Growth has to pay for growth," Goodman said in his radio and television commercials, using his fast-talking West Philadelphia accent to get his message across. "Either it has to come from taxpayers or developers.... It's time to make developers pay impact fees when they build new homes."

Goodman argued that even a modest impact fee would bring in more than $15 million a year to help pay for sprawl costs and revitalize Las Vegas' downtown. He recommended a fee of roughly $2,000 for each new home, and said he would use it to revitalize neglected downtown neighborhoods. Such a scheme would require a change in state law—and a big fight with the real estate lobby—but Goodman promised to take on both the legislature and the builders. Polls showed that 80 percent of Las Vegans supported higher development fees.

Goodman was widely perceived in Las Vegas as a "Jesse Ventura candidate"—a celebrity from another field with a populist message and enough personal wealth to assure that he would not be bought by special interests. His opponents essentially played into Goodman's hands. Adamsen, his opponent in the June runoff, insisted that the fees would merely be passed on to homebuyers. "Development fees are very popular with the public because they want someone else to pay for it," he said. Adamsen focused on Goodman's criminal connections, saying they would harm the city's new, family-oriented image. If anybody in town knew about extracting money from a community through nefarious means, he charged, it was Goodman, not the developers.

In the end, though, Goodman's contacts with organized crime were a non-issue. In the Las Vegas of 1999, impact fees are a bigger issue than crime, organized or otherwise. What mattered to most voters was that the candidate didn't have any close ties to the real estate business. Goodman won 64 percent of the vote in his runoff with Adamsen. The Jesse Ventura approach worked so well that even many leading members of the Las Vegas political establishment embraced Goodman during the runoff.

On the last Monday in June, Oscar Goodman took the oath of office before an overflow crowd in the Las Vegas City Council chambers, then stepped outside for his first press conference as mayor. At 10:30 in the morning, it was 95 degrees, with a hot, dry desert wind blowing across the City Hall courtyard. Wearing a trademark double-breasted suit, Goodman was unfazed by what was, for Las Vegas, a typical June morning.

And he also seemed unfazed by his quick introduction to the real world of municipal politics in America's most transient city. In the three short weeks between the election and the inauguration, one of Goodman's campaign aides was accused of attempting to charge $150,000 to arrange an appointment to the City Council. The new mayor was faced with the question of what to do about renewing the contract of the city's waste hauler, who had contributed heavily to his campaign. And, of course, during the Goodman interregnum, 4,000 new residents had arrived in Las Vegas.

Goodman insists he will take his plan for a $2,000-a-house impact fee to the legislature. But the next regular session does not begin until January of 2001, and even with 18 months to build support, his task will not be easy. Current state law prevents any such scheme from being implemented—as it does many other development fees—and the real estate lobby has vowed to oppose it.

Meanwhile, says Goodman with his typical glib charm, "I've sat down with the developers, and I've told them we're going to run an efficient City Hall operation that will meet their day-to-day needs. Those guys are going to *want* impact fees to pay us back for all the good things we're going to do for them in the next two years."

Even Goodman sympathizers agree that Goodman—like that other populist, Jesse Ventura—faces a difficult challenge in learning how to govern on the job. "Growth is a tough issue, and he's going to have to hit the ground running on it," says former Mayor Jan Jones. "It's going to be a big learning curve." In particular, Goodman will have to cultivate regional agencies and suburban governments in Las Vegas—with which Jones often clashed—in order to curb growth on the metropolitan fringe and encourage renewal of older areas in Las Vegas itself.

Increasingly, however, there is agreement among both politicians and developers that the future of Las Vegas—like the future of most American communities—involves developers paying for the impact of growth somehow: There simply is no other way to finance the infrastructure costs. Whatever Goodman does, the Southern Nevada Water Authority is planning to levy a hookup fee of several thousand dollars a unit to help pay for the second straw from Lake Mead. That will only increase the average home price in the metro area, which is already up to $142,000.

Oscar Goodman may or may not succeed as mayor, but it's clear that the political sentiment that he tapped into is here to stay. In wide-open Las Vegas—as in so many of the more conventional cities across the country—the crucial votes no longer lie with the upwardly mobile families desperate for a place to live. They lie with the established middle-class residents who are tired of paying the bill.

From *Governing*, August 1999, pp. 16–21. © 1999 by William Fulton and Paul Shigley, editors of the California Planning & Development Report. Reprinted by permission.

# A MODEST PROPOSAL TO STOP GLOBAL WARMING

**While evidence continues to mount that humans are heating the globe, the world's nations squabble over a complex fix too timid to solve the problem. But we can stop global warming—by calling an end to the Carbon Age.**

by Ross Gelbspan

**The United States is constantly warning** against the danger posed by "rogue states" like Iraq or North Korea. But last November we behaved very much like an outlaw nation ourselves by unilaterally scuttling climate talks at The Hague, Netherlands. More than half of the world's industrial nations declared their willingness to cut their consumption of fossil fuels to forestall global warming, but when the United States would commit to nothing more than planting a few trees and buying up cheap pollution allowances from poor countries, the talks collapsed.

The meeting was probably irrelevant anyway. As the three years of frustrating negotiations fell apart, the United Nations–sponsored Intergovernmental Panel on Climate Change (IPCC), which had previously projected an increase in average global temperatures of 3 to 7 degrees Fahrenheit this century, raised its upper estimate to 10.4 degrees. To restabilize the climate, declared the 2,000 eminent climatologists and other scientists, humanity needs to cut its greenhouse-gas emissions ten times more than the 5.2 percent reductions discussed at The Hague.

As heat records continue to be broken and extreme weather events intensify around the world, the reality of global warming is sinking in—everywhere, it seems, except on Capitol Hill. At the 1998 World Economic Forum in Davos, Switzerland, the CEOs of the world's 1,000 biggest corporations surprised organizers by voting climate change the most critical problem facing humanity. European countries are planning drastic reductions in their $CO_2$ emissions, while growing numbers of corporate leaders are realizing that the necessary transition to highly efficient and renewable energy sources could trigger an unprecedented worldwide economic boom.

This growing international consensus may show us the way to a workable global solution. Instead of The Hague's torturous haggling over the complex minutiae of virtually meaningless goals, the earth's nations could jointly initiate an aggressive worldwide effort to halt and turn back the ominous heating of the globe—and come out stronger, safer, and richer.

**The alternative is dismal and frightening.** A recent report from the National Climatic Data Center predicts ever harsher droughts, floods, heat waves, and tropical storms as the atmosphere continues to warm. "We found that extreme weather events have had increasing impact on human health, welfare, and finances," said the Center's David Easterling. "This trend is likely to become more intense as the climate continues to change and society becomes more vulnerable to weather and climate extremes."

This vulnerability is underscored by a financial forecast from the world's sixth-largest insurance company. Previous reports from property insurers had emphasized the financial risks to the industry itself, but last November Dr. Andrew Dlugolecki, an executive of the United Kingdom's CGNU, released a study projecting that infrastructure and other property damage, bank and insurance industry losses, crop failures, and other costs of unchecked climate change could bankrupt the global economy by 2065.

And the coming changes will occur 50 percent faster than previously thought, say researchers at the Hadley Center, the UK's leading climate-research agency. Previous estimates of the rate of climate change have been based on projections of the earth's capacity—at current temperatures—to absorb carbon dioxide through its vegetation and, to a lesser extent, its oceans. For the last 10,000 years, these natural "carbon sinks" have maintained atmospheric carbon levels of about 280 parts per million. Since the late 19th century, however, human use of coal and oil has escalated dramatically, leading to our present atmospheric carbon level of about 360 parts per million—a level not experienced in 420,000 years. In a blow to the United States'

## Article 18. A MODEST PROPOSAL TO STOP GLOBAL WARMING

hope that planting forests in developing countries could absolve it of the need to conserve energy, Hadley's researchers found that photosynthesis slows as the climate warms. Plants' absorption of $CO_2$ diminishes, and soils begin to release more carbon than they absorb, turning what had been carbon sinks into carbon sources.

---

### Life in a Warmer World

#### by Paul Rauber

Unless we take drastic action—and soon—global warming threatens to plunge the world into a series of climatic crises. The following are some of the impacts foreseen by the Intergovernmental Panel on Climate Change, from a report published in February:

- Average temperatures will increase by 10.4°F over the next 100 years.
- Heavier flooding, especially in coastal cities, will affect 200 million people.
- Deserts will expand, particularly in Asia and Africa. Crop yields will decline, and droughts will grow more severe.
- The Gulf Stream may slow down, which would result in a dramatically colder northern Europe.
- The Greenland or West Antarctic ice sheets will shrink significantly, raising sea level by almost 3 feet this century and causing the inundation of low-lying islands like Samoa, the Maldives, Mauritius, and the Marshalls. Over the course of the millennium, sea level could rise by 20 feet.
- Endangered species will disappear as habitat dwindles. Coastal ecosystems will flood, freshwater fish will be unable to migrate to cooler regions, and animals already adapted to cold may be left with nowhere to go.
- Incidence of heat-related deaths and infectious diseases like malaria and dengue fever will increase, spreading beyond what are now the tropics.

---

Similarly, a team led by Sydney Levitus, head of the National Oceanic and Atmospheric Administration's Ocean Climate Laboratory, found that while oceans absorb heat, that effect can be temporary. During the 1950s and '60s, the group found, subsurface temperatures in the Atlantic, Pacific, and Indian Oceans rose substantially while atmospheric temperatures remained fairly constant. But in the 1970s atmospheric temperatures trended upward—driven, in part, by warmth released from deep water. "[O]cean heat content may be an early indicator of the warming of surface, air, and sea surface temperatures more than a decade in advance," said Levitus. Later this century, his researchers predicted, the oceans may release even more heat into an already warming atmosphere.

That grim prediction was echoed by a report from the International Geosphere-Biosphere Programme, which cast doubt on the ability of farmland or forests to soak up the vast amounts of $CO_2$ that humanity is pumping into the atmosphere. "There is no natural 'savior' waiting to assimilate all the anthropogenically produced $CO_2$ in the coming century," the report concluded.

The inadequacy of the percentage goals haggled over at The Hague was underscored by a research team led by Tom M. W. Wigley of the National Center for Atmospheric Research, which estimated that the world must generate about half its power from wind, sun, and other noncarbon sources by the year 2018 to avoid a quadrupling of traditional atmospheric carbon levels, which would almost certainly trigger catastrophic consequences. Writing in the journal *Nature*, Wigley's team recommended "researching, developing, and commercializing carbon-free primary power technologies … with the urgency of the Manhattan Project or the Apollo space program."

**Far from recognizing that urgency,** the United States' official position seems to be to minimize the severity of global warming. This recalcitrance can be traced to a relentless disinformation campaign by the fossil-fuel lobby to dismiss or downplay the climate crisis. For years, coal and oil interests have funded a handful of scientists known as "greenhouse skeptics" who cast doubt on the implications and even existence of global warming. Enormous amounts of money spent by their corporate sponsors have amplified the skeptics' voices out of all proportion to their standing in the scientific community, giving them undue influence on legislators, policymakers, and the media.

But with the skeptics being marginalized by the increasingly united and alarming findings of mainstream science, industry PR campaigns have taken to exaggerating the economic impacts of cutting back on fossil fuels. On the other side are more than 2,500 economists, including 8 Nobel laureates, who proclaimed in a 1997 statement coordinated by the think-tank Redefining Progress that the U.S. economy can weather the change, and even improve productivity in the long run. Industry is also attacking the diplomatic foundations of the Kyoto Protocol—the international agreement The Hague meeting was meant to implement—claiming that the United States would suffer unfairly because developing countries were exempted from the first round of emissions cuts. Yet the rationale for this exemption—that since the industrial nations created the problem, they should be the first to begin to address it—was ratified by President George H. W. Bush himself when he signed the 1992 Rio Treaty.

**The average American is responsible for about 25 times more $CO_2$ than the average Indian.**

The central mechanism of the Kyoto Protocol, as promoted by the United States, is "emissions trading." That system was intended to find the cheapest way to reduce global carbon levels. It allocated a certain number of carbon-emission "credits" to each country, and then permitted nations with greater emissions to buy unused credits from other countries—for example, by financing the planting of trees in Costa Rica.

But international carbon trading turned out to be a shell game. Carbon is burned in far too many places—vehicles, factories, homes, fields—to effectively track even if there were an international monitoring system. Trading also became a huge source of contention between industrial and developing countries. In allocating emission "rights," for instance, all countries were given their 1990 emission levels as a baseline, but the developing nations argue that this would lock in the advantages of the already-industrialized First World. Many developing countries advocate what they claim is a far more democratic, "per capita" basis for allocating emissions, which would grant every American the same quantity of emissions as, say, every resident of India. (Currently, the average American is responsible for about 25 times more $CO_2$ than the average Indian.)

A second level of inequity embedded in emissions trading is that industrialized countries could buy as many credits from poor countries as they want, banking those big, relatively cheap reductions indefinitely into the future. So when developing countries are eventually obliged to cut their emissions, they will be left with only the most expensive options, such as financing the production of fuel cells or solar installations.

Finally, carbon trading in itself can only go so far; its optimal use would be as a fine-tuning mechanism to help countries achieve the last 10 to 15 percent of their obligations. Measured against what it would take to actually cool the planet, emissions trading is ultimately a form of institutional denial.

**Despite U.S. obstructionism,** several European countries are now setting more ambitious goals. The United Kingdom last year committed to reductions of 12.5 percent by 2010, and a royal commission is calling for 60 percent cuts by 2050. Germany is also considering 50 percent cuts. Holland—a country at particular risk from rising sea levels—just completed a plan to slash its emissions by 80 percent in the next 40 years. It will meet those goals through an ambitious program of wind-generated electricity, low-emission vehicles, photovoltaic and solar installations, and other noncarbon energy sources. And a number of developing countries are voluntarily installing solar, wind, and small-scale hydro projects, despite their exemption under the Kyoto Protocol from the first round of cuts.

Some major industrial players are also reading the handwriting on the wall. British Petroleum, despite its attempts to drill in the Arctic National Wildlife Refuge, is investing substantial resources in solar power. The company, which now promotes itself as "Beyond Petroleum," anticipates doing $1 billion a year in solar commerce by the end of the decade. Shell has created a $500 million renewable-energy company. In fact, most of the major oil companies—with the notable exception of ExxonMobil—now acknowledge the reality of climate change. In the automotive arena, Ford and DaimlerChrysler have invested $1 billion in a joint venture to put fuel-cell-powered cars on the market in 2004. And William Clay Ford recently declared "an end to the 100-year reign of the internal combustion engine."

While some environmentalists dismiss these initiatives as "greenwashing," they mark an enormous change in industry's public posture. Only a year or two ago, working through such groups as the Western Fuels Association and the Global Climate Coalition, the oil and coal companies sought to dismiss the reality of climate change and cast doubt on the findings of the IPCC. Today, with these arguments largely discredited, the Global Climate Coalition has essentially collapsed. Oil and auto executives are beginning to choose a new approach: to position their firms as prominent players in the coming new-energy economy. (This doesn't preclude backsliding. In March, conservatives' complaints persuaded Bush to break a campaign promise to regulate $CO_2$ emissions from power plants—thus hanging out to dry EPA chief Christie Todd Whitman, who had widely promoted the idea, and Treasury Secretary Paul O'Neill, who has called for a crash program to deal with climate change.)

U.S. labor unions are also facing up to the future, working with environmentalists on an agenda to increase jobs while reducing emissions—witness the recent call by AFL-CIO president John Sweeney and Sierra Club executive director Carl Pope for a "package of worker-friendly domestic carbon-emission reduction measures." Building and maintaining the necessary new energy facilities will take an army of skilled workers, which organized labor can provide.

---

### How to End—and Reverse—Global Warming

Turning down the earth's thermostat will take a 70 percent reduction in our current global level of carbon emissions. That's a big job—but not impossible. Here's one blueprint to get there:

- Redirect the $300 billion the world currently spends on subsidies for fossil fuels to renewable power: solar systems, wind farms, geothermal, and fuel cells.
- Require every country, whether developed or not, to commit to specific reductions in carbon emissions every year—say, 5 percent—until the goal of 70 percent is met.
- Fund renewable-energy projects in the developing world through a tax on international currency speculation. Such a tax could raise $300 billion a year to help other nations avoid the industrial world's mistakes. Combined with the progressive reduction schedule above, it would provide a monetary incentive for innovation and introduction of renewable technologies.

This isn't the only possible formula, but any approach to reversing the warming of the earth's atmosphere must be at least as ambitious. The transformation will be dramatic, but not necessarily painful—unless we delay too long. —*P.R.*

---

**Despite these encouraging developments,** the United States continues to obstruct rather than lead the world in addressing climate change. Former president Clinton blamed the media, saying that until the public knows more about the threat there will not be sufficient popular support to address the issue in a meaningful way. George W. Bush and Dick Cheney, oilmen both, are more inclined to protect the petroleum industry's short-term profitability than to promote its inevitable transformation.

Thus, the public debate is still stuck in the ineffective Kyoto framework. So two years ago, a small group of energy executives, economists, energy-policy specialists, and others (including the author) fashioned a bundle of strategies designed to cut carbon emissions by 70 percent, while at the same time creating a surge of new jobs, especially in developing countries.

At present, the United States spends $20 billion a year to subsidize fossil fuels and another $10 billion to subsidize nuclear power. Globally, subsidies for fossil fuels have been estimated at $300 billion a year. If that money were put behind renewable technologies, oil companies would have the incentive to aggressively develop fuel cells, wind farms, and solar systems. (A portion of those subsidies should be used to retrain coal miners and to construct clean-energy manufacturing plants in poor mining regions.)

The strategy also calls on all nations to replace emissions trading with an equitable fossil-fuel efficiency standard. Every country would commit to improving its energy efficiency by a specified amount—say 5 percent—every year until the global 70 percent reduction is attained. By drawing progressively more energy from noncarbon sources, countries would create the mass markets for renewables that would bring down their prices and make them competitive with coal and oil. This approach would be easy to negotiate and easy to monitor: A nation's progress could be measured simply by calculating the annual change in the ratio of its carbon fuel use to its gross domestic product.

A global energy transition will cost a great deal of money (although not nearly as much as ignoring the problem). Until clean-energy infrastructures take root, providing clean energy to poor countries would cost several hundred billion dollars a year, say researchers at the Tellus Institute, an energy-policy think tank in Boston. A prime source for that funding would be a "Tobin tax" on international currency transactions, named after its developer, Nobel prize-winning economist Dr. James Tobin. Every day, speculators trade $1.5 trillion in the world's currency; a tax of a quarter-penny on the dollar would net about $300 billion a year for projects like wind farms in India, fuel-cell factories in South Africa, solar assemblies in El Salvador, and vast, solar-powered hydrogen farms in the Middle East. Unlike a North-South giveaway, the fund is a transfer of resources from the finance sector—in the form of speculative transactions—to the industrial sector-in the form of productive, wealth-generating investments. Banks would be paid a small percentage fee to administer the fund, partly offsetting their loss of income from the contraction of currency trading. Creation of a fund of this magnitude would follow the kind of thinking that gave rise to the Marshall Plan after World War II. Without that investment, the nations of Europe could be a collection of impoverished, squabbling states instead of the fruitful and prosperous trading partners we have today.

This approach has another precedent, in the Montreal Protocol, the treaty that ended the production of ozone-destroying chemicals. It was successful because the same companies that made the destructive chemicals were able to produce their substitutes. The energy industry can be reconfigured in the same way. Several oil executives have said in private conversations that they can, in an orderly fashion, decarbonize their energy supplies. But they need the governments of the world to regulate the process so that all companies can make the transition simultaneously, without losing market share to competitors.

## The very act of addressing the crisis would acknowledge that we are living on a finite planet and foster a new ethic of sustainability.

The plan would be driven by two engines: The progressive-efficiency standard would create the regulatory drive for all nations to transform their energy diets, and the tax generating $300 billion a year for developing countries would create a vast market for clean-energy technologies. It has been endorsed by a number of national delegations—India, Bangladesh, Germany, Mexico, and Britain, among others. While the plan will require refinement, it is of a scale appropriate to the magnitude of the problem.

**Ultimately, the climate crisis** provides an extraordinary opportunity to help us calibrate competition and cooperation in the global economy, harnessing the world's technical ingenuity and the power of the market within a regulatory framework that reflects a consensus of the world's citizens. The very act of addressing the crisis would acknowledge that we are living on a finite planet and foster a new ethic of sustainability that would permeate our institutions and policies in ways unimaginable today. It could subordinate our current infatuation with commerce and materialism to a restored connection to our natural home, ending the exploitative relationship between our civilization and the planet that supports it.

Angry nature is holding a gun to our heads. Drought-driven wildfires last summer consumed 6 million acres in the western United States. Last fall, the United Kingdom experienced its worst flooding since record-keeping began 273 years ago. In Iceland, Europe's biggest glacier is disintegrating. And the sea ice in the Arctic has thinned by 40 percent in the last 40 years.

We have a very small window of opportunity. The choice is clear. The time is short.

---

ROSS GELBSPAN *is author of* The Heat Is On: The Climate Crisis, the Cover-up, the Prescription *(Perseus Books, 1998). He maintains the Web site* www.heatisonline.org, *a project of the Green House Network.*

# A greener, or browner, Mexico?

**NAFTA purports to be the world's first environmentally friendly trade treaty, but its critics claim it has made Mexico dirtier. There is evidence on both sides**

CIUDAD JUAREZ

GIVEN the industrial invasion that the North American Free-Trade Agreement has brought to the cities that line Mexico's border with the United States, one might expect the skies of Ciudad Juarez to be brown with pollution, and its watercourses solid with toxic sludge. But no. The centre of Ciudad Juarez looks like a poorer version of El Paso, Texas, its cross-border neighbour: flat, dull and full of shopping malls. Since NAFTA took effect on January 1st 1994, Mexico has passed environmental laws similar to those of the United States and Canada, its NAFTA partners; has set up a fully-fledged environment ministry; and has started to benefit from several two- and three-country schemes designed to fulfil "side accords" on the environment—the first such provisions in a trade agreement.

But the debate over NAFTA's impact on Mexico's environment remains polarised. Certainly NAFTA has had the hoped-for result of encouraging industries to move to Mexico. Since 1994, the number of *maquiladoras* (factories using imported raw materials to make goods for re-export, usually to the United States) has doubled, and half of the new factories are in the 100km-wide (60 mile) northern border region.

But that has put pressure on water supplies. The influx of migrant workers is slowly drying out cities like Juarez, which shares its only water source, an underground aquifer, with El Paso (which has other water supplies). Growth is aggravating old deficiencies: 18% of Mexican border towns have no drinking water, 30% no sewage treatment, and 43% inadequate rubbish disposal, according to Franco Barreno, head of the Border Environment Co-operation Commission (BECC), a bilateral agency set up under NAFTA. Putting that right, which is the BECC's job, will cost $2 billion–3 billion.

The problem of toxic waste is more serious still, and is being ignored. Mexico has just one landfill site for hazardous muck, which can take only 12% of the estimated waste from non-*maquila* industry; the rest is dumped illegally. Mexican law says *maquiladoras* must send their toxic waste back to the country the raw materials came from but, with enforcement weak and repatriation expensive, many are probably ignoring the law, says Victoriano Garza, of the Autonomous University of Ciudad Juarez.

Mexico has long been a dumping-ground for unwanted rubbish from the United States. But in fact less than 1% of the country's toxic gunge is generated in the border region. After 1994, despite the growth in *maquiladoras*, Mexico's exports of toxic waste to the United States dropped sharply—suggesting that tougher rules were making companies adopt greener manufacturing policies. But then, in 1997, waste exports mysteriously shot up again (see chart).

Even greens are split about NAFTA. "In 15 years as an environmentalist, I haven't noticed any change in policy," laments Homero Aridjis, a writer. Quite the contrary, says Tlahoga Ruge, of the North American Centre for Environmental Information and Communication: "NAFTA has brought the environment into the mainstream in Mexico, as something to be taken seriously."

Such is the complexity of the issue, only in June did the Commission for Environmental Co-operation (CEC), a Montreal-based body created in 1994 to implement the environmental side-agreement, publish its methodology for assessing NAFTA's impact. An accompanying case study, on maize farming in Mexico, illustrates the difficulties. Lower protective tariffs are forcing Mexico's subsistence maize farmers to modernise, change their

**Bubbling** Mexico's *maquiladora* plants: employment; transfer of toxic waste* (US tons '000), 1991–97. *to the United States. Sources: Mexican trade and industry ministry; United States Environmental Protection Agency

crops or look for other work; each choice has potential impacts, such as soil deterioration, depletion of the maize gene pool, or migration to cities. Social and environmental impacts go hand-in-hand.

Critics say that, as well as being slow, the CEC has too few powers. It can investigate citizens' complaints about breaches of each NAFTA country's national environmental laws, but governments are free to ignore its recommendations. Moreover, the side-agreement's gentle urgings are outgunned by the main accord's firm admonition, in its Chapter 11, to protect foreign investors from uncertainty. A recent report by the International Institute for Sustainable Development, a group based in Winnipeg, Canada, warns that this provision is increasingly being used by business to the detriment of environmental protection. And unlike the CEC, the arbitration bodies that settle Chapter 11 disputes make binding decisions.

Companies have challenged environmental laws in all three NAFTA countries, but Mexico is particularly vulnerable. Its laws are new, its institutions untested and its environmental culture undeveloped. To these handicaps was added the collapse of Mexico's peso in 1995, from which it is still trying to recover. Beefing up environmental institutions has not been its top priority. In Juarez, for example, there are only 15 federal environmental inspectors (and they cover the whole of Chihuahua state). Mexico also lags in data-collection: next week the CEC is due to publish a report on industrial emissions and air pollution, but in the United States and Canada only, because Mexican factories do not have to report their emissions.

These things need to change. NAFTA is supposed to lead to completely free trade after 15 years; its full impact has yet to be felt. In 2001, the *maquiladoras* will lose some of their tax breaks. Increasingly, American firms may opt to set up ordinary factories in Mexico, which would not have to send their waste back home. Mexico badly needs somewhere to put this stuff, but so far attempts to find sites for new toxic-waste landfills have been scuppered by not-in-my-backyard opposition (and, ironically, by the strict stipulations of the new environmental laws).

But not all is murk. Besides better laws and institutions, Mexico has seen a rapid rise in non-governmental organisations, themselves partly an import from the north, which work on the education and awareness-raising that they say the government is neglecting. Ciudad Juarez may be running out of water to drink, but Mexico City's aquifer is so low that the whole place is slowly sinking. And, unlike the capital, Juarez benefits from all kinds of two-country efforts to fix its problem, thanks mainly to NAFTA.

---

Reprinted with permission from *The Economist*, August 7, 1999, pp. 26–27. © 1999 by The Economist, Ltd. Distributed by The New York Times Special Features.

# UNIT 3
# The Region

## Unit Selections

20. **The Rise of the Region State**, Kenichi Ohmae
21. **Continental Divide**, Torsten Wohlert
22. **A Dragon With Core Values**, *The Economist*
23. **The Late Great Wall**, Melinda Liu
24. **A Continent in Peril**, *Time*
25. **AIDS Has Arrived in India and China**, Ann Hwang
26. **Greenville: From Back Country to Forefront**, Eugene A. Kennedy
27. **Death of a Small Town**, Dirk Johnson
28. **The Rio Grande: Beloved River Faces Rough Waters Ahead**, Steve Larese
29. **"You Call That Damn Thing a Boat?"** Charles W. Ebeling

## Key Points to Consider

- To what regions do you belong?
- Why are maps and atlases so important in discussing and studying regions?
- What major regions in the world are experiencing change? Which ones seem not to change at all? What are some reasons for the differences?
- What regions in the world are experiencing tensions? What are the reasons behind these tensions? How can the tensions be eased?
- Why are regions in Africa suffering so greatly?
- How will East Asia change as China and Taiwan expand relationships?
- Discuss whether or not the nation-state system is an anachronism.
- Why is regional study important?

 Links: www.dushkin.com/online/
These sites are annotated in the World Wide Web pages.

**AS at UVA Yellow Pages: Regional Studies**
http://xroads.virginia.edu/~YP/regional.html

**Can Cities Save the Future?**
http://www.huduser.org/publications/econdev/habitat/prep2.html

**IISDnet**
http://iisd.ca

**NewsPage**
http://www.individual.com

**Treaty on Urbanization**
http://www.geocities.com/atlas/urb/tretyurb.html

**Virtual Seminar in Global Political Economy/Global Cities & Social Movements**
http://csf.colorado.edu/gpe/gpe95b/resources.html

**World Regions & Nation States**
http://www.worldcapitalforum.com/worregstat.html

The region is one of the most important concepts in geography. The term has special significance for the geographer, and it has been used as a kind of area classification system in the discipline.

Two of the regional types most used in geography are "uniform" and "nodal." A uniform region is one in which a distinct set of features is present. The distinctiveness of the combination of features marks the region as being different from others. These features include climate type, soil type, prominent languages, resource deposits, and virtually any other identifiable phenomenon having a spatial dimension.

The nodal region reflects the zone of influence of a city or other nodal place. Imagine a rural town in which a farm-implement service center is located. Now imagine lines drawn on a map linking this service center with every farm within the area that uses it. Finally, imagine a single line enclosing the entire area in which the individual farms are located. The enclosed area is defined as a nodal region. The nodal region implies interaction. Regions of this type are defined on the basis of banking linkages, newspaper circulation, and telephone traffic, among other things.

This unit presents examples of a number of regional themes. These selections can provide only a hint of the scope and diversity of the region in geography. There is no limit to the number of regions; there are as many as the researcher sets out to define.

"The Rise of the Region State" suggests that the nation-state is an unnatural and even dysfunctional unit for organizing human activity. "Continental Divide" deals with changing geopolitical situations. The next two articles consider regional changes in China. A time map of AIDS in Africa and an article on AIDS in India and China follow. The devastation of this disease is a global problem. The continuing rise of Greenville, South Carolina, is documented in the next article. "Death of a Small Town" deals with population declines in the Great Plains. The hard-pressed Rio Grande River is discussed next. The final article, a study in historical geography, reviews the era of "whaleback" tankers on the Great Lakes.

# THE RISE OF THE REGION STATE

Kenichi Ohmae

**The Nation State Is Dysfunctional**

THE NATION STATE has become an unnatural, even dysfunctional, unit for organizing human activity and managing economic endeavor in a borderless world. It represents no genuine, shared community of economic interests; it defines no meaningful flows of economic activity. In fact, it overlooks the true linkages and synergies that exist among often disparate populations by combining important measures of human activity at the wrong level of analysis.

For example, to think of Italy as a single economic entity ignores the reality of an industrial north and a rural south, each vastly different in its ability to contribute and in its need to receive. Treating Italy as a single economic unit forces one—as a private sector manager or a public sector official—to operate on the basis of false, implausible and nonexistent averages. Italy is a country with great disparities in industry and income across regions.

On the global economic map the lines that now matter are those defining what may be called "region states." The boundaries of the region state are not imposed by political fiat. They are drawn by the deft but invisible hand of the global market for goods and services. They follow, rather than precede, real flows of human activity, creating nothing new but ratifying existing patterns manifest in countless individual decisions. They represent no threat to the political borders of any nation, and they have no call on any taxpayer's money to finance military forces to defend such borders.

Region states are natural economic zones. They may or may not fall within the geographic limits of a particular nation—whether they do is an accident of history. Sometimes these distinct economic units are formed by parts of states, such as those in northern Italy, Wales, Catalonia, Alsace-Lorraine or Baden-Württemberg. At other times they may be formed by economic patterns that overlap existing national boundaries, such as those between San Diego and Tijuana, Hong Kong and southern China, or the "growth triangle" of Singapore and its neighboring Indonesian islands. In today's borderless world these are natural economic zones and what matters is that each possesses, in one or another combination, the key ingredients for successful participation in the global economy.

Look, for example, at what is happening in Southeast Asia. The Hong Kong economy has gradually extended its influence throughout the Pearl River Delta. The radiating effect of these linkages has made Hong Kong, where GNP per capita is $12,000, the driving force of economic life in Shenzhen, boosting the per capital GNP of that city's residents to $5,695, as compared to $317 for China as a whole. These links extend to Zhuhai, Amoy and Guangzhou as well. By the year 2000 this cross-border region state will have raised the living standard of more than 11 million people over the $5,000 level. Meanwhile, Guangdong province, with a population of more than 65 million and its capital at Hong Kong, will emerge as a newly industrialized economy in its own right, even though China's per capita GNP may still hover at about $1,000. Unlike in Eastern Europe, where nations try to convert entire socialist economies over to the market, the Asian model is first to convert limited economic zones—the region states—into free enterprise havens. So far the results have been reassuring.

These developments and others like them are coming just in time for Asia. As Europe perfects its single market and as the United States, Canada and Mexico begin to explore the benefits of the North American Free Trade Agreement (NAFTA), the combined economies of Asia and Japan lag behind those of the other parts of the globe's economic triad by about $2 trillion—roughly the aggregate size of some 20 additional region states. In other words, for Asia to keep pace existing regions must continue to grow at current rates throughout the next decade, giving birth to 20 additional Singapores.

Many of these new region states are already beginning to emerge. China has expanded to 14 other areas—many of them inland—the special economic zones that have worked so well for Shenzhen and Shanghai. One such project at Yunnan will become a cross-border economic zone encompassing parts of Laos and Vietnam. In Vietnam itself Ho Chi Minh City (Saigon) has launched a similar "sepzone" to attract foreign capital. Inspired in part by Singapore's "growth triangle," the governments of Indonesia, Malaysia and Thailand in 1992 unveiled a larger triangle across the Strait of Malacca to link Medan, Penang and Phuket. These developments are not, of course, limited to the developing economies in Asia. In economic terms the United States has never been a single nation. It is a collection of region states: northern and southern California, the "power corridor" along the East Coast between Boston and Washington, the Northeast, the Midwest, the Sun Belt, and so on.

## What Makes a Region State

THE PRIMARY linkages of region states tend to be with the global economy and not with their host nations. Region states make such effective points of entry into the global economy because the very characteristics that define them are shaped by the demands of that economy. Region states tend to have between five million and 20 million people. The range is broad, but the extremes are clear: not half a million, not 50 or 100 million. A region state must be small enough for its citizens to share certain economic and consumer interests but of adequate size to justify the infrastructure—communication and transportation links and quality professional services—necessary to participate economically on a global scale.

It must, for example, have at least one international airport and, more than likely, one good harbor with international-class freight-handling facilities. A region state must also be large enough to provide an attractive market for the brand development of leading consumer products. In other words, region states are not defined by their economies of scale in production (which, after all, can be leveraged from a base of any size through exports to the rest of the world) but rather by their having reached efficient economies of scale in their consumption, infrastructure and professional services.

For example, as the reach of television networks expands, advertising becomes more efficient. Although trying to introduce a consumer brand throughout all of Japan or Indonesia may still prove prohibitively expensive, establishing it firmly in the Osaka or Jakarta region is far more affordable—and far more likely to generate handsome returns. Much the same is true with sales and service networks, customer satisfaction programs, market surveys and management information systems: efficient scale is at the regional, not national, level. This fact matters because, on balance, modern marketing techniques and technologies shape the economies of region states.

Where true economies of service exist, religious, ethnic and racial distinctions are not important—or, at least, only as important as human nature requires. Singapore is 70 percent ethnic Chinese, but its 30 percent minority is not much of a problem because commercial prosperity creates sufficient affluence for all. Nor are ethnic differences a source of concern for potential investors looking for consumers.

Indonesia—an archipelago with 500 or so different tribal groups, 18,000 islands and 170 million people—would logically seem to defy effective organization within a single mode of political government. Yet Jakarta has traditionally attempted to impose just such a central control by applying fictional averages to the entire nation. They do not work. If, however, economies of service allowed two or three Singapore-sized region states to be created within Indonesia, they could be managed. And they would ameliorate, rather than exacerbate, the country's internal social divisions. This holds as well for India and Brazil.

## The New Multinational Corporation

WHEN VIEWING the globe through the lens of the region state, senior corporate managers think differently about the geographical expansion of their businesses. In the past the primary aspiration of multinational corporations was to create, in effect, clones of the parent organization in each of the dozens of countries in which they operated. The goal of this system was to stick yet another pin in the global map to mark an increasing number of subsidiaries around the world.

More recently, however, when Nestlé and Procter & Gamble wanted to expand their business in Japan from an already strong position, they did not view the effort as just another pin-sticking exercise. Nor did they treat the country as a single coherent market to be gained at once, or try as most Western companies do to establish a foothold first in the Tokyo area, Japan's most tumultuous and overcrowded market. Instead, they wisely focused on the Kansai region around Osaka and Kobe, whose 22 million residents are nearly as affluent as those in Tokyo but where competition is far less intense. Once they had on-the-ground experience on how best to reach the Japanese consumer, they branched out into other regions of the country.

Much of the difficulty Western companies face in trying to enter Japan stems directly from trying to shoulder their way in through Tokyo. This instinct often proves difficult and costly. Even if it works, it may also prove a trap; it is hard to "see" Japan once one is bottled up in the particular dynamics of the Tokyo marketplace. Moreover, entering the country through a different regional doorway has great economic appeal. Measured by aggregate GNP the Kansai region is the seventh-largest economy in the world, just behind the United Kingdom.

Given the variations among local markets and the value of learning through real-world experimentation, an incremental region-based approach to market entry makes excellent sense. And not just in Japan. Building an effective presence across a landmass the size of China is of course a daunting prospect. Serving the people in and around Nagoya City, however, is not.

If one wants a presence in Thailand, why start by building a network over the entire extended landmass? Instead focus, at least initially, on the region around Bangkok, which represents the lion's share of the total potential market. The same strategy applies to the United States. To introduce a new top-of-the-line car into the U.S. market, why replicate up front an exhaustive coast-to-coast dealership network? Of the country's 3,000 statistical metropolitan areas, 80 percent of luxury car buyers can be reached by establishing a presence in only 125 of these.

## The Challenges for Government

TRADITIONAL ISSUES of foreign policy, security and defense remain the province of nation states. So, too, are macroeconomic and monetary policies—the taxation and public investment needed to provide the necessary infrastructure and incentives for region-based activities. The government will also remain responsible for the broad requirements of educating and training citizens so that they can participate fully in the global economy.

Governments are likely to resist giving up the power to intervene in the economic realm or to relinquish their impulses for protectionism. The illusion of control is soothing. Yet hard evidence proves the contrary. No manipulation of exchange rates by central bankers or political appointees has ever "corrected" the trade imbalances between the United States and Japan. Nor has any trade talk between the two governments. Whatever cosmetic actions these negotiations may have prompted, they rescued no industry and revived no economic sector. Textiles, semiconductors, autos, consumer electronics—the competitive situation in these industries did not develop according to the whims of policymakers but only in response to the deeper logic of the competitive marketplace. If U.S. market share has dwindled, it is not because government policy failed but because individual consumers decided to buy elsewhere. If U.S. capacity has migrated to Mexico or Asia, it is only because individual managers made decisions about cost and efficiency.

The implications of region states are not welcome news to established seats of political power, be they politicians or lobbyists. Nation states by definition require a domestic political focus, while region states are ensconced in the global economy. Region states that sit within the frontiers of a particular nation share its political goals and aspirations. However, region states welcome foreign investment and ownership—whatever allows them to employ people productively or to improve the quality of life. They want their people to have access to the best and cheapest products. And they want whatever surplus accrues from these activities to ratchet up the local quality of life still further and not to support distant regions or to prop up distressed industries elsewhere in the name of national interest or sovereignty.

When a region prospers, that prosperity spills over into the adjacent regions within the same political confederation. Industry in the area immediately in and around Bangkok has prompted investors to explore options elsewhere in Thailand. Much the same is true of Kuala Lumpur in Malaysia, Jakarta in Indonesia, or Singapore, which is rapidly becoming the unofficial capital of the Association of Southeast Asian Nations. São Paulo, too, could well emerge as a genuine region state, someday entering the ranks of the Organization of Economic Cooperation and Development. Yet if Brazil's central government does not allow the São Paulo region state finally to enter the global economy, the country as a whole may soon fall off the roster of the newly industrialized economies.

Unlike those at the political center, the leaders of region states—interested chief executive officers, heads of local unions, politicians at city and state levels—often welcome and encourage foreign capital investment. They do not go abroad to attract new plants and factories only to appear back home on television vowing to protect local companies at any cost. These leaders tend to possess an international outlook that can help defuse many of the usual kinds of social tensions arising over issues of "foreign" versus "domestic" inputs to production.

In the United States, for example, the Japanese have already established about 120 "transplant" auto factories throughout the Mississippi Valley. More are on the way. As their share of the U.S. auto industry's production grows, people in that region who look to these plants for their livelihoods and for the tax revenues needed to support local communities will stop caring whether the plants belong to U.S.- or Japanese-based companies. All they will care about are the regional economic benefits of having them there. In effect, as members of the Mississippi Valley region state, they will have leveraged the contribution of these plants to help their region become an active participant in the global economy.

Region states need not be the enemies of central governments. Handled gently, region states can provide the opportunity for eventual prosperity for all areas within a nation's traditional political control. When political and industrial leaders accept and act on these realities, they help build prosperity. When they do not—falling back under the spell of the nationalist economic illusion—they may actually destroy it.

Consider the fate of Silicon Valley, that great early engine of much of America's microelectronics industry. In the beginning it was an extremely open and entrepreneurial environment. Of late, however, it has become notably protectionist—creating industry associations, establishing a polished lobbying presence in Washington and turning to "competitiveness" studies as a way to get more federal funding for research and development. It has also begun to discourage, and even to bar, foreign investment, let alone foreign takeovers. The result is that Boise and Denver now prosper in electronics; Japan is developing a Silicon Island on Kyushu; Taiwan is trying to create a Silicon Island of its own; and Korea is nurturing a Silicon Peninsula. This is the worst of all possible worlds: no new money in California and a host of newly energized and well-funded competitors.

Elsewhere in California, not far from Silicon Valley, the story is quite different. When Hollywood recognized that it faced a severe capital shortage, it did not throw up protectionist barriers against foreign money. Instead, it invited Rupert Murdoch into 20th Century Fox, C. Itoh and Toshiba into Time-Warner, Sony into Columbia, and Matsushita into MCA. The result: a $10 billion infusion of new capital and, equally important, $10 billion less for Japan or anyone else to set up a new Hollywood of their own.

Political leaders, however reluctantly, must adjust to the reality of economic regional entities if they are to nurture real economic flows. Resistant governments will be left to reign over traditional political territories as all meaningful participation in the global economy migrates beyond their well-preserved frontiers.

Canada, as an example, is wrongly focusing on Quebec and national language tensions as its core economic and even political issue. It does so to the point of still wrestling with the teaching of French and English in British Columbia, when that province's economic future is tied to Asia. Furthermore, as NAFTA takes shape the "vertical" relationships between Canadian and U.S. regions—Vancouver and Seattle (the Pacific Northwest region state); Toronto, Detroit and Cleveland (the Great Lakes region state)—will become increasingly important. How Canadian leaders deal with these new entities will be critical to the continuance of Canada as a political nation.

In developing economies, history suggests that when GNP per capita reaches about $5,000, discretionary income crosses an invisible threshold. Above that level people begin wondering whether they have reasonable access to the best and cheapest available products and whether they have an adequate quality of life. More troubling for those in political control, citizens also begin to consider whether their government is doing as well by them as it might.

Such a performance review is likely to be unpleasant. When governments control information—and in large measure because they do—it is all too easy for them to believe that they "own" their people. Governments begin restricting access to certain kinds of goods or services or pricing them far higher than pure economic logic would dictate. If market-driven levels of consumption conflict with a government's pet policy or general desire for control, the obvious response is to restrict consumption. So what if the people would choose otherwise if given the opportunity? Not only does the government withhold that opportunity but it also does not even let the people know that it is being withheld.

Regimes that exercise strong central control either fall on hard times or begin to decompose. In a borderless world the deck is stacked against them. The irony, of course, is that in the name of safeguarding the integrity and identity of the center, they often prove unwilling or unable to give up the illusion of power in order to seek a better quality of life for their people. There is at the center an understandable fear of letting go and losing control. As a result, the center often ends up protecting weak and unproductive industries and then passing along the high costs to its people—precisely the opposite of what a government should do.

**The Goal is to Raise Living Standards**

THE CLINTON administration faces a stark choice as it organizes itself to address the country's economic issues. It can develop policy within the framework of the badly dated assumption that success in the global economy means pitting one nation's industries against another's. Or it can define policy with the awareness that the economic dynamics of a borderless world do not flow from such contrived head-to-head confrontations, but rather from the participation of specific regions in a global nexus of information, skill, trade and investment.

If the goal is to raise living standards by promoting regional participation in the borderless economy, then the less Washington constrains these regions, the better off they will be. By contrast, the more Washington intervenes, the more citizens will pay for automobiles, steel, semiconductors, white wine, textiles or consumer electronics—all in the name of "protecting" America. Aggregating economic policy at the national level—or worse, at the continent-wide level as in Europe—inevitably results in special interest groups and vote-conscious governments putting their own interests first.

The less Washington interacts with specific regions, however, the less it perceives itself as "representing" them. It does not feel right. When learning to ski, one of the toughest and most counterintuitive principles to accept is that one gains better control by leaning down toward the valley, not back against the hill. Letting go is difficult. For governments region-based participation in the borderless economy is fine, except where it threatens current jobs, industries or interests. In Japan, a nation with plenty of farmers, food is far more expensive than in Hong Kong or Singapore, where there are no farmers. That is because Hong Kong and Singapore are open to what Australia and China can produce far more cheaply than they could themselves. They have opened themselves to the global economy, thrown their weight forward, as it were, and their people have reaped the benefits.

For the Clinton administration, the irony is that Washington today finds itself in the same relation to those region states that lie entirely or partially within its borders as was London with its North American colonies centuries ago. Neither central power could genuinely understand the shape or magnitude of the new flows of information, people and economic activity in the regions nominally under its control. Nor could it understand how counterproductive it would be to try to arrest or distort these flows in the service of nation-defined interests. Now as then, only relaxed central control can allow the flexibility needed to maintain the links to regions gripped by an inexorable drive for prosperity.

---

Kenichi Ohmae is Chairman of the offices of McKinsey & Company in Japan.

---

From *Foreign Affairs*, Spring 1993, pp. 78–87. © 1993 by Kenichi Ohmae. Reprinted by permission.

# Article 21

# Continental Divide

With 13 countries on the applicant list, the European Union confronts the dilemma of expansion in 2000. But are the advantages worth the price of admission? Even as leaders of Bulgaria, Cyprus, the Czech Republic, Estonia, Hungary, Latvia, Lithuania, Malta, Poland, Romania, the Slovak Republic, Slovenia, and Turkey maneuver to bring their policies in line with EU law, there is dissension within member countries over EU standards and the uniformity of their enforcement. And judging from the recent electoral successes of far-right political parties in Western Europe, nationalist sentiment and distaste for open borders will lend controversy to upcoming EU decisions on such hot-button issues as asylum and immigration policy.

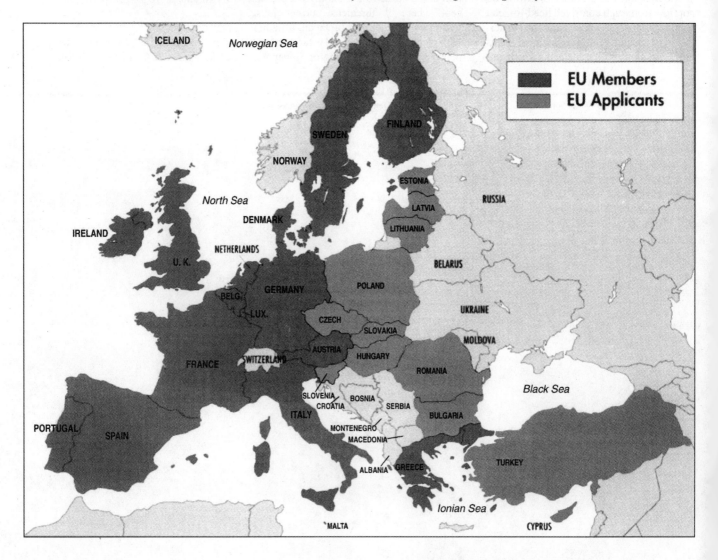

## DESPERATELY SEEKING ADMISSION

### Limitless Growth?

Eighteen months ago the European Union opened admission negotiations with Estonia, Poland, Slovenia, the Czech Republic, Hungary, and Cyprus. In December this group should expand by six additional nations. If Bulgaria, Latvia, Lithuania, Malta, Romania, and Slovakia are included, the 15-member EU will be dealing with 12 candidates. In addition, Turkey, Albania, and the successor states to Yugoslavia have sought admission.

The candidates are pressing for a quick response, but there is no way to say when the EU will act, since it is not clear when the candidates will meet the performance prerequisites. The only exception is Cyprus. The divided island republic faces political hurdles. Economically, Cyprus has long since met all the criteria. For another reason, it is not at all clear that the EU is ready for any expansion.

Even in its present makeup, it finds it very difficult to function. An EU of 20 or even more members would be practically ungovernable. By 2002 the EU wants, therefore, to finish its internal reforms.

But where the European giant ought to be going, just how large the EU ought to be, and what the political and geographic borders of Europe ought to be are unclear. And not by chance. It is only because the future political shape of the Union has been left vague that the integration process has gone as far as it has. Had it, from the beginning, been based on one of the many suggested models—a federative state, a confederation, a union—the European train would have never left the station.

Now, however, with more and more cars hitched on the train, there is increasing pressure for reform. The EU can scarcely continue with the luxury of unanimous decisions and lengthy negotiations preceding any changes. This is true for the first and until now most important pillar of the EU, the internal market, with a common agricultural policy and the economic and currency union. And it will be all the more true for the second pillar, the common foreign and security policy, and the third, domestic and legal policies.

The EU special summit meeting in Tampere, Finland, [in October] made this dilemma clear. It makes sense, in a Europe that is increasingly closely knit, to have a single set of rules governing legal rights, penal codes, border security, and asylum policies. At the same time, this raises the danger that national achievements will be destroyed.

Therefore, it would be reasonable to begin by moving toward a European constitution that would define basic rights and obligations before tackling agreements on the detailed questions. But how should this work, when up till now both the geographical borders and the political form of this entity have been left vague? So, in Tampere there was an agreement to work out a Charter of Basic Rights for Europe by 2000. Long sought by the European left, the charter would in fact represent remarkable progress.

The Council of Sages created by European Commission President Romano Prodi is also aimed at a quasi-constitution for Europe. Belgium's former Prime Minister Jean-Luc Dehaene; Lord Simon, the former British minister to Europe; and former German President Richard von Weizsäcker advise the EU to restructure its legal framework. The Sages' goal is unmistakable: An expanded EU should, and must, abandon the right to veto and replace it with democratic decisions based on majority votes. That would increase the power of the European Parliament. It would also mean that individual nation-states would have to give up even more of their national sovereignty.

But even this proposed reform is a compromise that would consciously leave the future of European integration open. The candidates for membership are of course profiting from this open character—for otherwise the EU would have long since become a closed club.

At the same time, it means living with an undecided and unfinished integration. Deputies to the European Parliament are elected nationally, EU officials are chosen by national governments, and the first concerns of the European Union's heads of government are always the next election back home, and only then European politics. Common interests always take second priority, and when action is taken, fear of what national interests might be affected often leads to stasis or even retreat at the European level.

Europe has put itself under pressure to integrate in order to preserve peace and its own prosperity. That drives the EU onward, but also feeds the tendency of members to keep what they've got. The result is great pressure for conformity. It turns candidates into petitioners who cannot predict the real results of their negotiations or wishes.

The EU's agricultural policy is the best example of this contradiction. It favors intensive farming, aimed at world markets, the kind that damages the environment. To demand of the new Central European nations that they conform to this recipe is hopeless. It would not serve the EU, or Eastern farmers, or the environment, let alone Third World farmers.

The way out lies in the reform of the EU agriculture policy. But that flies in the face of narrow-minded yet understandable national interests. The only way out of this dilemma lies in forced democratization of the EU, creating a public receptive to new ideas, and winning supportive majorities. Without stronger citizen participation, the Union will remain a distant, bureaucratic Moloch, or a cash cow that each state will, according to its power, seek to milk without regard for the interests of the entire European Community.

—*Torsten Wohlert*, Freitag *(leftist weekly),*
*Berlin, Oct. 22, 1999.*

The branding of Hong Kong

# A dragon with core values

## Stick that in your lapel

HONG KONG

ON HANDOVER day in 1997, Donald Tsang, then Hong Kong's finance secretary, pinned a little emblem on to his lapel. It was a double flag—Communist China's joined to Hong Kong's *bauhinia* flower—that stood for "one country, two systems". For more than three years Mr Tsang was rarely seen without it, and it became, along with his bow ties, his trademark.

Last year—by when he was chief secretary—Mr Tsang replaced the emblem with a little dragon, the fruit of three years of research by international brand consultants. Besides cosmopolitanism, says Kerry McGlynn, the government public-relations director behind the project, the dragon projects five "core values". These are three adjectives—"progressive", "free" and "stable"—and two nouns, "opportunity" and "quality".

The visual link, according to the government, is self-evident: Hong Kong stands for "East meets West". So the dragon is composed of two parts that could, if you twist it, stand for the letters H and K, as well as the Chinese characters *Heung* and *Gong*. Combined into a dragon, an ancient Chinese metaphor for energy, the strokes represent Hong Kong's legendary dynamism.

The dragon appeared on brochures, buses and much else last summer, and within days Hong Kong's people were naming it. Expatriates saw it mostly as a "flying fox", while Cantonese speakers—usually more creative in such matters—settled on "shocked chicken". Those appraised of the consultants' fees called it "the HK$9m dragon".

Perhaps the most perplexing thing about the dragon, however, is that it took Hong Kong so long to get one. Canada branded itself in 1970, and New York ("the Big Apple") a decade later. Besides, their fetish for brands is one of the few core values that most Hong Kong residents agree on. As one long-term resident puts it, "When the going gets tough, Hong Kong goes shopping."

# The Late Great Wall

A wonder of the world is vanishing, unable to resist the destructive forces of nature and economics. What can be done to save it? A tour of the ruins.

**BY MELINDA LIU**

THE GREAT WALL OF CHINA can't quite match the myths that have grown up around it. Still, the truth is astonishing enough. The Chinese call it the Long Wall of 10,000 Miles—an exaggeration, even though its actual length would stretch from Miami to Seattle. The wall wasn't built 2,000 years ago, as some sources claim, and yet a few parts are centuries older. In fact, it's really not a single wall at all, but a tangle of parallel and proximate fortifications. The pieces weren't organized into a unified system until the Ming dynasty, which lasted from 1368 to 1644. And one more quibble: it's not visible from the moon.

> "This is the largest single cultural-relics challenge in the world. It's difficult to protect because there's so much of it."
> —WILLIAM LINDESAY, head of International Friends of the Great Wall

The sad part is, less and less of it is visible from earth. The Great Wall is vanishing, unable to withstand the destructive forces of nature and economics as deserts, development and tourists spread across China. This year the New York-based World Monuments Fund added the wall to its "most endangered sites" list. "It's harder for really well-known sites to be selected because there's skepticism as to whether they really need help," observes Bonnie Burnham, the group's president. Truth is, the wall needs urgent help—but where to start? "It's difficult to protect because there's so much of it," says William Lindesay, a British preservationist who is trying to rescue at least part of the untouched "wild wall" and its spectacular natural landscape near Beijing. He calls it "the largest single cultural-relics-protection challenge in the world."

The upcoming 2008 Olympics have made cultural preservation a particularly hot issue in Beijing. China desperately wants to put on its best face for the occasion. Unfortunately, Chinese authorities often think the way to look good is by tearing down old buildings and putting up shiny new ones. Nearly two decades ago China's then paramount leader Deng Xiaoping launched a national campaign under the slogan "Love your country, rebuild the Great Wall." By that point, the local press estimated, two thirds of the vast national symbol had been reduced to rubble by centuries of war, weather and peasant farmers mining its bricks to build homes and pigsties. Some Chinese think the rebuilders should have left bad enough alone.

The first stretch of wall to be rebuilt was at Badaling, in the hills roughly 45 miles northwest of Beijing. Zhang Jianxin, an official of the National Bureau of Cultural Relics, recalls how unspoiled it was in 1979, when he took a weeklong bike tour nearby and encountered wolves. Today the site is part theme park, part carnival and part shopping mall, managed by a corporation that is listed on the Hong Kong Stock Exchange. The area around the wall is packed with tour buses, T-shirt vendors, souvenir "ride a camel" photo stands and a huge, grinning likeness of Colonel Sanders clutching an oversized bucket of fried chicken. Zhang tries not to go anywhere near the place now. "It's lost its sense of history," he says.

The mandarins of Beijing didn't seem to mind. The place was a money machine. Next they renovated another section of "tourist wall" 60 miles northeast of Beijing at Mutianyu. Sightseers can ride a cable car to the wall's crest, more than 3,000 feet up, and swoosh back down the grassy hillside on a toboggan. Not surprisingly, when Harvard University's president, Larry Summers, visited a stretch of the wall near Beijing in May, he sounded more than a little concerned. "Go-kart rides, Disneyland-type scenes and golden arches," he said with a sigh to NEWSWEEK. "Is this good?"

Good or bad, modern times have hit the wall—and not only around Beijing. Some 200 miles northeast of the capital, the wall's eastern terminus, the Old Dragon's Head, rises from the sea. You can still see a few bits of the wall's ancient foundation there, enshrined in weather-beaten glass cases atop the rebuilt wall. What stands on the site now is actually a reconstruction, erected in the late 1980s. The original Dragon's Head was demolished by European expeditionary forces in 1900. These days you run a gantlet of aggressive hawkers brandishing trinkets and offering to take your picture dressed as an emperor or a modern Chinese Army general. On the grounds of the Old Dragon's Head, passengers ride "the Dragon Boat," an amusement-park attraction that rocks back and forth at increasingly sharp angles until the keel is perpendicular to the ground.

## "We no longer know how [to repair it]," says... one of China's foremost wall experts. "We just cannot meet the old standards."

But tacky tourism isn't the most serious threat besieging the wall. It's indifference—that of impoverished locals who seek to eke out a living from hikers and "wall walkers," and that of county authorities who are always willing to take a bribe to look the other way when locals violate the few existing preservation laws. In fact, most of the wall is unrestored "wild wall," as Lindesay and other preservationists call it. Imperial history still resonates through the crumbling bricks, tangled undergrowth and pristine natural settings of these dilapidated but majestic sites. The question is how much longer they can survive. Wherever hikers stop on the wall, they are increasingly likely to find litter, graffiti and peasant-operated tourist traps. One of the most spectacular sites, roughly 40 miles north of Beijing, is the village of Huanghuacheng, where a crumbling 500-year-old watchtower now houses a soft-drink stand.

On a recent summer afternoon, villagers stood on the roof of the tower setting off firecrackers and cherry bombs. Selling fireworks is a favorite way to coax money from wall walkers. After a series of earsplitting explosions, the men found they had ignited some dry grass growing on the ancient structure. They danced around giggling, stamping out the flames on top of the ancient tower. A little farther on, farmers have set up unauthorized "ticket booths" and ladders to extract entrance fees. One local man has bolted a crude metal door to a tower's archway, creating a private room where he can rest when he's not taking admissions.

Still, the damage is relatively minor around Huanghuacheng. In the backcountry, far from Beijing's oversight, progress is the only priority. Three years ago, in Inner Mongolia, highway builders demolished part of a sentry-post wall dating back more than 2,200 years. To the west, some parts of the wall have entirely disappeared beneath the sands of the expanding Gobi Desert. The parched wasteland is advancing all across northern China, thanks to decades of overgrazing and reckless land use.

No one has any magic recipes for saving the wall. Beijing has some cultural-relics regulations to protect the roughly 400 miles in its direct jurisdiction, but no one seems to enforce them. The fact that all commercial structures are banned within a quarter mile or so of the wall has not kept entrepreneurs in Huanghuacheng from putting up several restaurants, a modern hotel complex and even a mobile-phone repeater station right on top of an ancient watchtower. What's Beijing doing about it? Writing more laws.

Some of the wall's problems are beyond human legislation and modern technology. One of its oldest standing fragments is a rammed-earth barricade some 50 yards long and 12 feet high at Yumengyuan, Inner Mongolia, not far from the wall's western tip. The ancient builders used a kind of adobe made from soil, straw, tamarisk, egg yolk and rice paste. Now it's disintegrating, and no one can repair it. "We no longer know how," says Luo Zhewen, one of China's foremost wall experts. "We just cannot meet the old standards."

## To many Chinese the wall stands for feudal oppression as much as it represents cultural pride.

Nearby, the 630-year-old fortress at Jiayuguan rises out of the desert like a mirage against the snowcapped Qilian Mountains. Caretakers thought they knew a way to patch its crumbling brick walls to make them stronger than ever. "We thought cement was good because it was a modern invention," says a local tour guide. "But it was too heavy for the original materials. The repaired portion collapsed." Perhaps to distract visitors from the damage, someone has draped the walls in Christmas lights.

Despite all the obstacles, Lindesay is determined to save the wild wall at least. He calls it "the world's largest open-air museum, without a curator." As head of the International Friends of the Great Wall, he organizes regular cleanup drives and educational campaigns. Just last week he signed an agreement with Beijing municipal authorities and UNESCO to help protect the wild wall and its natural setting. By designating special protection zones, he hopes to convince local officials that the wall is not just a structure but a unique landscape, requiring careful management and what he calls "stewardship."

It's a tough sell. Most Chinese see the wall merely as the country's biggest tourist attraction, while others remain profoundly ambivalent about their national treasure. To them the wall stands for feudal oppression as much as it represents cultural pride. Tradition says China's first emperor, the despotic Qin

Shihuang, worked laborers to death by the tens of thousands in erecting a barricade against the "barbarians" of present-day northern China. Some members of today's older generation saw him as a model for the tyranny of Mao Zedong. Others have never forgotten the popular lullaby about Meng Jiang, a Han-dynasty woman whose husband died of hunger while working on the wall. After he was buried beneath its ramparts, the song says, she cried until it collapsed.

The preservationists hope young Chinese will eventually learn to love the wall. Rightly or not, nothing else in China inspires such awe in the eyes of Westerners. China likes to pretend it doesn't care what foreigners think. Didn't the ancient emperors build the wall in order keep out the meddlesome barbarians? It didn't work. Wave after wave of invaders, from the Mongols to the Manchus, swept past the wall as if it didn't exist. Today people fly halfway around the world just to see it. China might also benefit from taking a fresh look.

With PAUL MOONEY and JUNE SHIH

From *Newsweek*, July 29, 2002, pp. 40-43. © 2002 by Newsweek, Inc. All rights reserved. Reprinted by permission.

*Article 24*

# A CONTINENT IN PERIL

**17 million Africans have died since the AIDS epidemic began in the late 1970s, more than 3.7 million of them children. An additional 12 million children have been orphaned by AIDS. An estimated 8.8% of adults in Africa are infected with HIV/AIDS, and in the following seven countries, at least 1 adult in 5 is living with HIV**

## HOT SPOT

Of the 36 million adults and children in the world living with HIV/AIDS in 2000, more than 70% were in sub-Saharan Africa. 3.8 million Africans were newly infected last year

## 1. Botswana

Though it has the highest per capita GDP, it also has the highest estimated adult infection rate—**36%**. 24,000 die each year. 66,000 children have lost their mother or both parents to the disease

## 2. Swaziland

Swaziland More than **25%** of adults have HIV/AIDS in this small country. 12,000 children have been orphaned, and 7,100 adults and children die each year

## 3. Zimbabwe

**One-quarter** of the adult population is infected here. 160,000 adults and children died in 1999, and 900,000 children have been orphaned. Because of AIDS, life expectancy is 43

## 4. Lesotho

**24%** of adults are infected with HIV/AIDS. 35,000 children have been orphaned, and 16,000 adults and children die each year

## 5. Zambia

**20%** of the adult population is infected, 1 in 4 adults in the cities. 650,00 children have been orphaned, and 99,000 Zambians died in 1999

## 6. South Africa

This country has the largest number of people living with HIV/AIDS, about **20%** of its adult population, up from 13% in 1997. 420,000 children have been orphaned, and 250,000 people die each year from the disease.

## 7. Namibia

**19.5%** of the adult population is living with HIV. 57% of the infected are women. 67,000 children are AIDS orphans, and 18,000 adults and children die each year

Article 24. A CONTINENT IN PERIL

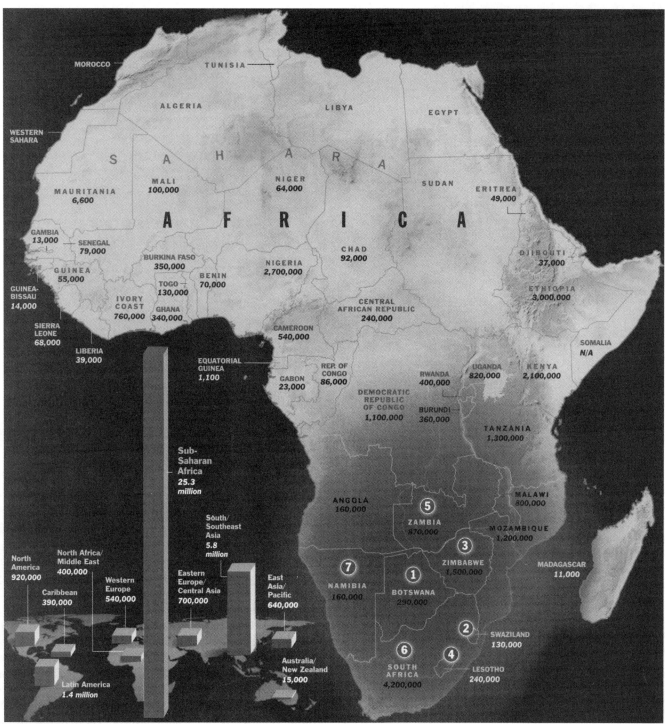

Source: UNAIDS

TIME Graphic by Lon Tweeten

From *Time*, February 12, 2001, pp. 38-39. © 2001 by Time, Inc. Magazine Company. Reprinted by permission.

# AIDS Has Arrived in India and China

*How will the world's two most populous countries cope with the pandemic?*

by Ann Hwang

In 1348, the Black Death arrived in Europe from its probable home in Central Asia, and over the next couple of years it is believed to have killed 25 million people. Sometime soon, mortality from AIDS will exceed the death toll of that worst outbreak of bubonic plague. Since the start of the AIDS pandemic roughly 20 years ago, 20 million people have died and over 50 million have been infected. Every 11 seconds, someone dies from AIDS. According to statistics compiled by the World Health Organization, AIDS is now killing more people each year than any other infectious disease. AIDS has become one of the greatest epidemics in the history of our species.

The AIDS epidemic took much longer to build momentum than did the Black Death, but AIDS has far more staying power. For all their intensity, the bubonic plague epidemics were relatively short: *Yersinia pestis*, the plague bacterium, tends to burn itself out quickly. And in any case, *Y. pestis* and *Homo sapiens* are no longer caught up in an intense epidemic cycle. Plague still kills people in various parts of the world, but it does not spark epidemics on a continental scale. Even if it did, antibiotics have made it far less deadly than it was 650 years ago. But HIV, the virus that causes AIDS, shows no sign of releasing us from its grip. Indeed, it has evolved into several new forms, even as it continues to burn through humanity. And although there are now drugs that can prolong the lives of its victims—or at least, those who can afford treatment—there is no cure for the disease and no vaccine for it. (See box, "An AIDS Vaccine?")

Within the AIDS pandemic, sub-Saharan Africa has become the equivalent of mid-14th century Europe. Ignorance of the disease, poverty, war, and frequently, a rather relaxed attitude toward sexual activity (especially when it comes to men)—such factors have allowed HIV to explode through some African societies. In 1996, the Joint United Nations Programme on HIV/AIDS (UNAIDS) predicted that by 2000, over 9 million Africans would be infected with HIV. The actual number turned out to be 25 million. Though Africa is home to less than 9 percent of the world's adults, it has more than two-thirds of adult HIV infections. In Botswana, the county with the world's highest infection rate, one in three adults is now infected. And as the infected continue to die, places like Botswana may become increasingly unstable for lack of farmers, teachers, community leaders, even parents.

But in large measure, the course of the pandemic will depend on what happens not in Africa but in Asia, the continent that is home to nearly 60 percent of the world's people. AIDS is already well established in Asia, although no one knows precisely when or where it first arrived. By the mid-1980s, however, infections were beginning to appear in several Asian counties, including Thailand and India. A few years later, it was obvious that HIV infection was increasing dramatically among two of the best known "high risk populations"—prostitutes and users of injection drugs. As its incidence increased, the disease began to travel the highways of Asia's drug trade, radiating outward from the opium-producing "Golden Triangle," where Myanmar (Burma), Laos, and Thailand converge. The infecting of the world's most populous continent had begun.

That process may now be reaching a kind of critical mass. AIDS has arrived in the two most heavily populated countries in the world: India and China. With populations of 1 billion and 1.3 billion respectively, these countries are home to over a third of the world's population and nearly 70 percent of Asians people. Thus far, neither has suffered the kind of explosive epidemic that has ravaged sub-Saharan Africa. Each still has important opportunities to stem the epidemic. What will the giant societies of Asia make of those opportunities? This is one of the greatest social and ethical issues of our era.

# Mapping the Epidemic

## Reported Risk of HIV infection in India and China, 1999

**Caveat lector:** the data in this map, which derive from official country sources, do not give a complete picture of the epidemic. In particular, information on high risk groups is incomplete. The high risk data do not include homosexuals in either India or China, or blood sellers in China. Because the latter group is omitted, the map does not accurately portray the epidemic in central China.

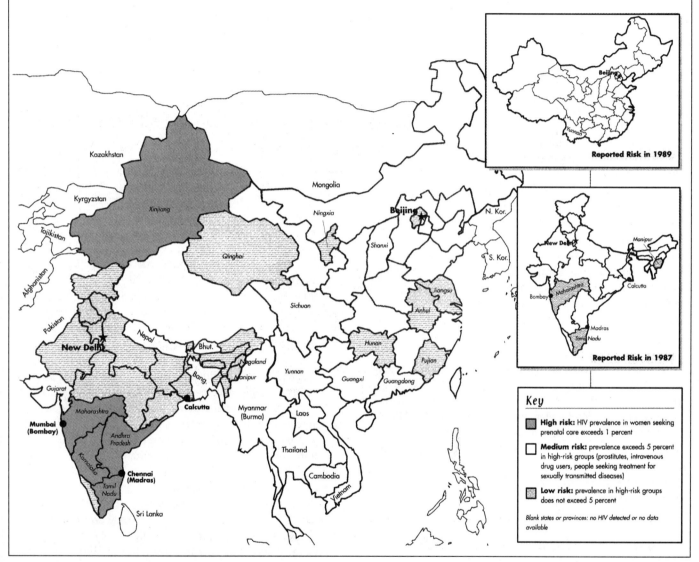

*Sources:* Indian Ministry of Health and Family Welfare, National AIDS Control Organization; UNAIDS/WHO Epidemiological Fact Sheet: China, 2000 update; UNAIDS Country Profile: China

## Four Million Infected in India

India is home to an estimated 4 million people with HIV—more than any other country in the world. Because of India's huge population, the level of infection as a national average is very low—just 0.4 percent, close to the U.S. national level of 0.3 percent. But this apparently comfortable average masks huge regional disparities: in some of India's states, particularly in the extreme northeast, near the Myanmar border, and in much of the south, the rate of infection among adults has reached 2 percent or more—five times the national rate and more than enough to kindle a widespread epidemic.

Among these more heavily infected regions, there is another kind of disparity as well, in the way the virus is spreading. In southern India, AIDS fits the standard profile of a sexually transmitted disease (STD), with particularly high infection rates among prostitutes. Sex is big business in India, generating revenues of $8.7 billion each year, according to the Centre of Concern for Child Labour, a Delhi-based non-profit. Mumbai (Bombay), the country's largest west coast city, has twice the population of New York yet almost 20 times the number of prostitutes. By 1997, over 70 percent of those prostitutes were HIV positive. The prostitutes' clients, in addition to risking infection themselves, put their wives or other sex partners in jeopardy, thereby creating a bridge that allows the virus to spread from a high-risk enclave to the general population.

In some segments of Indian society, that bridge is now very broad. Long-distance truck drivers, for example, are usually away from home for long periods and many visit prostitutes en route. For one study, published in the *British Medical Journal* in 1999, nearly 6,000 long-distance truckers were interviewed and nine out of ten married drivers described themselves as "sexually promiscuous," defined as having frequent and indiscriminate change of sexual partners. Not surprisingly, HIV incidence is now rising among married Indian women. A study from 1993 to 1996 found that over 10 percent of female patients at STD clinics in Pune, near Mumbai, were HIV positive; over 90 percent of these women were married and had had sexual contact only with their husbands. (See box, "Increasingly, A Women's Disease.")

In India's northeast, the epidemic has a very different character. This region has an extensive drug culture—which is hardly surprising, given its proximity to the Golden Triangle. Here, the epidemic has been driven by intravenous drug use, particularly among young unemployed men and students. By sharing contaminated needles, addicts are injecting the virus into their bloodstreams. Data are scarce, but according to government estimates, there are 1 million heroin users in India, and roughly 100,000 of them reside in the comparatively small states that make up the northeast.

In the northeastern state of Manipur, on the Myanmar border, HIV among intravenous drug users and their sexual partners increased from virtually nothing in 1988 to over 70 percent four years later. By 1999, 2.2 percent of pregnant women attending prenatal care clinics in Manipur tested positive for HIV. Because the infection risk in women seeking prenatal care should be roughly representative of the general population, epidemiologists often use this group to estimate trends in the general population. In the northeast, as in the south, HIV is apparently moving into mainstream society.

## Perhaps One Million Infected in China

In China, the shadow of AIDS is at present just barely discernable. Current estimates put the number of HIV infections at 500,000 to 1 million. In a country of 1.3 billion, that works out to an infinitesimal national level of infection: eight one-hundredths of a percent at most. But even though the virus is very thinly spread, it seems to be present nearly everywhere: all of the country's 31 provinces have reported AIDS cases.

As with India, the character of this incipient epidemic differs greatly from one region to the next. China's original HIV hotspot is in the south: Yunnan province, which borders Laos and Myanmar, had almost 90 percent of the country's HIV cases in 1990. Yunnan lies on the periphery of the Golden Triangle and is home to a large (but not readily definable) proportion of China's intravenous drug users. Today, however, the virus has moved well beyond Yunnan, in part because of a surge in the popularity of injection drugs. By the middle of the 1990s, half of new infections in intravenous drug users were occurring outside Yunnan, mostly in other southern provinces. Guangxi province, which borders Yunnan to the east, saw infection levels in surveyed drug users climb from zero in 1993 to 40 percent by 1997.

## An AIDS Vaccine? No Magic Bullet

"People expect a magic bullet," says Chris Collins, president of the board of the AIDS Vaccine Advocacy Coalition, a network of U.S. activists that seeks to increase funding for HIV vaccine research. But he cautions, "the AIDS vaccine probably isn't going to be that."

It is true that vaccine researchers have made substantial progress over the past few years. A California-based company known as VaxGen is now conducting the first ever large-scale tests in humans of a possible vaccine. An interim analysis of the tests, which involve 8,000 volunteers on three continents, is scheduled for November 2001. Many experts believe that such efforts will eventually pay off, but the results are not likely to compare with the smallpox vaccine, which eventually eliminated that earlier global pandemic.

One big obstacle is the virus's mutation rate. Mutations appear to occur in at least one of the virus's genes each time it replicates, once every 8 hours. In HIV, as in any other organism, most mutations prove to be evolutionary dead ends. But not all of them: the virus has already spawned more than a dozen different subtypes around the globe, and it is unclear whether a single vaccine would be effective against every subtype. China in particular has a very heterogeneous epidemic, with nearly all known subtypes represented. This global mosaic of subtypes may exacerbate the medical North-South divide. How much industrialized-country R&D will be invested in developing vaccines for strains that predominate in developing countries?

Even when a viable vaccine is discovered, researchers are likely to face formidable challenges in determining its use. Suppose, for example, that a vaccine is only 50 percent effective: should it be licensed, given the possibility that people receiving it may be less inclined to have safe sex or use clean needles? Assuming that a strong general case could be made for the use of such a vaccine, who is going to pay for the inoculation of the developing world's high-risk populations? Vaccine researchers may find the sociology of the epidemic as difficult to deal with as the biology of the virus itself. No doubt, an effective vaccine will be a valuable tool against the pandemic, but it is not likely to replace any of the other tools already in use.

Last year, China's official count of registered intravenous drug users reached 600,000—more than double the number in 1992. And as the number of users has grown, so has the custom

## Increasingly, A Women's Disease

In the developing world, women now account for more than half of HIV infections, and there is growing evidence that the position of women in developing societies will be a critical factor in shaping the course of the AIDS pandemic. In general, greater gender inequality tends to correlate with higher levels of HIV infection, according to the World Bank researchers who track literacy rates and other general indicators of social well-being.

As in the AIDS-ravaged countries of sub-Saharan Africa, India and China offer women far fewer social opportunities than men. Both countries score in the lower half of the "Gender-Related Development Index," a measure of gender equity developed by the United Nations Development Programme.

Double sexual standards that demand female virginity while condoning male promiscuity put many women at risk. Studies in India and Thailand, by the Washington D.C.-based International Center for Research on Women (ICRW), have found that young, single women are expected not only to be virgins but also to be ignorant of sexual matters. As a result, young women lack basic knowledge about their bodies and are poorly prepared to insist on the use of condoms to protect themselves from HIV or other sexually transmitted diseases (STDs).

Even within marriage, women may have little influence over sex. "A woman does not have much say in the house," said one Indian woman participating in an ICRW focus group. "He is the husband. How long can we go against his wish?" Without adequate legal protection or opportunities for economic independence, such women may have little choice but to remain in abusive marriages and follow their husbands' dictates. Of 600 women living in a slum in Chennai (Madras), a major city on India's east coast, 90 percent said they had no bargaining power with their spouses about sex and couldn't convince them to use condoms. And 95 percent of these women were financially dependent on their husbands.

Women's risk is compounded by biological factors. During vaginal intercourse without a condom, transmission of HIV from an infected man to a woman is two to four times more likely than transmission in the opposite direction. The two key factors appear to be the surface area of exposed tissue and the viral load. Women lose on both counts: the virus concentrates in semen, and the surface area of the vagina is relatively large and subject to injury during sex. Tears in the lining of the vagina or cervix may admit the virus more readily. Women suffer another biological disadvantage as well. In general, STDs are harder to detect in women because the symptoms are more likely to be internal. Lesions from unrecognized STDs can increase a woman's susceptibility to HIV.

Once infected, women are less likely to be treated. In couples where both partners are infected with HIV but where treatment can be afforded only for one, it is the husband who almost invariably gets the drugs. Subhash Hira, director of Bombay's AIDS Research and Control Center, explained it this way to an AP reporter: "It is the woman who is stepping back. She thinks of herself as expendable." A 1991–93 study in Kagera, Tanzania found that in AIDS-afflicted households, more than twice as much, on average, was spent caring for the male victims than for the female victims: $80 versus only $38.

The stigma of infection also seems to fall more heavily upon women. Unease over female sexuality appears to translate readily into a tendency to see infection in women as punishment for sexual promiscuity. Women are sometimes even blamed for being the source of the disease. Suneeta Krishnan, an expert on AIDS in southern India, notes that the local languages contain few words for STDs, but the most commonly used formula is "diseases that come from women." One man explained the term to her: "The man may be the transmitter of the disease, but the source is the woman. She is the one who is blamed. For example, if a well is poisoned, and a man drinks from it and falls ill, people do not blame him. They blame the well. In the same way, people blame women for sexually transmitted diseases."

—*Ann Hwang*

---

of sharing needles. Information on this habit is hard to come by, but based on the most recent data the government has provided, UNAIDS estimated that 60 percent of users shared needles in 1998, up from 25 percent the year before.

In many parts of China, and particularly in the countryside of the central provinces, the virus is spreading through a very different form of injection. Selling one's own blood is a common way for poor people to make a little extra money, but it puts them at high risk for HIV infection. The government banned blood sales in 1998 (the blood supply in China is supposed to come from voluntary donations). But growing demand for blood virtually ensures the continuation of the practice. In some illegal collection centers, blood of the same blood type may be pooled, the plasma extracted to make valuable clotting and immune factors, and the remaining cells re-injected into the sellers. (Re-injection shortens the recovery period, allowing people to sell their blood more frequently.) The needles and other collection equipment are often reused as well. A January 2000 raid on one such center in Shanxi province, southwest of Beijing, turned up 64 bags of plasma, all of which tested positive for HIV and hepatitis B.

The extent of the black market in blood is unknown. China's news media are banned from reporting on the topic, outside researchers have been prevented from studying it, and government officials won't discuss it. But it's a good bet that the system is not about to be weaned off black market blood anytime soon; official donations are apparently inadequate even though their "voluntary" character is already badly strained. Inland from Hong Kong, for instance, in the city of Guangzhou, work groups are fined if they do not meet their blood donation quotas. Workers sometimes avoid the fines without donating by hiring "professional donors" to take their place. One could argue that such quotas still work, albeit in a somewhat indirect and callous way. But the system is riddled with flaws. The general cultural reluctance to give blood in China has been exacerbated by a widespread perception that donation is dangerous. And unfortunately, that perception is probably justified, since even official blood collection centers may reuse needles and tubing. (Such reuse is not necessarily intentional; sometimes unscrupulous dealers collect used equipment, repackage it, and sell it as new.) Another unfortunate consequence follows when the blood is actually used: apart from the larger urban hospitals, the Chinese blood supply is probably not adequately screened for HIV or other diseases, and "professional donors" have much higher levels of infection than the general public.

In the major cities and especially along China's highly developed southeastern coast, AIDS is primarily an STD. At least in the cities, sexual mores appear to have loosened considerably over the past couple of decades. Not surprisingly, prostitution is becoming more common. For the country as a whole, prostitution arrests now number about 500,000 annually; China's Public Security Department estimates the number of prostitutes to be between 3 and 4 million, a figure that has been increasing since the 1980s. STDs, such as syphilis and gonorrhea, were virtually eradicated in the 1960s under an aggressive public health campaign, but have returned with a vengeance. Infection rates are increasing by 30 to 40 percent each year, according to the Ministry of Health. That portends an increase in AIDS, not only because of what it suggests about the growing sexual permissiveness, but also because the genital sores caused by other STDs make people more vulnerable to HIV.

Sexual contact, intravenous drug use, blood selling—in many parts of the country, these and perhaps other modes of transmission are increasingly likely to "overlap" as the virus spreads. The results may be difficult to anticipate, or to counter. For example, in 1998, the most recent year for which statistics were available, the province reporting the largest number of new HIV infections was not Yunnan or Guangxi, but the remote Xinjiang, in China's arid and lightly populated northwest. Why? In part, the answer appears to be drugs. Despite its apparent isolation, Xinjiang is enmeshed in the opium trade. Some studies have found infection levels of about 80 percent among the province's intravenous drug users. Local prostitutes seem to be heavily infected as well. And HIV has begun to appear in women coming to clinics for prenatal care—a strong indication that the virus is starting to leak into the province's general population. But despite the fact that it has become an HIV hotspot, Xinjiang has attracted little official attention, and that suggests another reason for the province's plight. Most of Xinjiang's inhabitants are Uigur, a people of Turkish descent. (The area is sometimes called "Chinese Turkestan.") Like some of China's other ethnic minorities, the Uigur suffer disproportionately from HIV. The country's AIDS prevention and education programs, very small to begin with, may be even less effective among ethnic minorities. Lack of official interest in minorities may be a factor Xinjiang's epidemic; perhaps also there is some sort of cultural "communications gap."

In early 2000, a group of concerned Chinese scientists—including some members of the Chinese Academy of Sciences—submitted a report to the government that warned, "The spread of AIDS is accelerating rapidly and we face the prospect of remaining inert against the threat." Without decisive action, according to China's National Center for AIDS Prevention and Control (NCAIDS), 10 million people in China could be infected with HIV by 2010.

## *Death on the Margins*

In China and most of India, AIDS is still concentrated among socially marginal high-risk groups—groups engaged in activities that attract mainstream disapproval and that are often illegal. One of the most obscure of these groups is male homosexuals. Despite the prominence of homosexuality in the AIDS controversies of the industrialized countries, very little is known about gay life in China or India. But studies in Chennai (Madras), the largest city on the southeast coast, reveal one ominous characteristic of the Indian homosexual underground: most participants do not appear to be exclusively homosexual. Most are married.

Gay men who are married, heterosexual men who patronize prostitutes, intravenous drug users and their sexual partners: AIDS may still be a disease of the social margins, but in both India and China there are several major bridges between the margins and mainstream society. It's possible that the virus will tend to cross those bridges relatively slowly. If it remains largely in the fringe populations, it should be easier to control. But even this scenario entails serious risk, since it could encourage callousness towards the victims and complacency towards the disease.

Take the complacency potential first: if AIDS is portrayed as a disease of marginalized groups, people who are not in those groups may be reluctant to acknowledge their own vulnerability. Suneeta Krishnan, a researcher at the University of California at Berkeley, has studied HIV for the past three years in southern India and seen this reluctance first hand. "The perception is that AIDS is only a problem of female commercial prostitutes sitting in Bombay," she said. "It's only a problem for us if we have sex with them." Such attitudes could easily heighten the risk of contagion.

The "us-them" mentality can also greatly increase the suffering of those who are already infected. One effect of stigmatizing AIDS-prone minorities is that *all* AIDS sufferers tend to end up stigmatized. Rajesh Vedanthan, one of the founders of Swasthya, a nonprofit that provides HIV counseling to women in the southern Indian state of Karnataka, recalls the story of a pregnant woman who sought care at a hospital for profuse vaginal bleeding. Without her consent or knowledge, she was tested for HIV and found to be infected. The hospital doctor—without informing her of her HIV status—placed gauze to soak up the blood, discharged her from the hospital without treatment, and told her never to return. By the time she came to Swasthya, she had a raging infection. Such inhumanity can greatly compound the contagion of the disease itself.

## *"Avoiding Unnecessary Agony"*

In Beijing, the streets are swept clean by women wielding brooms made from twigs. Licensed taxis queue at the airport waiting for uniformed guards to assign them passengers. But you needn't go far from China's capital before all the taxis have inexplicably broken meters, and beggars crowd the trash-covered streets. A similar duality is apparent in the country's efforts to deal with AIDS. As a totalitarian state with a strong tradition of public health and social services, China would appear to be in good shape to control the AIDS epidemic. But China spends only about seven tenths of 1 percent of its GDP on health care. (The United States is at the other end of the spectrum, with public health care expenditures amounting to 6 percent of its GDP.) China's anti-AIDS efforts thus far have amounted to little more than crackdowns on prostitution, drug use, and blood sales—strong-arm tactics that have had negligible effect.

Public education about the epidemic has been stalled by censorship. The language of official AIDS announcements reflects a deep awkwardness in discussing sexual issues. "The government calls to the attention of its citizens whether their words and deeds conform to the standards of the Chinese nation," explained one official declaration dating from the beginning of the epidemic. The announcement added, with muffled urgency, that citizens should "know what to do and what not to do when making sexual decisions and avoiding unnecessary agony." Though times are changing, China's first nationally televised advertisement promoting condom use to prevent AIDS was taken off the air in December 1999, after just two days of broadcast, because it violated a ban on ads for sex products.

Technical infrastructure for treating and tracking the epidemic is also in short supply. In its most recent report, released in 1997, NCAIDS noted that China had only 400 labs capable of testing for HIV, or roughly one for every 3 million people. There is also a shortage of medical personnel trained to treat people infected with HIV or other STDs. When workers at STD clinics in the southern city of Shenzen were tested on their medical knowledge, only 23 percent passed, according to Xinhua, the official Chinese news agency. According to Zeng Yi, an AIDS researcher and member of the Chinese Academy of Sciences, local officials in various parts of the country are reluctant to collect data on HIV, for fear that their province will be blackballed as a highly infected area. Even more alarming is the apparent drop in resources committed to fighting the epidemic. Following budget cuts of 40 percent, the number of HIV screening tests in disease surveillance programs fell from 3.4 million in 1997 to 1.3 million in 1998.

India, the world's largest democracy, has little reason for complacency either. Early in the epidemic, some Indian politicians were calling for banning sex with foreigners, isolating HIV-positive people, and urging a return to traditional values—cries that were being heard in other countries as well. The proposal to ban sex with foreigners was put forth in 1988 by A.S. Paintal, the government's chief medical researcher, but was scuttled immediately under a barrage of domestic and international criticism. In Goa state, on India's west coast, a law permitting the resting and isolation of anyone suspected of being HIV-positive was overturned only after repeated protests. On the federal level, an unsuccessful 1989 "AIDS Prevention Bill" called for the forcible testing and detention of any HIV-positive person or anyone suspected of being HIV-positive.

In 1992, India's Ministry of Health and Family Welfare established the National AIDS Control Organisation (NACO) to carry out AIDS prevention and education. NACO has put into place a surveillance system to monitor disease trends, but limited resources have hampered prevention and made treatment impossible. Anti-retroviral therapy, the "drug cocktail" that can slow the progression of AIDS, costs $270 to $450 per month. The country's average per capita income is only $444 per year. Even among India's rapidly expanding middle class, the average per capita income is only about $4,800 per year—roughly the same as a year's worth of the cocktail. Nor is there preventive care for the many opportunistic infections that ultimately kill people whose immune systems are ravaged by AIDS, even though in industrialized countries, these infections can usually be held at bay for years with relatively inexpensive medications. Like China, India spends only about 1 percent of its GDP on health care—a number that Jeffrey Sachs, a professor of international trade at Harvard University, calls "shockingly low."

## *The Sonagachi Prostitutes and the Future of AIDS*

In 1989, when surveys of Thai brothels turned up rising levels of HIV infection among prostitutes, the Thai government collaborated with several non-governmental organizations to launch a massive public information campaign urging condom use. The "100% Condom Program" distributed condoms to brothels and massage parlors, and enforced use by tracking the contacts of men who sought care for STDs. Over the course of the next three years, condom use in brothels increased from 14 to 90 percent. By 1995, the number of men treated at government clinics for new sexually transmitted infections had dropped tenfold. A year later, HIV prevalence among conscripts to the army had dropped below 2 percent—less than half of what it had been in mid-1993.

The lesson from the early stages of the epidemic in Thailand is clear: it's worth dispensing with moral scruples to give people a clear sense of the medical issues. Public education works, at least when it's backed by some degree of enforcement. There's no reason to think this approach would be any less effective in China or India. Consider the prostitutes of Calcutta's Sonagachi district. Against considerable odds, these women have managed, not just to inform themselves about AIDS, but to organize themselves. The over 30,000 dues-paying members in their informal union have improved working conditions, educated other prostitutes about AIDS, started reading classes, and reduced the number of child prostitutes. They understand the need for condoms and have even threatened collective action against brothel owners reluctant to require condom use. As a result, their HIV infection levels remain at 5 percent—very low compared to the 60 or 70 percent levels typical of prostitutes in other Indian cities.

Frank talk about condoms and safe sex is of course just a start. An effective AIDS program must also have a reliable, confidential, and voluntary HIV testing program. It must protect the rights of infected people and secure treatment for them. But perhaps the greatest challenge of all is the need to build some form of long-term support for those marginalized, high-risk groups—support that invites the kind of initiative shown by the Sonagachi prostitutes. As Suneeta Krishnan puts it, "HIV is intimately linked to social and economic inequality and deprivations. As long as these problems persist, HIV is going to persist."

That is perhaps one of the lessons from the latter stages of the Thai epidemic. The Asian financial crisis dried up funding for Thailand's AIDS programs. Spending fell from $90 million in 1997 to $30 million in 1998 before rebounding somewhat, to $40 million in 1999 and 2000. The drop in funds has made the weak points in the Thai approach more apparent. Among populations other than brothel workers and their clients, the epidemic has proceeded largely unchecked. Male homosexuals have not generally been included in the program. Neither have intravenous drug users—a group whose infection level has passed 40 percent. And the worst news of all is that the infection level among women receiving prenatal care is now climbing.

In India and China as in many other places, prostitutes, homosexuals, and drug addicts are frequently the objects of contempt and legal sanction. But these are the people who should be top priorities for any serious AIDS program, for both practical and humanitarian reasons. How much of an investment are we really willing to make in the egalitarian principles upon which every public health program is built? AIDS is an acid test of our humanity. Over and over again, the virus teaches its terrifying lesson. There is no such thing as an expendable person.

---

Ann Hwang is a medical student at the University of California, San Francisco and a former intern at the Worldwatch Institute.

From *World Watch*, January/February 2001, pp. 12-20. © by Ann Hwang, the Worldwatch Institute. Reprinted by permission.

# GREENVILLE: FROM BACK COUNTRY to FOREFRONT

Eugene A. Kennedy

What factors are crucial in determining the success or failure of an area? This article explores the past and present and glimpses what may be the future of one area which is experiencing great success. The success story of Greenville County, S.C. is no longer a secret. This article seeks to find the factors which led to its success and whether they will provide a type of yardstick to measure the future.

PHOTO: E. KENNEDY
An open courtyard off Main Street, downtown Greenville, S.C.

The physical geography of this area is explored, as well as the economic factors, history, transportation, energy costs, labor costs and new incentive packages designed to lure new industries and company headquarters to the area.

## Physical geography: advantageous

Greenville County is situated in the northwest corner of South Carolina on the upper edge of the Piedmont region. The land consists of a rolling landscape butted against the foothills of the Appalachian Mountains. Monadnocks, extremely hard rock structures which have resisted millions of years of erosion, rise above the surface in many places indicating that the surface level was once much higher than today. Rivers run across the Piedmont carving valleys between the plateaus. The cities, farms, highways and rail lines are located on the broad, flat tops of the rolling hills.

Climatologically, the area is in a transition zone between the humid coastal plains and the cooler temperatures of the mountains, resulting in a relatively mild climate with a long agricultural growing season. The average annual precipitation for Greenville County is 50.53 inches at an altitude of 1040 feet above sea level. The soil is classified as being a Utisoil. This type of soil has a high clay base and is usually found to be a reddish color due to the thousands of years of erosion which has leached many of the minerals out of the soil, leaving a reddish residue of iron oxide. This soil will produce good crops if lime and fertilizer containing the eroded minerals are added. Without fertilizers, these soils could sustain crops on freshly cleared areas for only two to three years before the nutrients were exhausted and new fields were needed. This kept large plantations from being created in the Greenville area. The climate and land are such that nearly anything could be cultivated with the proper soil modification. Physical potential, although a limiting factor, is not the only determining factor in the success of an area. As Preston James, one of the fathers of modern geography pointed out, the culture of the population which comes to inhabit the area greatly determines the response to that particular physical environment.

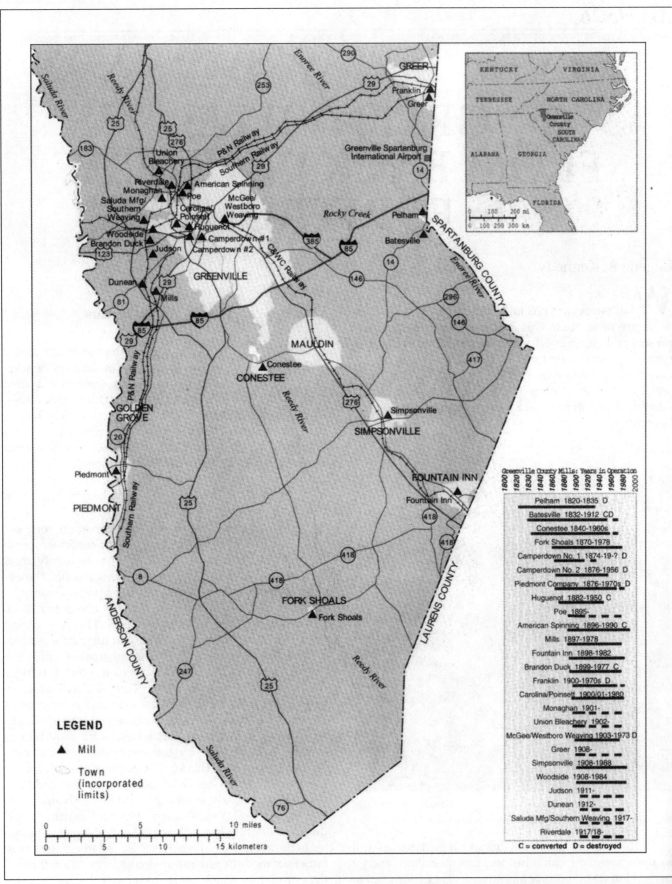

## Article 26. GREENVILLE: FROM BACK COUNTRY to FOREFRONT

### From European settlement through the textile era

An Englishman named Richard Pearis was the first to begin to recognize the potential of what was then known as the "Back Country." The area was off limits to white settlers through a treaty between the British and the Indians. In order to get around the law, Pearis married a native American and opened a trading post in 1768 at the falls of the Reedy River. He soon built a grist mill and used the waterfalls for power. Pearis prospered until the end of the Revolutionary War. He had remained loyal to the British, lost all his property when the new nation was established and left the country.

Others soon realized the potential of what was to become Greenville County. Isaac Green built a grist mill and became the area's most prominent citizen. In 1786, the area became a county in South Carolina and was named for Isaac Green. The Saluda, Reedy and Enoree Rivers along with several smaller streams had many waterfalls, making them excellent locations for mills during the era of water power. During the Antebellum period, Greenville County's 789 square miles was inhabited by immigrants with small farms and also served as a resort area for Low Country planters who sought to escape the intense summer heat and the disease carrying mosquitoes which flourished in the flooded rice fields and swampy low country of the coastal plains. Most of the permanent residents farmed and a few mills were built to process the grains grown in the area.

Although very little cotton was ever actually grown in Greenville County, cotton became the driving force behind its early industrialization. Beginning with William Bates in 1820, entrepreneurs saw this area's plentiful rivers and waterfalls as a potential source of energy to harness. William Bates built the first textile mill in the county sometime between 1830 and 1832 on Rocky Creek near the Enoree River. This was known as the Batesville Mill. Water power dictated the location of the early southern textile mills, patterned after the mills built in New England. Mill owners purchased the cotton from farms and hauled it to their mills, but lack of easy transportation severely limited their efforts until 1852. That year, the Columbia and Greenville Railroad finally reached Greenville County. Only the interruption of the Civil War kept the local textile industry from becoming a national force during the 1850s and 1860s.

The area missed most of the fighting of the Civil War and escaped relatively unscathed. This provided the area with an advantage over those whose mills and facilities had been destroyed during the war. The 1870 census reported a total $351,875 in textiles produced in the county. This success encouraged others to locate in Greenville. Ten years later, with numerous mills being added each year, the total reached $1,413,556.11. William Bates' son-in-law, Colonel H. P. Hammett, was owner of the Piedmont Company which was the county's largest producer. Shortly after the construction of the water powered Huguenot Mill in the downtown area of the City of Greenville County in 1882, the manufacture of cotton yarn would no longer be controlled by the geography of water power.

PHOTO: E. KENNEDY

F.W. Poe Manufacturing Co., Old Buncombe Road, Greenville, S.C. Built in 1895, purchased by Burlington Industries and closed in 1997. Palmetto State Dyeing and Finishing Co. opened in 1987. The company employs approximately 110 people.

The development of the steam engine created a revolution in the textile industry. No longer was the location of the mill tied to a fast moving stream, to turn a wheel that moved machinery. Large amounts of water were still needed but the dependence upon the waterfalls was severed. Between 1890 and 1920, four textile plants were built in the county outside the current city limits of the City of Greenville. At least thirteen large mills were built near the city to take advantage of the rail system, as shown on the map, "Greenville County Mills." Thus, with cheaper and more efficient steam power, transportation costs became a deciding factor. These mills built large boiler rooms adjacent to their plants and dug holding ponds for water.

Another drastic change took place in the textile industry around 1900. This change would provide even greater flexibility for the mill owners. A hydroelectric dam was constructed on the Saluda River, five miles west of the city of Greenville. It was completed in 1902 and would provide cheap electricity for the county. John Woodside, a local mill owner who foresaw electricity as the next step in the evolution of the industry, built what was then the largest textile mill in the world in the city of Greenville that same year. He located it further from a water source than previously thought acceptable. However, John had done some primitive locational analysis and chose the new site well. It was located just beyond the city boundary to limit his tax liability and directly between the lines of two competing rail companies—the Piedmont Railroad and the Norfolk and Western (now known as Norfolk and Southern). John Woodside's mill proved to be a

PHOTO: E. KENNEDY

Panorama of downtown Greenville, S.C.

tremendous success. With water no longer a key factor of location, the owners identified transportation as the key factor of location. Others began to build near rail lines.

The textile industry made Greenville County very prosperous. The mills needed workers and shortly outstripped the area's available labor supply. Also, many did not want to work in the hot, poorly ventilated, dangerous conditions found in the mills. When most of the mills were still built of wood, the cotton fibers floating in the air made fire a very real danger. Many businesses sprang up to service the needs of the workers and the textile mill owners. Farmers, sharecroppers, former slaves and children of former slaves were recruited to work in the mills. Housing soon became scarce and the infrastructure wasn't equipped to handle the influx of new workers. To alleviate the problem, the mill owners built housing for their workers. These were very similar to the coal camps of Appalachia and other factory owned housing in the north. They were very simple dwellings built close to the mill so the workers could easily walk to and from work. They also provided company-owned stores, doctors and organized recreational activities for their employees, creating mill communities. Many people who worked for the mills would have told you they lived at Poe Mill or Woodside, the names of their mill communities, rather than Greenville.

In the 1960s, rail transportation of textiles was a cost the owners wished to lower. They found a cheaper, more versatile form of transportation in the trucking industry. The interstate highway system was now well developed and provided a means of keeping costs down for the operators. In the 1970s, owners began to identify wages and benefits as a major factor in their cost of operation and many firms relocated in foreign countries, which offered workers at a fraction of the wages paid in the United States and requiring few if any benefits.

## Meeting the challenge of economic diversification

Greenville County used its natural physical advantage to become the "Textile Capital of the World." Many of the other businesses were tied directly or indirectly to the textile industry. These ranged from engineering companies who designed and built textile machinery to companies which cleaned or repaired textile machines. Employment in the textile industry in Greenville County peaked in 1954 with 18,964 workers directly employed in the mills. As the industry began to decline, the leaders of the industry along with local and state leaders showed great foresight by combining their efforts into an aggressive move to transform Greenville County into a production and headquarters oriented economy. A state sponsored system of technical schools greatly facilitated this effort. Workers could get the training they needed to pursue almost any vocation at these centers. This system still is a factor in Greenville County's success.

The group emphasized the ability to make a profit in Greenville County. The focus of their efforts was turned to creating a sound technical education network along with the flexibility to negotiate packages of incentives to lure large employers. Incentives included negotiable tax and utility rates, plus a strong record of worker reliability due to South Carolina's nonunion tradition, with very few work stoppages. The foresight of this group has paid off handsomely. The majority of the textile mills which provided the backbone of the economy of Greenville County are no longer in business. Many of the old buildings still stand. Ten of the mills built before 1920 now are used in other capacities. American Spinning was built in 1896 and now is used as a warehouse, office space and light manufacturing all under one roof. Most of the mills are used for warehouse space or light manufacturing such as the Brandon Duck Mills, which operated between 1899 and 1977 as a cotton mill. It now houses two small factories which assemble golf clubs and part of the mill is used as a distribution center. The low lease cost (from $1 to $15 per square foot) is an enticement for other businesses to locate in these old buildings.

The old Huguenot Mill, the last water powered mill built in the county, was recently gutted and has been rebuilt as offices for the new 35 million dollar Peace Center entertainment complex in downtown Greenville. The Batesville Mill, the first in the county, was built of wood. It burned and was rebuilt in brick in 1881. It closed its

## Article 26. GREENVILLE: FROM BACK COUNTRY to FOREFRONT

doors in 1912 because the water-powered mill was not competitive. The mill was purchased by a husband and wife in 1983, converted into a restaurant, and was the cornerstone and headquarters of a chain of FATZ Restaurants until it burned again in 1997. So, in considering diversification, one of the first steps was to look for other uses for the facilities which already existed.

Other efforts also met with great success. As businesses began to look south during the 1970s for relocation sites, Greenville began to use its natural advantages to gather some impressive companies into its list of residents. By 1992, the combination of these efforts made Greenville County the wealthiest county in the state of South Carolina. Twenty-five companies have their corporate headquarters in the county, as shown in Table 1.

Forty-nine others have divisional headquarters in the county, as shown in Table 2. This constitutes a sizable investment for the area, yet even this list does not include a 150 million dollar investment by G. E. Gas Turbines in 1992 for expansion of their facility. This was the largest recent investment until 1993.

> **What ultimately swayed the automaker to choose Greenville? One of the main reasons was physical location.**

Along with American companies, foreign investment was sought as well. Companies such as Lucas, Bosch, Michelin, Mita and Hitachi have made major investments in the county. Great effort has been put into reshaping the face of Main Street in Greenville as well. The city is trying to make a place where people want to live and shop. Many specialty stores have opened replacing empty buildings left by such long time mainstays as Woolworths. The Plaza Bergamo was created to encourage people to spend time downtown. The Peace Center Complex provides an array of entertainment choices not usually found in a city the size of Greenville. The Memorial Auditorium, which provided everything from basketball games, to rodeo, concerts, high school graduations and truck pulls has closed its doors and was demolished in 1997 to make way for a new 15,000 seat complex which will be named for its corporate sponsor. It will be called the Bi-Lo Center. Bi-Lo is a grocery store chain and a division of the Dutch Company, Ahold. A new parking garage is being built for this center and two other garages have recently been added to improve the infrastructure of the city. City leaders have traveled to cities such as Portland, Oregon, to study how they have handled and managed growth and yet kept the city friendly to its inhabitants.

### Table 1
### CORPORATE HEADQUARTERS IN GREENVILLE COUNTY

1. American Leprosy Mission International
2. American Federal Bank
3. Baby Superstores, Inc.
4. Bowater Inc.
5. Builder Marts of America Inc.
6. Carolina First Bank
7. Delta Woodside Industries Inc.
8. Ellcon National Inc.
9. First Savings Bank
10. Heckler Manufacturing and Investment Group
11. Henderson Advertising
12. Herbert-Yeargin, Inc.
13. JPS Textile Group Inc.
14. Kemet Electronics Corp.
15. Leslie Advertising
16. Liberty Corp.
17. Mount Vernon Mills Inc.
18. Multimedia Inc.
19. Ryan's Family Steakhouses
20. Span America
21. Steel Heddle Manufacturing Co.
22. Stone Manufacturing Co.
23. Stone International.
24. TNS Mills Inc.
25. Woven Electronics Corp.

The largest gamble for Greenville County came in early 1989. The automaker BMW announced that it was considering building a factory in the United States. Greenville County and the state of South Carolina competed against several other sites in the midwest and southeast for nearly two years. On June 23, 1992, the German automaker chose to locate in the Greenville-Spartanburg area. Although the plant is located in Spartanburg County, the headquarters are in Greenville and both counties will profit greatly. When the announcement was made, the question was: what ultimately swayed the automaker to choose Greenville? One of the main reasons was physical location. The site is only a four hour drive from the deepwater harbor of Charleston, SC. Interstates 26 and 85 are close by for easy transportation of parts to the assembly plant. The Greenville-Spartanburg Airport is being upgraded so that BMW can send fully loaded Boeing 747 cargo planes and have them land within five miles of the factory. Plus, the airport is already designated as a U.S. Customs Port of Entry and the flights from Germany can fly directly to Greenville without having to stop at Customs when entering the country. Other incentives in the form of tax breaks, negotiated utility rates, worker training and state purchased land helped BMW choose the 900 acre site where it will build automobiles.

## Table 2
### DIVISIONAL HEADQUARTERS IN GREENVILLE COUNTY

1. Ahold, Bi-Lo.
2. BB&T, BB&T of South Carolina.
3. Bell Atlantic Mobile
4. Canal Insurance Company
5. Coats and Clark, Consumer Sewing Products Division.
6. Cryovac—Div. of W. R. Grace & Co.
7. Dana Corp., Mobile Fluid Products Division
8. DataStream Systems, Inc.
9. Dodge Reliance Electric
10. Dunlop Slazenger International, Dunlop Slazenger Corp.
11. EuroKera North America, Inc.
12. Fluor Daniel Inc.
13. Fulfillment of America
14. Frank Nott Co.
15. Gates/Arrow Inc.
16. General Nutrition Inc., General Nutrition Products Corp.
17. Gerber Products Co., Gerber Childrenswear, Inc.
18. GMAC
19. Goddard Technology Corp.
20. Greenville Glass
21. Hitachi, Hitachi Electronic Devices (USA)
22. Holzstoff Holding, Fiberweb North America Inc.
23. IBANK Systems, Inc.
24. Insignia Financial Group, Inc.
25. Jacobs-Sirrine Engineering
26. Kaepa, Inc.
27. Kvaerner, John Brown Engineering Corp.
28. Lawrence and Allen
29. LCI Communications
30. Lockheed Martin Aircraft Logistics Center Inc.
31. Manhattan Bagel Co.
32. Mariplast North America, Inc.
33. Michelin Group, Michelin North America
34. Mita South Carolina, Inc.
35. Moovies, Inc.
36. Munaco Packing and Rubber Company
37. National Electrical Carbon Corp.
38. O'Neal Engineering
39. Personal Communication Services Dev.
40. Phillips and Goot
41. Pierburg
42. Rust Environment and Infrastructure
43. SC Teleco Federal Credit Union
44. Sodotel
45. South Trust Bank
46. Sterling Diagnostic Imaging, Inc.
47. Umbro, Inc.
48. United Parcel Service
49. Walter Alfmeier GmbH & Co.

The large incentive packages might appear self-defeating but BMW's initial investment was scheduled to be between 350 and 400 million dollars. The majority of the companies supplying parts for BMW also looked for sites close enough to satisfy BMW's just-in-time manufacturing needs. The fact that Michelin already made tires for their cars here, Bosch could supply brake and electrical parts from factories already here and J.P. Stevens and others could supply fabrics for automobile carpets and other needs readily from a few miles away also was a factor.

## A combination of physical, environmental and cultural factors greatly influence the location of businesses.

One major BMW supplier, Magna International, which makes body parts for the BMW Roadster and parts for other car manufacturers, located its stamping plant in Greenville County. Magna invested $50 million and will invest $35 million more as BMW expands. Magna needed 100 acres of flat land without any wetlands and large rock formations. This land needed to be close enough to provide delivery to BMW. After studying several sites, Magna chose South Donaldson Industrial Park, formerly an Air Force base, just south of the city of Greenville. The county and state will help prepare the location for their newest employer.

PHOTO: E. KENNEDY

Huguenot Mill (lower left). Built in 1882, on Broad Street, Greenville, S.C., it is the last waterpowered mill in Greenville County. It is being refurbished to become part of the Peace Center Complex at right.

Road improvements, the addition of a rail spur and an updating of water and sewer facilities will all be provided to Magna in this agreement. Also, Magna will receive a reduced 20 year fixed tax rate along with other incentives for each worker hired. These incentive packages may seem unreasonable but they have proven to be necessary in the 1990s when large organizations are deciding where to locate.

## The future: location, location, and location

From the time of the earliest European settlers, the natural advantages of Greenville County helped bring it to prosperity. The cultural background of the settlers was one of industry and a propensity for changing the physical environment to maximize its industrial potential. Nature provided the swift running rivers and beautiful waterfalls. The cultural background of the settlers caused them to look at these natural resources and see economic potential.

The people worked together to create an environment which led Greenville County to be given the title of "Textile Center of the World" in the 1920s. Then, again taking advantage of transportation opportunities and economic advantages, the area retained its textile center longer than the majority of textile centers.

Today, after 30 years of diversification, economic factors now are normally the deciding factor in the location of a new business or industry. Greenville County with its availability of land, reasonable housing costs, low taxes, willingness to negotiate incentive packages, and positive history of labor relations helped make it a desirable location for business. Proximity to interstate transportation, rail and air transport availability help keep costs low. The county's physical location about half way between the mega-growth centers of Charlotte, North Carolina, and Atlanta, Georgia, places it in what many experts call the mega-growth center of the next two decades. Now, with BMW as a cornerstone industry for the 1990s and beyond, Greenville County looks to be one of the areas with tremendous growth potential.

Thus, a combination of physical, environmental and cultural factors greatly influence the location of businesses. Transportation costs, wage and benefit packages and technical education availability are all interconnected.

The newest variable involves incentive packages of tax, utility reduction, worker training, site leasing and state and local investment into improving the infrastructure for attracting employers. The equation grows more and more complex with no one factor outweighing another; however, economic costs of plant or office facilities, wages and benefits and transportation seem to be paramount. Greenville County is blessed with everything it needs for success. It will definitely be one the places "to be" in the coming years.

## References and further readings

DuPlessis, Jim. 1991. Many Mills Standing 60 Years After Textile Heydays. *Greenville News-Piedmont* July 8. pp. 1c–2c.
Greater Greenville Chamber of Commerce, 1990. *1990 Guide to Greenville.*
*Greenville News-Piedmont.* 1991–1993. *Fact Book 1991; 1992; 1993.*
Patterson, J. H. 1989. *North America.* Oxford University Press: N.Y. Eighth edition.
Scott, Robert. 1993. Upstate Business. *Greenville News-Piedmont.* 15 August. pp. 2–3.
Shaw, Martha Angelette. 1964. *The Textile Industry in Greenville County.* University of Tennessee Master's Thesis.
Strahler, Arthur. 1989. *Elements of Physical Geography.* John Wiley and Sons: N.Y. Fourth edition.

---

Eugene A. Kennedy is a native of West Virginia who attended Bluefield State College, Bluefield, West Virginia, and received an M.A. in geography from Marshall University in Huntington. He is currently a public educator in the Greenville County School system, Greenville S.C. He was awarded a "Golden Apple" by Greenville television station WYFF in 1997, has been a presenter at the South Carolina Science Conference, and a consultant to the South Carolina State Department of Education. He can be reached at GEOGEAK@aol.com.

---

From *Focus*, Spring 1998, pp. 1–6. © 1998 by the American Geographical Society. Reprinted by permission.

# Death of a Small Town

Even the mayor is moving on in Bisbee, North Dakota, the next 'rural ghetto' victim

BY DIRK JOHNSON

BOB WELTIN, A BURLY AND BEARDED North Dakota homeboy with the oil-stained hands of a working man, squeezed tight on a cup of coffee in the Chocolate Shop on Main Street in tiny Bisbee, N.D. So much in his life has been slipping away. Weltin, 43, is the mayor of Bisbee, a town that gave him his childhood: movies at Pettsinger's Theater, root-beer floats at Brannon's Drug and Soda Fountain, groceries at Dick's Red Owl. All that is gone now. There isn't a doctor or lawyer in Bisbee anymore, or a plumber. The priest at Holy Rosary Catholic Church left town; parishioners can no longer afford to support a pastor. Even the police department closed. The population dropped more than 30 percent in the last decade. There are now but 227 hearty souls hanging on for life in Bisbee.

But Mayor Weltin won't be one of them. He is stepping down from his post in September and moving away. There reaches a point in the decline of a town, he explained apologetically, "where it just doesn't work anymore." Besides everything else, Weltin says he wants to move closer to a medical center. His 13-year-old boy, Brendan, has heart troubles. His 18-year-old son, Jordan, died of cerebral palsy in June. Weltin stared away for a moment, then put down his coffee cup and started talking about the weather. The winds can howl wickedly across the northern plains. Even a strong man can stand for only so long.

Hundreds of little towns on the Great Plains are teetering on collapse. About half the counties on the plains lost population in the '90s—some to levels lower than during frontier days in the 1800s, when pioneers rolled westward in covered wagons. Outside these isolated little places, it's hard to know whether many people really care. In the suburban ethos that prevails in 21st-century America, the death of so many hick towns elicits barely more than a shrug.

The land of white picket fences was never idyllic. But as these towns become further hollowed out, the spine of America today is becoming older and poorer. Economist Karl Stauber told a conference sponsored by the Federal Reserve Bank that rural poverty is generally higher than urban—15.9 percent, compared with 12.6 percent. Nearly 23 percent of rural children live in poverty, compared with 19.2 percent of urban children. "The rural ghetto, if it is allowed to continue and expand, will be a powerful symbol of failure in America and of American culture," said Stauber. "It will mean that America accepts the idea that success and prosperity should be allocated based on race and location, rather than being available to all." He and other rural advocates fault policymakers in Washington for focusing only on farmers—direct subsidies to agriculture last year totaled more than $25 billion—even though most people in the countryside do not farm. These advocates say the federal government should aid rural businesses, bolster state and community colleges and promote immigration to counties that are begging for warm bodies.

The Bush administration has shown little interest in crafting a plan for the survival of withering little towns, said Mark Drabenstott, a vice president at the Federal Reserve Bank in Kansas City, Mo. "We write farm bills every four or five years, but we don't have a rural policy," said Drabenstott, who directs the Center for the Study of Rural America. "So many communities are asking: where is our next economic engine?"

Small towns that do prosper these days tend to be those with the stunning natural features of the Rockies—ranchettes in Montana go for millions—not the plain-Jane farm towns of the corn belt, said Calvin Beale, a demographer for the Agriculture Department. "Mountains and lakes," Beale said, "are seen as amenities."

Places like Bisbee have a different kind of beauty. E. B. White waxed lyrical about the flat fields and endless horizons in North Dakota. Lately, bird-watchers have been coming to the state to gaze at the geese. But they're not staying year-round and buying houses that now sell for as little as $2,000 in Bisbee and other towns in the state.

It has never been easy to live in a place that gets too hot in the summertime (over 100 degrees), too cold in the wintertime (gallows humor here: "40 below—it keeps out the riffraff"), and too flat all year round. Rand McNally once forgot to include North Dakota in an atlas of the 50 states. Aware of its image problem, some politicians are pushing a measure to change the name of the state. They urge dropping the "North" from North Dakota. "The thinking is that maybe it won't sound quite so cold," said Terry Jorde, a bank president in Cando, the seat of Towner County, which includes Bisbee.

As a way to try to survive, counties on the plains are putting aside rivalries and trying to cooperate, rather than compete. Since the high school in Bisbee is too small to field a football team, it forms a cooperative with other schools. So it's the Cando-Bisbee-Egeland-Starkweather team that plays the Adams-Edinburg-Edmore squad. But it means that teams must travel farther to find an opponent. The players from Bisbee sometimes journey 160 miles for a game. Wayne Lingen, who is the superintendent of school districts in Cando and Bisbee, said the makeup of the cooperatives is different for each sport. "And then it's still another combination for the girls' sports," said Lingen. "It can be very confusing to try to keep rivalries straight."

There is more than rivalry to worry about. Lingen points to enrollment figures that underscore the bleak future for his school. The high school, which includes grades seven to 12, has 69 students. But the grade school, kindergarten through sixth, has only 31. The decline is sharp and fast. And nobody here is fooling himself. There is some brave talk at Dutch's tavern, especially when it gets crowded on Tuesday nights for bingo. But for the most part, people are smart enough—and honest enough—to know that someday, maybe soon, their town is going to die.

The Chocolate Shop, the only restaurant in town, loses money every time it opens the door. It stays open as a gesture of goodwill by its owner, Sylvia Schmidt, an elderly, well-to-do woman who remembers another Bisbee, a place that bustled and preened and made a little noise on a Saturday night. "You can't imagine what it was like," says Schmidt, with eyes that still manage to sparkle.

---

From *Newsweek*, September 10, 2001. © 2001 by Newsweek, Inc. All rights reserved. Reprinted by permission.

# The Río Grande

## Beloved river faces rough waters ahead

### By Steve Larese

Corrales farmer Gus Wagner allows himself just a minute to admire a peach and honey sunrise pouring over the Sandía Mountains, its patchy light filtering through cottonwoods and bobbing on the Río Grande. Turning back to his work, his callused hands turn the wheel of his *compuerta* (floodgate), and the river's paced water soon sluices through the *acequia* to flood his blossoming apple orchard.

"The river's beautiful, *que no?*" he asks, stealing another moment from his busy day. "I thank God everyday for the Río Grande and how it allows my family to live here. The river is our lifeblood."

It's safe to say there's not a Land of Enchanter around who would disagree with Wagner. Three states and two countries lay claim to the Río Grande, but only New Mexico claims it as our soul. To New Mexicans past and present, the Río Grande has been a gentle miracle in a harsh land.

But our beloved river was recently named the seventh most endangered in the nation by the respected American Rivers, a national conservation organization (www.Amrivers.org). "The Río Grande is really on its deathbed," says Betsy Otto, American Rivers' director of river restoration finance. "Increased population, outdated irrigation techniques and misguided engineering are drying the river up. But that being said, the situation isn't hopeless. Rivers are resilient if you let them be rivers. But too many people think of the Río Grande only as an irrigation ditch, drinking fountain or a sewer."

Beginning life as a twist of alpine streams 12,500 feet above sea level in the San Juan Mountains of Colorado, the Río Grande courses through Colorado, New Mexico and forms the 1,000-mile border between Texas and Mexico. Its life ends 1,887 miles later when it merges with salt water in the Gulf of Mexico. Today, the river appears fairly linear through New Mexico, lined by the nation's largest continuous cottonwood forest called the *bosque*, or woods. But this hasn't always been the case, says Dr. Cliff Crawford, a University of New Mexico research professor emeritus who has extensively studied the Río Grande and its ecosystem.

"The river and *bosque* as we see them today look quite different from how they appeared just a few decades ago," he says.

Crawford explains that until about the mid-'50s, the Río Grande was a braided river, meaning several sizable rivulets would snake over one another with cottonwood islands between them. Instead of perfectly lining the river, cottonwoods, coyote willows and other native flora created a mosaic pattern, their seeds deposited on the whim of the river. Depending on the year's runoff, the rivulets would swell to make one river worthy of the name Río Grande.

This seasonal flooding would establish new groves of trees, clean away debris and keep in check non-native species such as salt cedar and Russian olive. During heavy runoff years, the river would occasionally spread past its average bounds and devastate villages and pueblos built within the flood plain. The force of such flooding could reroute the very course of the river, cutting new channels through former farm fields and irrigation ditches.

*Ciénegas* (marshes) and oxbow lakes remained after the flood subsided. For the most part, New Mexicans ac-

# Article 28. The Río Grande

STEVE LARESE

LAURENCE PARENT

*Top*—The river courses through the Río Grande Gorge, an impressive feature it carved over millions of years through the volcanic basalt of the Taos Plateau. The canyon starts as a ditch in Colorado's San Juan Hills and eventually plummets to depths of 1,000 feet. *Bottom*—The Río Grande as it leaves New Mexico and forms the Texas/Mexico border.

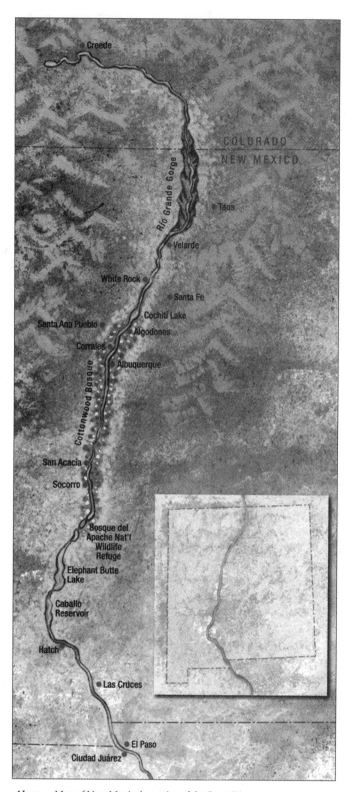

*Above*—Map of New Mexico's portion of the Great River.

cepted this fact of life, and many understood it was necessary. Like Egypt's Nile, with the floods came rich silt that was deposited on farmland. The infrequent inconvenience of rebuilding waterlogged adobes was worth the annual gifts provided by the river.

ANNUAL EDITIONS

STEVE LARESE

The Middle Río Grande supports the largest continuous cottonwood forest left in the United States, seen here north of Albuquerque. Scientists, citizens and environmentalists worry the forest (bosque) is dying out, partly because the river hasn't been able to flood in the past half century. Flooding is crucial in the establishment of new cottonwoods. Also, deadwood has been allowed to build up, which creates a severe fire hazard. The Catch 22 is that if flooding is allowed, there may not be enough water to meet water commitments to Texas, Chihuahua, Coahuila, Nuevo Leon and Tamaulipas, Mexico. As New Mexico enters an expected period of even drier weather, the growing cities of Albuquerque, El Paso and Ciudad Juárez are also becoming more dependent on the river.

But New Mexico's dynamics were changing. As more people moved to the state and communities—especially Albuquerque—grew, the price of flooding increased. Finally, after Albuquerque in part had to be evacuated because of the floods of 1941 and '42, state and federal governments decided the Great River needed to be controlled.

In the 1950s, the Southwest suffered through a devastating drought. The Río Grande dried up, and New Mexico was unable to meet its water obligations to Texas under the Río Grande Compact, a water-rights agreement signed by Colorado, New Mexico and Texas and approved by Congress in 1939. New Mexico avoided a lawsuit by aggressively pursuing channelization of the river to maximize flow downstream. Levees were built along the river's banks to contain flooding. "Jetty Jacks," the same type of crossed metal structures used to deter amphibious assaults during World War II, made the Río Grande look like Normandy Beach. Jetty Jacks lined the desired channels with other structures directing the water flow into the channels. The result was to trap silt, sand and debris, building up the banks and further confining the river. By narrowing the channel, more water was being delivered downstream, which also greatly reduced flooding.

The efforts worked perfectly. But altering the natural tendencies of the river on such a grand scale couldn't happen without affecting its very nature. Today, Crawford says, we are seeing those effects.

"The changes made to the river were very justified in people's minds at the time," he says. "But the hydrology of the system has completely changed. Unless we allow the system to come back to some level of how its ecosystem used to work, Albuquerque's *bosque* at least will be lost."

"You can't blame people for not wanting their houses washed away or their crops destroyed," says Rob Yaksich, an interpretive ranger at the Río Grande Nature Center State Park in Albuquerque. "But a relatively short time later, we're seeing the cottonwoods are certainly losing ground. Without flooding, they aren't regenerating. Most of the youngest trees we have were established in the floods of the '40s. Without some type of flooding, when they're gone in 40 or so years, that'll be it for the *bosque* as we know it."

Early last century, salt cedar and Russian olive trees were introduced to New Mexico as ornamental vegetation. Some were planted along the river to further stabilize its banks. These newcomers have done extremely

Realizing the threat, several governments and organizations have begun to take action to maintain and restore New Mexico's *bosque*.

Todd Caplan, department of natural resources director at Santa Ana Pueblo, watches as his crew plants new cottonwoods along part of the six miles of river that crosses the 74,000-acre pueblo. "What we're doing is restoring 7,000 acres of floodplain here," Caplan says. "The river through here used to average 1,200 feet wide. Now it's 300 feet. Cochití Dam has certainly reduced flooding, but now the *bosque* needs help."

Using a combination of state, federal and tribal funds, the Pueblo has painstakingly ripped out hundreds of acres of salt cedar and Russian olive that overran traditional cottonwood and willow habitat. The felled trees are cut and delivered to tribal elders for firewood. Towering piles of disassembled Jetty Jacks await new lives as fence posts. Below what will be the tribe's new Hyatt Regency Tamaya golf resort, 115 acres of native salt grass will be planted, and miles of nature trails will wind through the restored *bosque*, Caplan says. Already, the change is apparent.

CLAY MARTIN

STEVE LARESE

*Top*—Elephant Butte Dam near Truth or Consequences created the largest reservoir in the world upon the dam's completion in 1916. A major recreation destination, the reservoir impounds 2-million-acre-feet of water for irrigation in southern New Mexico. Cochití Dam, completed in 1975 north of Cochití Pueblo, is one of the largest earth-filled dams in the world at 5.5 miles long and 251 feet high. Constructed and managed by the U.S. Army Corps of Engineers, the no-wake lake is a favorite playground for windsurfers. *Bottom*—The Upper Río Grande near Pilar is an angler's paradise.

STEVE LARESE

The river provides a relaxing respite for New Mexicans and many species of birds that depend upon it during migrations. For more information about the river and its natural history, tour the Río Grande Nature Center State Park in Albuquerque, 2901 Candelaria NW (505) 344–7240 or visit www.unm.edu/~natrcent.

well, and have choked out native trees such as cottonwoods in areas. Salt cedars have increased the salinity of the soil, and they have also blocked the sun that young cottonwoods need, says Crawford. "New cottonwoods need to be established in the open with lots of sun and silt," he says. "But it's a matter of who wins the shade race."

"The best feeling I can have is when the elders come down and take a look at what we've done so far and say, 'This is what it looked like when we were growing up and playing down here.' What we're doing is for the health, recreation and enjoyment of the pueblo, but it will also positively impact the river downstream. The river is a living creature, and what is good for a part is good for the whole."

STEVE LARESE

*Having escaped the drought and hardships that befell northwestern New Mexico, many groups of ancestral Pueblo people eventually settled along the Río Grande, creating the oldest communities in the United States. Here, a pet helps out during San Juan Pueblo's deer dance.*

STEVE LARESE

*Glowing red in the late afternoon sun, salt cedar lines much of the river through New Mexico. The non-native, thirsty shrub was introduced from Eurasia as an ornamental plant early last century. Also called tamarisk, it is considered a problem along the Río Grande because it uses much water and chokes out native trees like cottonwoods and coyote willows.*

Bosque del Apache National Wildlife Refuge is also aggressively maintaining and restoring its 57,191 acres by tearing out salt cedar, planting cottonwoods and conducting controlled flooding this month, says ecologist Gina Dello Russo. "We have one of the nicest examples of what the river used to look like," she says. "We've had great success in bringing back native habitat."

STEVE LARESE

*Union and Confederate troops clashed along the Río Grande during the Civil War battle of Valverde near Socorro, which is re-enacted every February. Victorious Confederate troops rebuffed Union attacks from Fort Craig, captured six cannons and continued north to take Albuquerque and Santa Fe. For more information, log on to www2.cr.nps.gov/abpp/battles/nm001.htm.*

But conditions look like they're going to get harder before they get easier. A major drought is being predicted for New Mexico in the near future. The El Niño weather pattern gets much of the blame, but as old-timers, historians and scientists say, periods of unusually dry conditions aren't so unusual.

"The issues we're seeing with the Río Grande have always, always, always been true," says Steve Hansen, a Bureau of Reclamation hydrologist. "The Río Grande has always been in danger of drying up. It has a long history of water poverty."

True, historic accounts are filled with tales of drought and deluge. Coronado's conquistadores gave the river its impressive name after seeing what was described as a body of water many leagues wide. After seeing a trickle of water pick its way through a dusty riverbed, Will Rogers declared the Río Grande was the only river he'd ever seen in need of irrigation. Both accounts summarize the Great River, which has always been a paradox. It's the country's third-longest river next to the Mississippi and Missouri, yet some years you can step across it.

"We're coming out of a 20-year wet period," Hansen says. "It's going to get drier, which is normal. What has changed is that there's a lot more people depending on the Río Grande now, and we care about the environment a lot more. Fact is, we need more water but there's going to be less to go around. Everybody is going to have a compromise and be neighbors and partners in this."

Article 28. The Río Grande

*Above—This section of river north of Cochití Dam demonstrates the braided nature and oxbows once common for much of New Mexico's Río Grande. By damming and bank reinforcement, the river below Cochití Dam is now fairly straight. This has eliminated flooding and improved water availability downstream, but it has also destroyed much of the habitat needed by endangered species such as the silvery minnow and Southwest willow flycatcher.*
*Top left—Ron J. Sarracino places a beaver cage around a newly planted cottonwood as part of Santa Ana Pueblo's Río Grande Bosque Rehabilitation Project. Top right—Todd Caplan, Santa Ana's director of natural resources, explains how his employees have cleared out non-native salt cedar and Russian olive trees from the Pueblo's bosque, leaving cottonwoods and returning the area to how it appeared before the introduction of non-native species. Bottom top right—Rafters take advantage of the annual spring snow runoff that turns the Upper Río Grande into a national destination for thrill seekers.*

From *New Mexico* Magazine, June 2000, pp. 28–35. © 2000 by Steve Larese. Reprinted by permission.

# "You Call That Damn Thing a Boat?"

**MORE THAN A CENTURY AGO, SHIPS THAT LOOKED LIKE NUCLEAR SUBMARINES WERE EVERYWHERE ON THE GREAT LAKES**

BY CHARLES W. EBELING

***But for the cabins and smokestacks*** on their sterns, you might almost think the vessels shown on these pages were advanced underwater craft. Like submarines, they have hulls streamlined to minimize the resistance of the water they plow through. That hull design was conceived in the 1870s and 1880s by the Scottish-born Great Lakes captain Alexander McDougall, who had begun his career at 16 as a deck hand and porter and worked his way up to the command of ships by the time he was 25. He combined innovative thinking with all the knowledge he had gathered on the Great Lakes to come up with a truly radical new kind of ship.

Because of their unique appearance, having rounded tops and riding low in the water, McDougall's vessels were called "whalebacks." Nearly four dozen of them plied the Great Lakes (and occasionally other waterways) in the nineteenth and twentieth centuries. They typically carried bulk iron ore and wheat from ports on Lake Superior to manufacturing centers on the lower lakes and coal on the return journey. Before them, most of the bulk carriers on the lakes had been wooden schooners, which had very limited cargo space. Starting in the late 1860s, as increased shipping volumes created a need for bigger ships, iron vessels powered by coal-burning steam engines slowly became available, but their costs were often too high.

Captain McDougall envisioned a steel ship that could be built and operated as economically as any wooden one but would be better suited to handling bulk materials. His objective was to provide maximum cargo tonnage in a minimum depth of water along with improved stability in rough seas. The ship would be built from smooth plates and frames preformed at the rolling mill. The hull would have a convex top to let water roll off, straight sides, a conical bow, and a flat bottom, minimizing friction and letting the ship glide smoothly and rapidly through the sea. Cylindrical turrets would extend from the hull to support the cabin and working decks. As McDougall improved his idea during the 1880s, he devised hatches across the top deck to allow fast loading from chutes at ports and easy unloading by bucket conveyors.

## At first builders ridiculed the design, but McDougall persevered and had the prototype made at his own expense.

After patenting his earliest design in 1881, McDougall built a number of models and took them to shipbuilders in hope of gaining financial backing, but the whaleback concept was too advanced. McDougall persevered and, at his own expense, built a prototype on land he owned in Duluth, Minnesota. Before it was sheathed, the steel frame emanating from the keel looked like the rib cage of a giant carcass. McDougall finished the 178-foot-long craft, with hatches across its top, as a non-powered barge and named it the *101*. According to one story, McDougall chose the name because he had been offered 10-to-1 odds against the craft's success.

Upon completion, in June 1888, the *101* was launched and towed to a port near Duluth, where it took on a bulk load of 1,200 tons of iron ore bound for Cleveland. The vessel quickly came to be called a whaleback, but it also got another name, as McDougall explained in his autobiography: "I could not get anything from ship owners and from captains except comments such as: 'She will roll over, having no masts to hold her up,' or, '… Why, it looks more like a pig.'" Sailors took to calling whalebacks "pig boats." The *101* made it to Cleveland without incident, but shipbuilders still weren't ready to embrace the concept, so McDougall built a model for a whaleback that would be half again as big and he took the model to New York City.

There a group of businessmen, including John D. Rockefeller, agreed to back the concept. With their help, McDougall founded the American Steel Barge Company and began building a dry dock at West Superior, Wisconsin. While it was being built, he continued to operate the Duluth yard, which launched six more whalebacks (five barges and one steamer) before the company moved to West Superior in 1889. In 1890 the yard launched its first whaleback, and, over the next seven years, 34 followed. The strange-looking freighters began replacing wooden bulk-cargo ships on the lakes and occasionally showing up elsewhere.

One of the earliest, the *Charles W. Wetmore,* named after one of McDougall's backers, became the first to travel beyond North America. In 1891 it carried 95,000 bushels of wheat down the St. Lawrence and across the Atlantic to Liverpool, proving the seaworthiness of the design. Later that year the *Wetmore* steamed around the Horn, carrying building materials for the new city of Everett, Washington. The *City of Everett,* launched in 1894 in Everett, became the only whaleback to circumnavigate the globe, but it was lost in 1923 in the Gulf of Mexico en route from Cuba to New Orleans.

*When the 1893 Chicago world's fair,* known as the Columbian Exposition, was being planned as a celebration of the achievements of modern society in the arts, science, and technology, McDougall recognized a prime chance to publicize his unlikely design. He conceived a passenger whaleback as an elaborate ferryboat to carry visitors the six miles between downtown Chicago and the fairgrounds at Jackson Park. Its hull would contain compartments for freight or baggage, and its turrets would support a multiple-deck passenger compartment. The one-of-a-kind ship was built in 1892 at the West Superior yards in less than three months and named the *Christopher Columbus.* It had a ballast capacity of 4,000 tons and was built for 5,000 passengers but actually carried 7,500 people on its inaugural trip from West Superior into the lake and back.

When the *Christopher Columbus* arrived at the fair, on May 18, 1893, it was received with great enthusiasm. A souvenir booklet proclaimed it the greatest marine wonder of its time. Stretching 362 feet long and 42 feet wide at the beam, it could attain 17 knots, driven by two coal-burning triple-expansion steam engines producing 2,600 horsepower. Its low-riding hull carried 730 tons of water ballast in nine compartments. Seven cylindrical turrets supported its multilevel deck structure, and staircases inside the turrets provided access between decks and hull.

It was a spectacular hit with the fairgoers, transporting 1.7 million people during the exposition. When the fair was over, the *Christopher Columbus* went into regular passenger service on the 90-mile run between Chicago and Milwaukee. It was retired in 1931 and scrapped in 1936, but two of its anchors, specially designed by McDougall, survived and can be seen today at the Mariners' Museum, in Newport News, Virginia. They are as unorthodox as the whaleback design itself, with an odd-looking triangular shape.

With the *Christopher Columbus* such a success, McDougall drew up plans for a whaleback warship and presented them to the Secretary of the Navy. The Navy set up a commission to examine the scheme, but the commission rejected it, arguing that the ship's guns would have to ride too low in the water to be aimed effectively in rough seas.

Lake Superior was a graveyard for ships in the nineteenth and early twentieth centuries, some driven onto rocks, some foundering in violent storms, and some sinking after colliding with other ships. The whalebacks were no exception, and the wreck of the *Thomas Wilson,* in 1902, was the most poignant loss. The *Wilson* was one of the largest whalebacks, 308 feet long with a 38-foot beam and a 27-foot draft. It entered service in 1892, hauling wheat from Duluth to Buffalo and coal on the return trip. On June 7, 1902, it was leaving Duluth with a load of iron ore when it was rammed amidships by a wooden ship, the *George G. Hadley,* that had made a turn and couldn't stop. The *Wilson* sank within minutes, taking nine crew members with it. It came to rest in 70 feet of water, with its stern spar still protruding above the surface. The shallowness of the water and the wreck's proximity to shore led to a number of salvage attempts, but they all failed. In the 1970s, a team of scuba divers recovered the *Wilson*'s anchors, entered its hull through the hatches, and collected a number of artifacts that are now displayed at the S.S. Meteor Whaleback Ship and Maritime Museum, in Superior, Wisconsin. The ship's anchors, one of them of the odd triangular type, are at the Canal Park Marine Museum in Duluth.

The Meteor Museum is home to the longest-lived whaleback of all. The *Meteor* was launched at West Superior as the *Frank Rockefeller* in 1896. It was renamed the *South Park* in 1928, and a few years later, suitably refitted, it began carrying automobiles. In 1942 it ran onto rocks during a bad storm on Lake Michigan. The next year it was recovered, rebuilt as a tanker, and given its final name. It was retired in 1969, and in 1972 it was permanently berthed at Superior, where it now not only is a museum but is listed in the National Register of Historic Places.

Alexander McDougall died in 1923, when he was 78, some 25 years after his last whaleback had been launched. A few were still in service, but except for the *Meteor,* they all are gone now, fallen to the sea or the scrapyard. By the turn of the century, much larger cargo ships were needed on the Great Lakes; the whaleback, with its rounded decks and other eccentricities, couldn't be scaled up for them. The new bulk carriers were several times the size of the largest of McDougall's ships.

Fortunately for us, the preservation of the *Meteor* makes it still possible to step back in time and board a nineteenth-century whaleback. Guided tours are available daily from mid-May to mid-October, so that anyone can experience what more than a century ago led grave and responsible captions to exclaim, "You call that damn thing a boat?"

---

CHARLES W. EBELING wrote "Big Wheels" in the Winter 2001 issue.

---

From *American Heritage of Invention & Technology,* Fall 2001, pp. 24-27. © 2001 by American Heritage of Invention & Technology. Reprinted by permission.

# UNIT 4
# Spatial Interaction and Mapping

## Unit Selections

30. **Transportation and Urban Growth: The Shaping of the American Metropolis**, Peter O. Muller
31. **Internet GIS: Power to the People!** Bernardita Calinao and Candace Brennan
32. **ORNL and the Geographic Information Systems Revolution**, Jerome R. Dobson and Richard C. Durfee
33. **Mapping the Outcrop**, J. Douglas Walker and Ross A. Black
34. **Gaining Perspective**, Molly O'Meara Sheehan
35. **Counties With Cash**, John Fetto
36. **A City of 2 Million Without a Map**, Oakland Ross
37. **China Journal I**, Henry Petroski

## Key Points to Consider

- Describe the spatial form of the place in which you live. Do you live in a rural area, a town, or a city, and why was that particular location chosen?

- How does your hometown interact with its surrounding region? With other places in the state? With other states? With other places in the world?

- How are places "brought closer together" when transportation systems are improved?

- What problems occur when transportation systems are overloaded?

- How will public transportation be different in the future? Will there be more or fewer private autos in the next 25 years? Defend your answer.

- How good a map reader are you? Why are maps useful in studying a place?

 Links: www.dushkin.com/online/
These sites are annotated in the World Wide Web pages.

**Edinburgh Geographical Information Systems**
http://www.geo.ed.ac.uk/home/gishome.html

**Geography for GIS**
http://www.ncgia.ucsb.edu/cctp/units/geog_for_GIS/GC_index.html

**GIS Frequently Asked Questions and General Information**
http://www.census.gov/ftp/pub/geo/www/faq-index.html

**International Map Trade Association**
http://www.maptrade.org

**PSC Publications**
http://www.psc.isr.umich.edu/

**U.S. Geological Survey**
http://www.usgs.gov/research/gis/title.html

Geography is the study not only of places in their own right but also of the ways in which places interact. Highways, airline routes, telecommunication systems, and even thoughts connect places. These forms of spatial interaction are an important part of the work of geographers.

In "Transportation and Urban Growth: The Shaping of the American Metropolis," Peter Muller considers transportation systems, analyzing their impact on the growth of American cities. Two articles on GIS follow, one covering public input to airport expansion through a GIS Web site and the other dealing with the Oak Ridge National Laboratory and its role in the historical development of GIS products. Next, "Mapping the Outcrop" relates the use of digitized topographic maps and GIS in geology. An extensive analysis of satellite imagery and its applications follows. The next article illustrates the power of the choropleth map to tell its story. Managua, Nicaragua, devastated by an enormous earthquake in 1972, is a city literally without a map. "China Journal" discusses the Three Gorges project in China, including the spatial interaction aspects.

It is essential that geographers be able to describe the detailed spatial patterns of the world. Neither photographs nor words could do the job adequately, because they literally capture too much of the detail of a place. Therefore, maps seem to be the best way to present many of the topics analyzed in geography. Maps and geography go hand in hand. Although maps are used in other disciplines, their association with geography is the most highly developed.

A map is a graphic that presents a generalized and scaled-down view of particular occurrences or themes in an area. If a picture is worth a thousand words, then a map is worth a thousand (or more!) pictures. There is simply no better way to "view" a portion of Earth's surface or an associated pattern than with a map.

# Transportation and Urban Growth

## The shaping of the American metropolis

Peter O. Muller

In his monumental new work on the historical geography of transportation, James Vance states that geographic mobility is crucial to the successful functioning of any population cluster, and that "shifts in the availability of mobility provide, in all likelihood, the most powerful single process at work in transforming and evolving the human half of geography." Any adult urbanite who has watched the American metropolis turn inside-out over the past quarter-century can readily appreciate the significance of that maxim. In truth, the nation's largest single urban concentration today is not represented by the seven-plus million who agglomerate in New York City but rather by the 14 million who have settled in Gotham's vast, curvilinear outer city—a 50-mile-wide suburban band that stretches across Long Island, southwestern Connecticut, the Hudson Valley as far north as West Point, and most of New Jersey north of a line drawn from Trenton to Asbury Park. This latest episode of intrametropolitan deconcentration was fueled by the modern automobile and the interstate expressway. It is, however, merely the most recent of a series of evolutionary stages dating back to colonial times, wherein breakthroughs in transport technology unleashed forces that produced significant restructuring of the urban spatial form.

The emerging form and structure of the American metropolis has been traced within a framework of four transportation-related eras. Each successive growth stage is dominated by a particular movement technology and transport-network expansion process that shaped a distinctive pattern of intraurban spatial organization. The stages are the Walking/Horsecar Era (pre-1800–1890), the Electric Streetcar Era (1890–1920), the Recreational Automobile Era (1920–1945), and the Freeway Era (1945–present). As with all generalized models of this kind, there is a risk of oversimplification because the building processes of several simultaneously developing cities do not always fall into neat time-space compartments. Chicago's growth over the past 150 years, for example, reveals numerous irregularities, suggesting that the overall metropolitan growth pattern is more complex than a simple, continuous outward thrust. Yet even after developmental ebb and flow, leapfrogging, backfilling, and other departures from the idealized scheme are considered, there still remains an acceptable correspondence between the model and reality.

Before 1850 the American city was a highly compact settlement in which the dominant means of getting about was on foot, requiring people and activities to tightly agglomerate in close proximity to one another. This usually meant less than a 30-minute walk from the center of town to any given urban point—an accessibility radius later extended to 45 minutes when the pressures of industrial growth intensified after 1830. Within this pedestrian city, recognizable activity concentrations materialized as well as the beginnings of income-based residential congregations. The latter was particularly characteristic of the wealthy, who not only walled themselves off in their large homes near the city center but also took to the privacy of horse-drawn carriages for moving about town. Those of means

(LIBRARY OF THE BOSTON ATHENEUM)

*Horse-drawn trolleys in downtown Boston, circa 1885.*

also sought to escape the city's noise and frequent epidemics resulting from the lack of sanitary conditions. Horse-and-carriage transportation enabled the wealthy to reside in the nearby countryside for the disease-prone summer months. The arrival of the railroad in the 1830s provided the opportunity for year-round daily commuting, and by 1840 hundreds of affluent businessmen in Boston, New York, and Philadelphia were making round trips from exclusive new trackside suburbs every weekday.

As industrialization and its teeming concentrations of working-class housing increasingly engulfed the mid-nineteenth century city, the deteriorating physical and social environment reinforced the desires of middle-income residents to suburbanize as well. They were unable, however, to afford the cost and time of commuting by steam train, and with the walking city now stretched to its morphological limit, their aspirations intensified the pressures to improve intraurban transport technology. Early attempts involving stagecoach-like omnibuses, cablecar systems, and steam railroads proved impractical, but by 1852 the first meaningful transit breakthrough was finally introduced in Manhattan in the form of the horse-drawn trolley. Light street rails were easy to install, overcame the problems of muddy, unpaved roadways, and en-

(LIBRARY OF THE BOSTON ATHENEUM)
*Electric streetcar lines radiated outward from central cities, giving rise to star-shaped metropolises. Boston, circa 1915.*

abled horsecars to be hauled along them at speeds slightly (about five mph) faster than those of pedestrians. This modest improvement in mobility permitted the opening of a narrow belt of land at the city's edge for new home construction. Middle-income urbanites flocked to these "horsecar suburbs," which multiplied rapidly after the Civil War. Radial routes were the first to spawn such peripheral development, but the relentless demand for housing necessitated the building of cross-town horsecar lines, thereby filling in the interstices and preserving the generally circular shape of the city.

The less affluent majority of the urban population, however, was confined to the old pedestrian city and its bleak, high-density industrial appendages. With the massive immigration of unskilled laborers, (mostly of European origin after 1870) huge blue-collar communities sprang up around the factories. Because these newcomers to the city settled in the order in which they arrived—thereby denying them the small luxury of living in the immediate company of their fellow ethnics—social stress and conflict were repeatedly generated. With the immigrant tide continuing to pour into the nearly bursting industrial city throughout the late nineteenth century, pressures redoubled to further improve intraurban transit and open up more of the adjacent countryside. By the late 1880s that urgently needed mobility revolution was at last in the making, and when it came it swiftly transformed the compact city and its suburban periphery into the modern metropolis.

The key to this urban transport revolution was the invention by Frank Sprague of the electric traction motor, an often overlooked innovation that surely ranks among the most important in American history. The first electrified trolley line opened in Richmond in 1888, was adopted by two dozen other big cities within a year, and by the early 1890s swept across the nation to become the dominant mode of intraurban transit. The rapidity of this innovation's diffusion was enhanced by the immediate recognition of its ability to resolve the urban transportation problem of the day: motors could be attached to existing horsecars, converting them into self-propelled vehicles powered by easily constructed overhead wires. The tripling of average speeds (to over 15 mph) that resulted from this invention brought a large band of open land beyond the city's perimeter into trolley-commuting range.

# Article 30. Transportation and Urban Growth

Before 1850 the American city was a highly compact settlement in which the dominant means of getting around was on foot, requiring people and activities to tightly agglomerate in close proximity to one another.

The most dramatic geographic change of the Electric Streetcar Era was the swift residential development of those urban fringes, which transformed the emerging metropolis into a decidedly star-shaped spatial entity. This pattern was produced by radial streetcar corridors extending several miles beyond the compact city's limits. With so much new space available for homebuilding within walking distance of the trolley lines, there was no need to extend trackage laterally, and so the interstices remained undeveloped. The typical streetcar suburb of the turn of this century was a continuous axial corridor whose backbone was the road carrying the trolley line (usually lined with stores and other local commercial facilities), from which gridded residential streets fanned out for several blocks on both sides of the tracks. In general, the quality of housing and prosperity of streetcar subdivisions increased with distance from the edge of the central city. These suburban corridors were populated by the emerging, highly mobile middle class, which was already stratifying itself according to a plethora of minor income and status differences. With frequent upward (and local geographic) mobility the norm, community formation became an elusive goal, a process further retarded by the grid-settlement morphology and the reliance on the distant downtown for employment and most shopping.

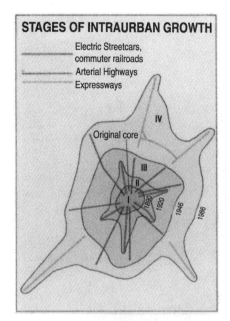

CARTOGRAPHY LAB. DEPT. OF GEOGRAPHY, UNIV. OF MINNESOTA

Within the city, too, the streetcar sparked a spatial transformation. The ready availability and low fare of the electric trolley now provided every resident with access to the intracity circulatory system, thereby introducing truly "mass" transit to urban America in the final years of the nineteenth century. For nonresidential activities this new ease of movement among the city's various parts quickly triggered the emergence of specialized land-use districts for commerce, manufacturing, and transportation, as well as the continued growth of the multipurpose central business district (CBD) that had formed after mid-century. But the greatest impact of the streetcar was on the central city's social geography, because it made possible the congregation of ethnic groups in their own neighborhoods. No longer were these moderate-income masses forced to reside in the heterogeneous jumble of row-houses and tenements that ringed the factories. The trolley brought them the opportunity to "live with their own kind," allowing the sorting of discrete groups into their own inner-city social territories within convenient and inexpensive traveling distance of the workplace.

By World War I, the electric trolleys had transformed the tracked city into a full-fledged metropolis whose streetcar suburbs, in the larger cases, spread out more than 20 miles from the metropolitan center. It was at this point in time that intrametropolitan transportation achieved its greatest level of efficiency—that the bustling industrial city really "worked." How much closer the American metropolis might have approached optimal workability for all its residents, however, will never be known because the next urban transport revolution was already beginning to assert itself through the increasingly popular automobile. Americans took to cars as wholeheartedly as anything in the nation's long cultural history. Although Lewis Mumford and other scholars vilified the car as the destroyer of the city, more balanced assessments of the role of the automobile recognize its overwhelming acceptance for what it was—the long-awaited attainment of private mass transportation that offered users the freedom to travel whenever and wherever they chose. As cars came to the metropolis in ever greater numbers throughout the interwar decades, their major influence was twofold: to accelerate the deconcentration of population through the development of interstices bypassed during the streetcar era, and to push the suburban frontier farther into the countryside, again producing a compact, regular-shaped urban entity.

While it certainly produced a dramatic impact on the urban fabric by the eve of World War II, the introduction of the automobile into the American metropolis during the 1920s and 1930s came at a leisurely pace. The earliest flurry of auto

(BOSTON PUBLIC LIBRARY)

*Afternoon commuters converge at the tunnel leading out of central Boston, 1948.*

adoptions had been in rural areas, where farmers badly needed better access to local service centers. In the cities, cars were initially used for weekend outings—hence the term "*Recreational* Auto Era"—and some of the earliest paved roadways were landscaped parkways along scenic water routes, such as New York's pioneering Bronx River Parkway and Chicago's Lake Shore Drive. But it was into the suburbs, where growth rates were now for the first time overtaking those of the central cities, that cars made a decisive penetration throughout the prosperous 1920s. In fact, the rapid expansion of automobile suburbia by 1930 so adversely affected the metropolitan public transportation system that, through significant diversions of streetcar and commuter-rail passengers, the large cities began to feel the negative effects of the car years before the auto's actual arrival in the urban center. By facilitating the opening of unbuilt areas lying between suburban rail axes, the automobile effectively lured residential developers away from densely populated traction-line corridors into the suddenly accessible interstices. Thus, the suburban homebuilding industry no longer found it necessary to subsidize privately-owned streetcar companies to provide low-fare access to trolley-line housing tracts. Without this financial underpinning, the modern urban transit crisis quickly began to surface.

The new recreational motorways also helped to intensify the decentralization of the population. Most were radial highways that penetrated deeply into the suburban ring and provided weekend motorists with easy access to this urban countryside. There they obviously were impressed by what they saw, and they soon responded in massive numbers to the sales pitches of suburban subdivision developers. The residential development of automobile suburbia followed a simple formula that was devised in the prewar years and greatly magnified in scale after 1945. The leading motivation was developer profit from the quick turnover of land, which was acquired in large parcels, subdivided, and auctioned off. Understandably, developers much preferred open areas at the metropolitan fringe, where

# Central City-Focused Rail Transit

The widely dispersed distribution of people and activities in today's metropolis makes rail transit that focuses in the central business district (CBD) an obsolete solution to the urban transportation problem. To be successful, any rail line must link places where travel origins and destinations are highly clustered. Even more important is the need to connect places where people really want to go, which in the metropolitan America of the late twentieth century means suburban shopping centers, freeway-oriented office complexes, and the airport. Yet a brief look at the rail systems that have been built in the last 20 years shows that transit planners cannot—or will not—recognize those travel demands, and insist on designing CBD-oriented systems as if we all still lived in the 1920s.

One of the newest urban transit systems is Metrorail in Miami and surrounding Dade County, Florida. It has been a resounding failure since its opening in 1984. The northern leg of this line connects downtown Miami to a number of low- and moderate-income black and Hispanic neighborhoods, yet it carries only about the same number of passengers that used to ride on parallel bus lines. The reason is that the high-skill, service economy of Miami's CBD is about as mismatched as it could possibly be to the modest employment skills and training levels possessed by residents of that Metrorail corridor. To the south, the prospects seemed far brighter because of the possibility of connecting the system to Coral Gables and Dadeland, two leading suburban activity centers. However, both central Coral Gables and the nearby International Airport complex were bypassed in favor of a cheaply available, abandoned railroad corridor alongside U.S. 1. Station locations were poorly planned, particularly at the University of Miami and at Dadeland—where terminal location necessitates a dangerous walk across a six-lane highway from the region's largest shopping mall. Not surprisingly, ridership levels have been shockingly below projections, averaging only about 21,000 trips per day in early 1986. While Dade County's worried officials will soon be called upon to decide the future of the system, the federal government is using the Miami experience as an excuse to withdraw from financially supporting all construction of new urban heavy-rail systems. Unfortunately, we will not be able to discover if a well-planned, high-speed rail system that is congruent with the travel demands of today's polycentric metropolis is capable of solving traffic congestion problems. Hopefully, transportation policy-makers across the nation will heed the lessons of Miami's textbook example of how not to plan a hub-and-spoke public transportation network in an urban era dominated by the multicentered city.

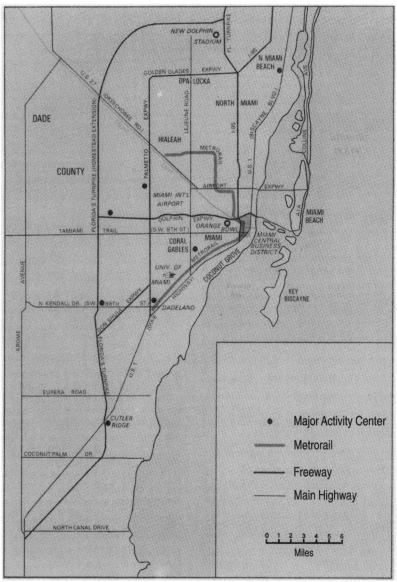

CARTOGRAPHY LAB. DEPT. OF GEOGRAPHY, UNIV OF MINNESOTA

large packages of cheap land could readily be assembled. Silently approving and underwriting this uncontrolled spread of residential suburbia were public policies at all levels of government: financing road construction, obligating lending institutions to invest in new homebuilding, insuring individual mortgages, and providing low-interest loans to FHA and VA clients.

> The ready availability and low fare of the electric trolley now provided every resident with access to the intracity circulatory system, thereby introducing truly "mass" transit to urban America.

Because automobility removed most of the pre-existing movement constraints, suburban social geography now became dominated by locally homogeneous income-group clusters that isolated themselves from dissimilar neighbors. Gone was the highly localized stratification of streetcar suburbia. In its place arose a far more dispersed, increasingly fragmented residential mosaic to which builders were only too eager to cater, helping shape a kaleidoscopic settlement pattern by shrewdly constructing the most expensive houses that could be sold in each locality. The continued partitioning of suburban society was further legitimized by the widespread adoption of zoning (legalized in 1916), which gave municipalities control over lot and building standards that, in turn, assured dwelling prices that would only attract newcomers whose incomes at least equaled those of the existing local population. Among the middle class, particularly, these exclusionary economic practices were enthusiastically supported, because such devices extended to them the ability of upper-income groups to maintain their social distance from people of lower socioeconomic status.

Nonresidential activities were also suburbanizing at an increasing rate during the Recreational Auto Era. Indeed, many large-scale manufacturers had decentralized during the streetcar era, choosing locations in suburban freight-rail corridors. These corridors rapidly spawned surrounding working-class towns that became important satellites of the central city in the emerging metropolitan constellation. During the interwar period, industrial employers accelerated their intraurban deconcentration, as more efficient horizontal fabrication methods replaced older techniques requiring multistoried plants-thereby generating greater space needs that were too expensive to satisfy in the high-density central city. Newly suburbanizing manufacturers, however, continued their affiliation with intercity freight-rail corridors, because motor trucks were not yet able to operate with their present-day efficiencies and because the highway network of the outer ring remained inadequate until the 1950s.

> Americans took to cars as wholeheartedly as anything in the nation's long cultural history.

The other major nonresidential activity of interwar suburbia was retailing. Clusters of automobile-oriented stores had first appeared in the urban fringes before World War I. By the early 1920s the roadside commercial strip had become a common sight in many southern California suburbs. Retail activities were also featured in dozens of planned automobile suburbs that sprang up after World War I—most notably in Kansas City's Country Club District, where the nation's first complete shopping center was opened in 1922. But these diversified retail centers spread slowly before the suburban highway improvements of the 1950s.

Unlike the two preceding eras, the postwar Freeway Era was not sparked by a revolution in urban transportation. Rather, it represented the coming of age of the now pervasive automobile culture, which coincided with the emergence of the U.S. from 15 years of economic depression and war. Suddenly the automobile was no longer a luxury or a recreational diversion: overnight it had become a necessity for commuting, shopping, and socializing, essential to the successful realization of personal opportunities for a rapidly expanding majority of the metropolitan population. People snapped up cars as fast as the reviving peacetime automobile industry could roll them off the assembly lines, and a prodigious highway-building effort was launched, spearheaded by high-speed, limited-access expressways. Given impetus by the 1956 Interstate Highway Act, these new freeways would soon reshape every corner of urban America, as the more distant suburbs they engendered represented nothing less than the turning inside-out of the historic metropolitan city.

The snowballing effect of these changes is expressed geographically in the sprawling metropolis of the postwar era. Most striking is the enormous band of growth that was added between 1945 and the 1980s, with freeway sectors pushing the metropolitan frontier deeply into the urban-rural fringe. By the late 1960s, the maturing expressway system began to underwrite a new suburban co-equality with the central city, because it was eliminating the metropolitanwide centrality advantage of the CBD. Now any location on the freeway network could easily be reached by motor vehicle, and intraurban accessibility had become a ubiquitous spatial good. Ironically, large cities had encouraged the construction of radial expressways in the 1950s and 1960s because they appeared to enable the downtown to remain accessible to the swiftly dispersing suburban population. However, as one economic activity

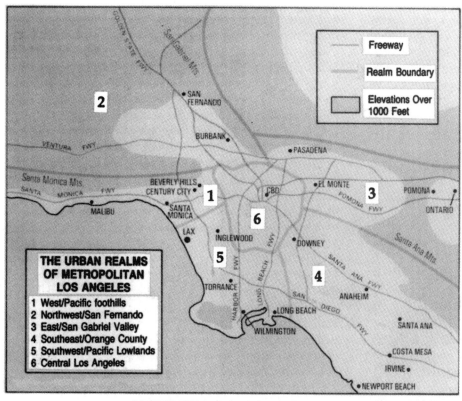

after another discovered its new locational flexibility within the freeway metropolis, nonresidential deconcentration sharply accelerated in the 1970s and 1980s. Moreover, as expressways expanded the radius of commuting to encompass the entire dispersed metropolis, residential location constraints relaxed as well. No longer were most urbanites required to live within a short distance of their job: the workplace had now become a locus of opportunity offering access to the best possible residence that an individual could afford anywhere in the urbanized area. Thus, the overall pattern of locally uniform, income-based clusters that had emerged in prewar automobile suburbia was greatly magnified in the Freeway Era, and such new social variables as age and lifestyle produced an ever more balkanized population mosaic.

> Retail activities were featured in dozens of planned automobile suburbs that sprang up after World War I—most notably Kansas City's Country Club District, where the nation's first complete shopping center was opened in 1922.

The revolutionary changes in movement and accessibility introduced during the four decades of the Freeway Era have resulted in nothing less than the complete geographic restructuring of the metropolis. The single-center urban structure of the past has been transformed into a polycentric metropolitan form in which several outlying activity concentrations rival the CBD. These new "suburban downtowns," consisting of vast orchestrations of retailing, office-based business, and light industry, have become common features near the highway interchanges that now encircle every large central city. As these emerging metropolitan-level cores achieve economic and geographic parity with each other, as well as with the CBD of the nearby central city, they provide the totality of urban goods and services to their surrounding populations. Thus each metropolitan sector becomes a self-sufficient functional entity, or *realm*. The application of this model to the Los Angeles region reveals six broad realms. Competition among several new suburban downtowns for dominance in the five outer realms is still occurring. In wealthy Orange County, for example, this rivalry is especially fierce, but Costa Mesa's burgeoning South Coast Metro is winning out as of early 1986.

> The new freeways would soon reshape every corner of urban America, as the more distant suburbs they engendered represented nothing less than the turning inside-out of the historic metropolitan city.

The legacy of more than two centuries of intraurban transportation innovations, and the development patterns they helped stamp on the landscape of metropolitan America, is suburbanization—the growth of the edges of the urbanized area at a rate faster than in the already-developed interior. Since the geographic extent of the built-up urban areas has, throughout history, exhibited a remarkably constant radius of about 45 minutes of travel from the center, each breakthrough in higher-speed transport technology extended that radius into a new outer zone of suburban residential opportunity. In the nineteenth century, commuter railroads, horse-drawn trolleys, and electric streetcars each created their own suburbs—and thereby also created the large industrial city, which could not have been formed without incorporating these new suburbs into the pre-existing compact urban center. But the suburbs that materialized in the early twentieth century began to assert their independence from the central cities, which were ever more perceived as undesirable. As the automobile greatly reinforced the dispersal trend of the metropolitan population, the distinction between central city and suburban ring grew as well. And as freeways eventually eliminated the friction effects of intrametropolitan distance for most urban functions, nonresidential activities deconcentrated to such an extent that by 1980 the emerging outer suburban city had become co-equal with the central city that spawned it.

As the transition to an information-dominated, postindustrial economy is completed, today's intraurban movement problems may be mitigated by the increasing substitution of communication for the physical movement of people. Thus, the city of the future is likely to be the "wired metropolis." Such a development would portend further deconcentration because activity centers would potentially be able to locate at any site offering access to global computer and satellite networks.

## Further Reading

Jackson, Kenneth T. 1985. *Crabgrass Frontier: The Suburbanization of the United States.* New York: Oxford University Press.

Muller, Peter O. 1981. *Contemporary Suburban America.* Englewood Cliffs, N.J.: Prentice-Hall.

Schaeffer, K. H. and Sclar, Elliot. 1975. *Access for All: Transportation and Urban Growth.* Baltimore: Penguin Books.

*Article 31*

# Internet GIS: Power to the People!

## A Web-based GIS provides a public-involvement tool for airport development

By Bernardita Calinao and Candace Brennan

A NEW AIRPORT RUNWAY CAN HAVE A MAJOR EFFECT ON A COMMUNITY. BUT HOW CAN CITIZENS BECOME MORE KNOWLEDGEABLE ABOUT SUCH ISSUES, AND HOW CAN THEY LET THEIR VOICES BE HEARD? IN ERIE, PA., C&S ENGINEERS USED A GIS-BASED WORLD WIDE WEB SITE TO ALLOW THE PUBLIC TO HELP CHOOSE WHICH RUNWAY EXTENSION ALTERNATIVES WORK BEST.

Erie's Internet GIS application is an analytical and public-involvement tool developed for an Environmental Assessment (EA) for the Proposed Runway 6-24 Extension at the Erie International Airport. The system is designed to help the public better understand the proposed project as well as its potential environmental and socio-economic effects. The process transcends the GIS from a tool for planners, managers and experts to its new function as a tool that enhances direct public participation for environmental decision making.

The basic elements of the Internet GIS reflect the contents of a standard EA, which include the following:

- Describe proposed alternatives.
- Present environmental feature maps within the project's area of influence.
- Delineate environmentally sensitive areas and present environmental consequences.
- Incorporate mitigation measures and action plans.
- The Internet GIS is consistent with Federal Aviation Administration (FAA) environmental guidelines, and it promotes the achievement of aviation-related environmental goals, including the following:
- Heighten objectivity in the EA process.
- Provide public information and input.
- Develop a more place-based approach to decision making.
- Ensure that the EA follows an iterative process.
- Comply with FAA's streamlining efforts.
- Develop new tools for interagency coordination.
- Create an effective platform for monitoring and environmental management.

The EA is developed through interagency coordination between the FAA and the Federal Highway Administration, which is currently overseeing the preparation of an EA for the relocation of Powell Ave., a road affected by the runway extension project.

## Project Specifics

The Erie Municipal Airport Authority proposed a runway extension for Runway 6-24, the primary runway at Erie International Airport. The existing runway is 6,500 feet long and 150 feet wide. The proposed project would extend the runway 1,900 feet to the northeast. Safety issues are a primary focus of the proposed extension.

Via the Internet-based GIS, community residents can view the proposed project in the context of their environment. Users can identify tax parcels and relate them to the proposed runway extension and existing land use.

The EA process is evaluating the proposed environmental and socio-economic impacts of all the alternatives identified in the "Master Plan" so alternatives are evaluated using technical and cost considerations as well as environmental and socio-cultural considerations.

A Web site (*http://www.erieairportprojects.org*) was developed to facilitate the link among the Internet mapping system and other related environmental information. The Web site's principal feature is a section on "Environmental Maps," which carries all the features of an Internet GIS. Because the project is developed in conjunction with the Powell Ave. project of the Pennsylvania Department of Transportation, the Internet GIS may likewise be accessed at http://www.airport-powellprojects.com under "Project Details, Airport Information."

## An Online Mapping System

The main objective for the Internet GIS is to provide public access to information used in developing the EA. Project engineers didn't want to spend much time processing and converting existing data. They also needed a product that could deliver a lightweight Internet application to the public as well as be accessed through a standard and readily accessible Internet client such as a Web browser.

### Application Requirements

Features such as runway alternatives, noise contours, runway safety areas and runway protection zones were created in a GIS format. Using a GIS as a data management tool is effective, because it can overlay and query spatial data. Government agencies already had data needed by the project. Tax parcel information was used along with details of the runway alternatives to calculate the number of residential properties affected by the proposed changes. Other information was added to the system, including watersheds, hazardous waste sites, air-quality monitoring data, roads, land use, neighborhoods and socio-economic census data.

By using an aerial photo as a background, a more realistic view of the potential impact areas is provided.

Airport environmental planners considered several different ways to make such information available. If they released the data on a CD-ROM, they would quickly lose control of updates and additions. The project requires data to be centralized, and updates need to be instantly displayed. If the engineers created a kiosk for the public to use at a specific location, they wouldn't be able to service multiple users at one time. An Internet application can handle multiple users from different locations, and the application can be reused and customized to meet the needs of future projects.

From the public's perspective, there are many issues to consider. The application should be quick, easy and convenient to use for someone inexperienced with computers and mapping applications. The project's Internet GIS doesn't require prior GIS knowledge, and there's no need to download or buy extra software. There are different choices for the type of viewer. Some require Active X or Java plug-ins, while others use basic HTML and Java Script, which are lightweight and provide access to all levels of Web users.

The main Web site serves as a platform and provides additional information about the project and the Internet GIS.

Content is another consideration. The site can be used to re-create and investigate maps seen in public meetings. It can be viewed by the public or accessed by people working on the project if they need quick information. The citizens living in the neighborhoods surrounding the airport are interested in evaluating the effect of runway extensions relative to their homes.

In addition, the Internet GIS will give insight to questions such as:

- Where are the extensions proposed?
- Will I have to be relocated?
- Will this change the noise levels around my home?
- Will I be able to receive benefits from the Sound Insulation Program?

### Internet GIS Configuration

The tool chosen to connect the public with the data was ESRI Inc.'s ArcIMS software, because most of the base maps and GIS layers were in ESRI's Shapefile format. The specific application directly accepts data that already have been collected and managed for other stages of the runway extension project.

There are three types of data presently included: 1) existing conditions, 2) potential impact areas and 3) cumulative effects. The existing conditions include all existing information such as roads, buildings, wetlands, floodplains, census block data, tax parcel information and airport pavement. Potential impact areas are created for each of the runway alternatives. The primary impact area is the predicted soil disturbance area and a buffer of approximately 500 feet around the runway safety area. The GIS also takes into account the runway protection zones defined for each alternative. Some cumulative areas will be affected by more than one source, such as a neighborhood block, for example, which will receive a significant increase in noise level and a reduction in air quality.

ArcIMS is a server-side software product that depends on a Web server and a Java servlet engine. Airport engineers already had a Microsoft Windows 2000 server with IIS 5.0 installed to serve company Web pages. There was enough room on the server to install ArcIMS, and the engineers networked another computer to serve the data. They installed The Apache Software Foundation's Jakarta Tomcat 3.2 product as a servlet engine as well as the ArcIMS application and spatial servers. The ArcIMS manager is installed on a separate computer used for development.

When a user views the Web site and sends a request for a map, the Web server sends the request through the Java servlet to the application server, which decides what to do with the request and sends it to the appropriate spatial server to handle the GIS computation. The spatial server then sends an output image of a new map to the user's Web server.

### Web Site Design

Citizens need to see more than a map when they visit the Web site. Airport engineers decided to create a site that will introduce the project and EA process. This site includes sections such as "About the Project," "Questions and Answers," "Completed Environmental Reports," "Environmental Maps," "Other Erie International Airport Projects," "News," "Glossary," "Links" and "Public Comments."

The "Environmental Maps" section provides an area where prepared static maps can be viewed using Adobe Acrobat. In addition, the metadata for the layers used in the GIS application are available.

A simple HTML viewer was chosen for the application, because it didn't require users to install plug-ins. It's also the most lightweight in terms of processing required by a client's computer. All the spatial processing is done on the server side, and images are returned to the client. The viewer is interactive—it's able to accept requests and send responses back to clients. Tools available for users include zoom, pan, identify, overview map, view legend, layer control (on and off) and print.

All the layers redisplay images according to user preferences. The setup also allows users to adjust detail levels. The application provides more functions, but airport engineers decided to keep the format simple and only display the necessary options.

## Project Concerns

Technical considerations are important when designing an Internet GIS. What type of audience is involved? What type of browsers will citizens have? Do they need an intuitive design or something more advanced? What type of equipment is available?

It's important to determine who is going to host the Web site. The airport project is served from two different locations. An Internet provider hosts the main site that includes the project descriptions, and the Internet GIS application is installed and hosted on a Web server at the C&S office.

The setup for the Internet GIS needs to be carefully planned before installation, and the existing infrastructure should be reviewed. It's important that network administrators work closely with GIS personnel to design an installation.

A significant issue that airport engineers faced when configuring their software was working around a firewall between the Web server and the rest of the C&S network. Due to some complications with strict firewall settings, the engineers decided to install all the ArcIMS components and data outside the firewall. Typically, this isn't the best installation, and the engineers are currently looking into alternative ways to configure the software.

## Lessons Learned

The Internet GIS recently has been completed and released for public use in the EA process of the proposed runway extension. Although positive responses have been received to date, results from the participation effort still aren't available. A few implementation lessons, however, have become apparent.

For example, Internet GIS is a new tool for public involvement in airport development. As such, there's a need to stir more enthusiasm among airport regulators and environmental specialists. Airport engineers also experienced that, unlike a standard GIS, the political concerns are more pronounced in the use of Internet GIS, perhaps because information is made more available to the public. Therefore, strategic planning and quality control for any Internet GIS effort are extremely vital to effective implementation.

Map requests are sent from the client to the server, where the Web Server and ArcIMS applications handle the request. The request's result is an output image, which is transferred back to the client.

Internet GIS isn't intended to replace other formats of public participation. It's an approach that complements existing formats, because it allows citizens to access and interact with mapping and GIS data to enhance their knowledge about proposed land development projects in their community and increase their participation in the overall environmental decision-making process.

Internet GIS revolutionizes the way environmental assessment is conducted. As a result, it empowers citizens by providing them with a more dynamic and interactive tool for improved participation.

---

Calinao is senior environmental planner, C&S Engineers Inc.;
e-mail: bcalinao@cscos.com.

---

Brennan is a GIS specialist, C&S Engineers Inc.;
e-mail: cbrennan@cscos.com.

---

From *GEO World*, June 2002. © 2002 by GEO World.

# ORNL and the Geographic Information Systems Revolution

*Explorers from competing teams race to find a mysterious lost city in the heart of Africa. The American team is continuously in touch with its Houston home base through satellite communications. In flight, team leader Karen Ross displays a map of Africa on her computer screen and notes the multicolored lines suggesting different routes from city to city and into the rain forest. Each pathway is accompanied by a precise estimate of travel time to the final destination. Zooming in on the target area, she switches to satellite images and interprets them in shades of blue, purple, and green. At each checkpoint, the team reports its progress and gets a revised estimate of arrival time.*

*Beset by difficulties, the explorers ask for a faster route, but the computer says the alternative is too dangerous. A simulation model with data representing geology, terrain, vegetation, weather, and many other geographic factors predicts local hazards, including the impending eruption of a nearby volcano. The Americans take the faster route anyway and beat the odds.*

### By Jerome E. Dobson and Richard C. Durfee

This fictional account of emerging geographic information system (GIS) technologies comes from Michael Crichton's 1980 novel *Congo*, which was made into a 1995 movie. The same technologies were highlighted in Clive Cussler's 1988 techno-thriller *Treasure*. In reality, GIS technology began more than a quarter of a century ago at key universities and government laboratories in the United States and Canada. Since 1969, Oak Ridge National Laboratory has been among the leading institutions in this diverse, now booming field. GIS has been evolving through new forms and applications ever since. Consider the following examples of GIS applications that rival and sometimes exceed Crichton's futuristic vision.

For the past three summers, ORNL geographers have monitored the potentially devastating effects of an Alaskan glacier with an annoying habit of rerouting whole river systems. We drive as far as the roads go or fly over roadless terrain with a color laptop computer that displays Ed Bright's interpretation of satellite images. A dot moves across the screen continuously showing our position on the image and thus on the ground calculated from Global Positioning System (GPS) signals from satellites. We've used the same system successfully in helicopters, boats, and even rental cars on the Oregon-Washington coast, the Gulf of Maine, and the North Slope of Alaska.

*Since 1969, ORNL has been a leading institution in developing and using GIS technologies.*

The roots of our remote sensing and GIS tradition started early at ORNL; more than 20 years ago, ORNL scientists studied some of the first satellite data from Landsat satellites (then called ERTS). By analyzing computer images of the Cumberland Mountains north of Oak Ridge, we were able to compute and display a three-dimensional perspective view of the

## Article 32. ORNL and the Geographic Information Systems Revolution

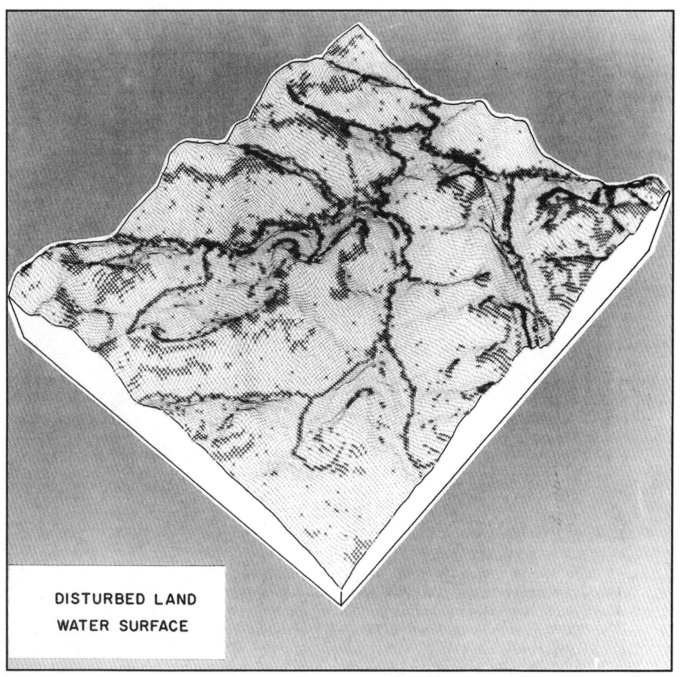

A three-dimensional perspective view of coal strip mines in the Cumberland Mountains north of ORNL. Streams are superimposed on the terrain. The spatial model computed the geographic relationship between disturbed land and nearby water surfaces, allowing assessment of environmental and visual impacts of strip mines.

coal strip mines in the area and superimpose the nearby streams on the terrain. After developing spatial models, we determined which streams were most likely to receive acid drainage from the strip mines. The visual impacts of strip mining on Oak Ridge residents were also predicted.

With or without satellite imagery, GIS is a powerful tool. In 1990, when the United States and other nations responded to Saddam Hussein's invasion of Kuwait, military leaders mounted the largest and most rapid deployment of military personnel and equipment ever attempted. The massive logistics were processed on the Airlift Deployment Analysis System (ADANS) developed at ORNL. ADANS, operating on networked computers, draws on a variety of logistic and spatial technologies to efficiently schedule the transport of U.S. military troops and equipment to trouble spots anywhere in the world. Since 1990, ADANS has been used to deploy military personnel and equipment not only to the Persian Gulf but also to Somalia, Rwanda, and Haiti.

In 1995, at ORNL's World Wide Web Showcase, Peter Pace showed a colorful high-resolution image of ORNL buildings and the roads, streams, and forested areas of the surrounding reservation. The view on his computer screen was constructed

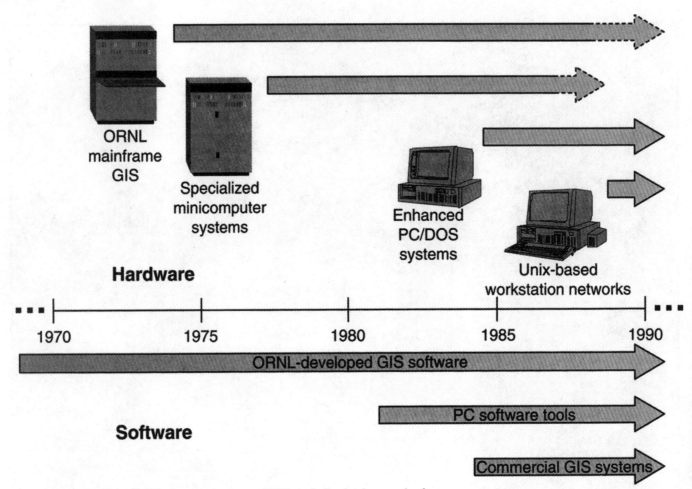

Historical development of GIS computer systems at ORNL, including hardware and software.

from a series of aerial photographs that had been scanned and converted to form a digital image. Various computer techniques were used to enhance and blend a series of images, eliminating unwanted elements and bringing out important details. Special photogrammetric techniques were used to remove distortions from the digitized photos. Each pixel (tiny rectangular element) on the screen represents 0.25 square meter ($m^2$) on the ground. Spatial registration of geographical features in the image is sufficiently accurate that a highly detailed map can be overlaid on the image. Pace zoomed in on a cooling tower and magnified it enough to see the blades of a fan. He printed out an image of the cooling tower alone. He and other ORNL researchers are preparing geographical data and imagery developed at Oak Ridge for distribution to selected users of the World Wide Web through Netscape, a navigational tool for accessing still and animated images as well as audio and text from the Internet.

Recent growth of GIS markets has been phenomenal. In 1994, GIS was listed under "Whole Systems" in the Whole Earth Catalog. Tens of thousands of people and organizations—universities, research centers, municipal planners, tax assessors, corporations, and resource managers—have come to depend on GIS for geographic data collection, analysis, and display. The commercial GIS industry, which started in the early 1980s, is now estimated to be worth $3.5 billion.

Today's rosy picture sharply contrasts with the situation in 1969 when GIS first began at ORNL. At that time only a few centers—principally Environment Canada, the U.S. Geological Survey, Harvard University, and ORNL—shared a common interest in solving the riddle of geographic analysis. Along with scientists from these centers and a few leading research universities, early members of ORNL's GIS and Computer Modeling (GCM) Group, led by Richard Durfee, contributed many of the developments that made the current boom possible.

*Dobson served on a committee that composed most of the new national Spatial Data Transfer Standard.*

These contributions include fundamental development of early geographic computational techniques that supported and accelerated the growth of a commercial industry; development and integration of key GIS data bases and methodologies; and use of geographic and spatial analysis to provide information to help policymakers make decisions on national issues, such as

# Article 32. ORNL and the Geographic Information Systems Revolution

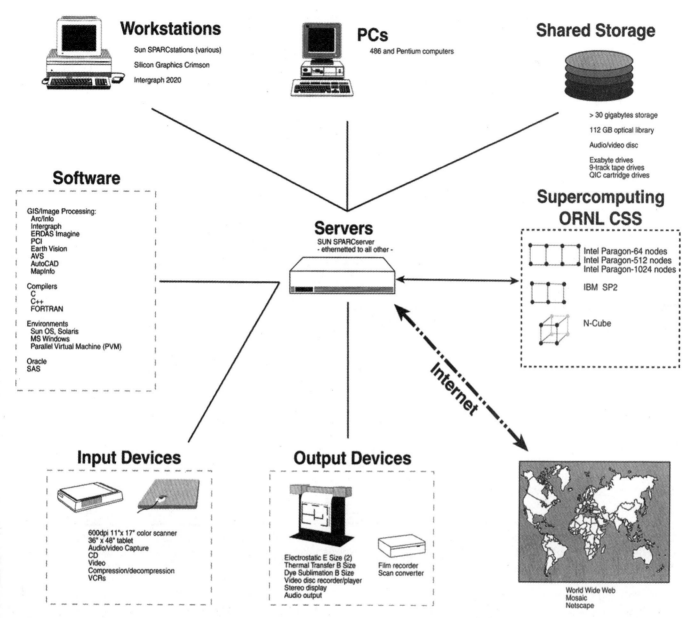

ORNL's GIS and Computer Modeling Group has a variety of computing resources used in support of a number of efforts ranging from natural resource assessments to environmental restoration.

development of energy sources and protection of water resources and fish populations, and to help government agencies assess natural resources and environmentally contaminated sites needing remediation.

Because of the increased use of GIS technology, a new national Spatial Data Transfer Standard (SDTS) has been established. Pioneering efforts by ORNL researchers Durfee, Bob Edwards, Phil Coleman, and Al Brooks helped build a foundation for the exchange of spatial data, and Jerry Dobson served on the Steering Committee of the National Committee for Digital Cartographic Data Standards, which composed most of SDTS. President Clinton's recently signed executive order requires all federal agencies to coordinate GIS data activities and make key data bases available to the public.

## What Is GIS?

Many people think of GIS as a computer tool for making maps. Actually, it is a complex technology beginning with the digital representation of landscapes captured by cameras, digitizers, or scanners, in some cases transmitted by satellite, and, with the help of computer systems, stored, checked, manipulated, enhanced, analyzed, and displayed as data referenced to the earth. This spatial information includes earth coordinates and geometric and topological configurations to portray spatial relationships between features such as streams, roads, cities, and mountains. GIS is "a digital representation of the landscape of a place (site, region, planet), structured to support analysis." Under this broad definition, GIS conceivably may include pro-

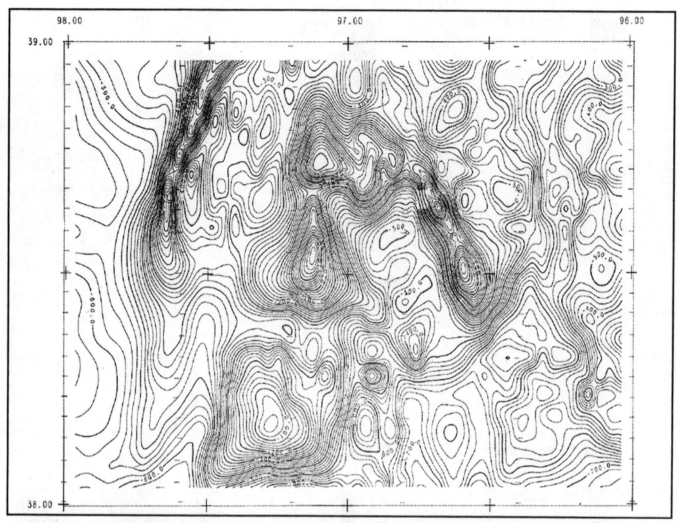

This map shows varying magnetic intensities in the Hutchinson, Kansas, region. ORNL helped develop a technique for converting one-dimensional flight line data into meaningful contour maps of regional magnetic data that could help identify anomalies potentially related to geologic deposits.

cess models and transport models as well as mapping and other spatial functions. The ability to integrate and analyze spatial data is what sets GIS apart from the multitude of graphics, computer-aided design and drafting, and mapping software systems.

Typical sources of geographic data for computer manipulation include digitized maps, field survey data, aerial photographs (including infrared photographs), and satellite imagery. Most image data are collected using remote sensing techniques. Aerial photographs are normally taken with special mapping cameras using photographic film. Most commercially available satellite imagery is collected using multispectral scanners, which record light intensities in different wavelengths in the spectrum—from infrared through visible light through ultraviolet light.

Spatial information can be represented in two distinctly different forms. Satellite images, for example, usually appear as raster data, a gridded matrix in which the position of each data point is indicated by its row and column numbers. Each position on a computer screen or map thus corresponds to the position on the ground measured by the satellite as it passes overhead. In contrast, cartographic features such as roads, boundaries, buildings, and contour lines usually are represented in vector form. In digitizing a lake, for example, the shoreline can be indicated as a series of points and line segments. In this case, each point is measured in Cartesian (X, Y) coordinates and each line segment is measured as a vector leading from one point to the next. The more points recorded, the more detailed the shoreline will be. Both forms, raster and vector, are essential to support environmental restoration projects on the ORNL reservation, for instance, and the software must be capable of rapid conversion from one form to the other.

## *GIS and remote sensing technology can detect changes in land features.*

For such geographic information to be meaningful, it must be accompanied by "metadata" documenting the source, description, specifications, accuracy, time of acquisition, and quality of each data element. As GIS technologies and multitudes of geographic data bases have spread to the desktop in the past decade,

## Article 32. ORNL and the Geographic Information Systems Revolution

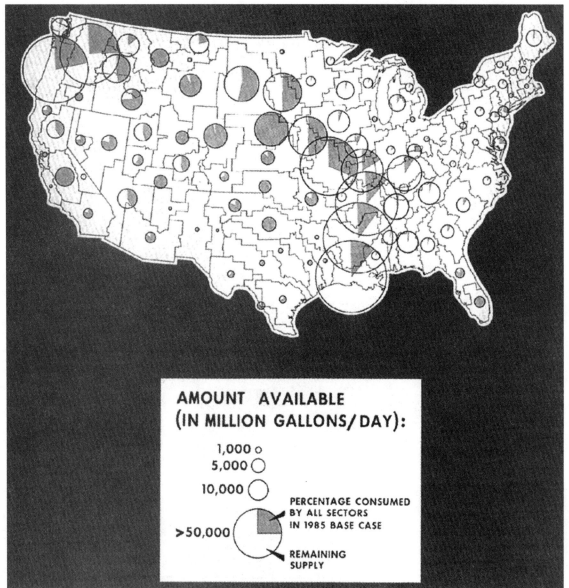

ORNL researchers used GIS and computer modeling techniques to predict the amount of water needed for national energy plans and compared it with the amount of water available in watershed basins across the United States.

metadata have become very important. Good metadata are essential in determining fitness of the geospatial data for each intended use—that is, determining which applications can be accomplished while ensuring the desired quality of results and decisions made from those data.

One of the most exciting applications of GIS combined with remote sensing technology is its ability to detect changes in features of large areas of land over many years by analyzing and comparing past and present landscape images. Each pixel can indicate a type of land cover, such as wetlands, forests, pastures, and developed areas. Such technology is now being used to monitor gains and losses in wetlands along the U.S. coast for assessing environmental impacts on U.S. fisheries. The technology has the potential for monitoring global change. For example, it is possible to detect increases in deforestation, which may alter the climate, or increases in desertification that may result from climate change.

In this article, we focus primarily on ORNL's role in the development and application of GIS to real-world problems over the past 25 years. Over this time, hundreds of projects and tasks involving GIS have been carried out by several organizations at ORNL involving a number of scientists, managers, and sponsors. It would be impossible to mention them all, but we do recognize and appreciate their significant contributions and collective vision for advancing GIS technologies over the years. In addition to the Computational Physics and Engineering Division, the examples of collaborating organizations within Martin Marietta Energy Systems have included the Energy Division, the Environmental Sciences Division, Chemical Technology Division, the Environmental Restoration Program, Biology Division, Data Systems Research and Development, and the Hazardous Waste Remedial Action Program. We highlight several of the larger efforts to illustrate the diversity of applications and tech-

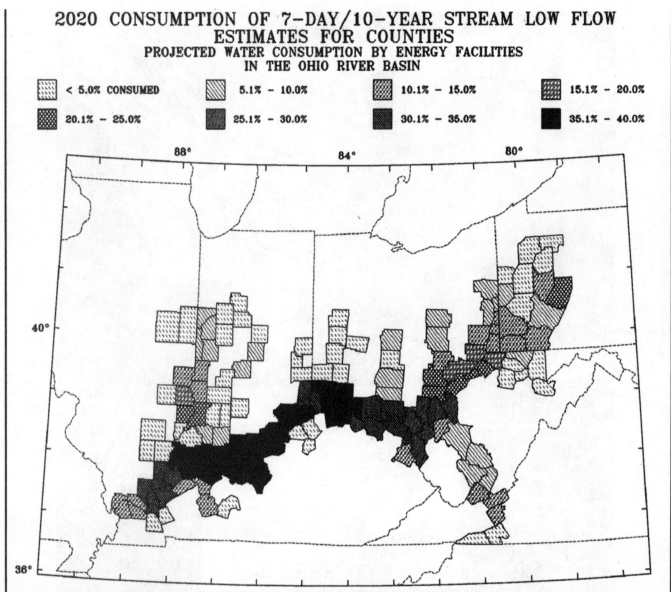

Results of one assessment of projected water consumption by energy facilities in the Ohio River Basin were shown to President Carter in a live presentation using a graphics station when he visited ORNL in 1978.

niques. We describe some of the early GIS developments and summarize some of the current systems capabilities. We offer examples in which GIS has proven useful in research and decision support.

## History of GIS Development

Actually, the term GIS, though first introduced in 1964, was not extensively used until the late 1970s. The first comprehensive geographic data management system—called the Oak Ridge Regional Modeling Information System (ORRMIS)—was developed in 1974 at ORNL by Durfee. Its purpose was to integrate and support the data management needs of a series of regional analytic models depicting and forecasting land-use, environmental, socioeconomic, and sociopolitical activities in the East Tennessee region.

Many early ORNL developments in GIS that are commonplace today are remarkable primarily because of their dates. Examples from the 1970s and early 1980s include perspective and isometric drawings of cartographic surfaces, integration of remote sensing and statistical techniques with GIS, raster-vector transformation, viewshed calculation, polygon intersections, transportation routing models, and true three-dimensional (3-D) imaging.

ORNL has a long heritage of GIS research, development, and application to complex problems ranging from national issues to site-specific impacts. After presenting an overview of GIS technology development in ORNL's computing environment, we discuss three eras of GIS history at ORNL—regional modeling and fundamental development (1969–1976), integrated assessments (1977–1985), and issue-oriented research and analysis (1986–1995).

# Article 32. ORNL and the Geographic Information Systems Revolution

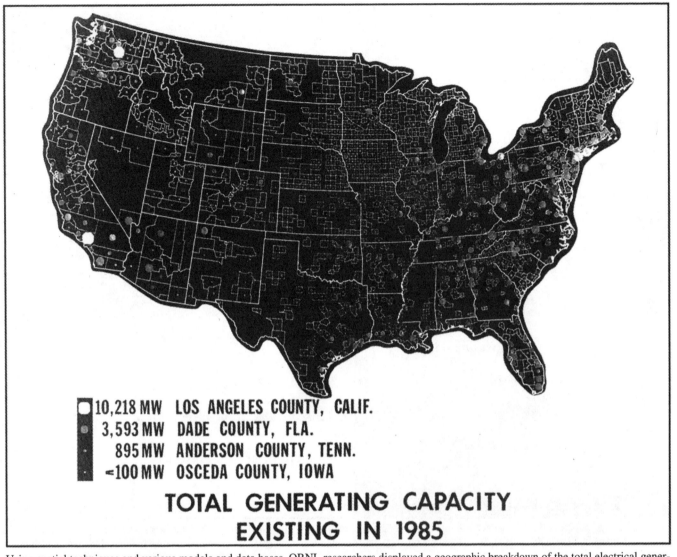

Using spatial techniques and various models and data bases, ORNL researchers displayed a geographic breakdown of the total electrical generating capacity in 1985 in the United States. The information aided utilities and energy agencies in the planning process for meeting projected demands for electricity.

## Evolution of GIS Technology at ORNL

In the past 25 years, GIS software development and applications have migrated from mainframe computers to minicomputers to personal computers (PCs) to networked UNIX workstations. GIS software is now being modified for use on parallel processors and supercomputers, such as the IBM SP2 and Intel Paragon X/PS machines at ORNL.

*Another technical development in the 1980s was integration of video information with digital data in the computer.*

In the very early 1970s, a technological feat was the development of a computer-generated 3-D perspective movie by Tom Tucker of ORNL. The movie simulated terrain and population changes over a 40-year period in the Norris, Tennessee, area as Norris Dam began operation as a hydroelectric facility.

Over the years, one of the benefits of these spatial technologies has been their applicability to many different types of problems. One example was the development in the early 1980s of electron microscope tomography for 3-D reconstruction of DNA chromosomes as a collaborative effort led by Don Olins and his colleagues in ORNL's Biology Division in cooperation with the GCM group at ORNL. Adaptation and development of hardware and software for a commercial remote sensing system, $I^2S$, on GCM minicomputers played a major role in the analysis of electron micrographs and display of chromosome structures. When it was determined that more sophisticated true 3-D displays were needed, a special varifocal mirror display was built. Depth visualization was provided by a vibrating mylar mirror synchronized with a monitor mounted above the mirror whose image was reflected to the operator. Data at greater depths were displayed when the mirror was at a greater

This GIS image of part of the Adirondack Mountains shows the geographical relationship between forested areas where trees were blown down by strong winds in 1950 and acidified lakes (numbers on lakes show acidity levels, or pH). The forest blowdown has been linked to acidification of the lakes.

deflection, thus varying the focal length to correspond to the appropriate depth. This occurred at a rate of 60 times per second, so the observer saw a continuous 3-D image.

Another ORNL breakthrough in GIS technology in the mid-1970s was the development of vector-based algorithms and their eventual integration with raster-based grid cell systems. The GCM group used these techniques for all types of water-resource and energy-related studies in collaboration with the Energy Division. In the late 1970s, ORNL developed transportation data bases and capabilities for routing hazardous wastes across the United States. Through use of GIS technology to match proposed routes with population density, the health and safety risks of hazardous waste transport could be estimated.

> *Development of new GIS hardware and software... improved our ability to solve old problems.*

Another technical development in the 1980s was integration of video information with digital data in the computer by Steve Margle and Ed Tinnel at ORNL. Raster digitization of video signals and the introduction of laser video discs opened up a whole new way of dealing with graphic and map data. Working in cooperation with the Data Systems Research and Development organization, the GCM group demonstrated the feasibility of using video from scanned map images recorded on laser video discs for simulations of war games as training exercises on a high-resolution workstation. In this technique implemented by Beverly Zygmunt, multiple video frames were located, computerized, and combined into large electronic maps that could be roamed and overlaid with other geographic and military information in real time.

Throughout the 1980s and into the 1990s, development of new GIS hardware and software technologies made new applications possible and improved our ability to solve old problems. It is interesting to note that some of our primary GIS applications in the first half of the 1990s have addressed a legacy of environmental problems, just as many initial applications in the early 1970s promoted GIS to help evaluate environmental impacts.

Some of the latest GIS research under way at ORNL involves developing software for use on parallel-processing supercomputers. Very recent work has shown that, by significant improvement of algorithms and by using parallel processors on ORNL supercomputers, the transformation and interpolation (estimation of values between data points) of large GIS data sets can be done 50 to 18,000 times faster than on smaller Sparc workstations. Because of the explosion in data collection from all types of earth sensor systems, workstations and supercomputers must be integrated to handle massive volumes of data.

*Real-time airborne GPS techniques have been used in aerial surveys of the Oak Ridge Reservation.*

We are also integrating portable GIS capabilities with GPS in which relative positions of objects on the earth can be pinpointed in real time by satellite sensors in communication with hand-held devices. As this technology becomes more commonplace, geospatial data will be collected at an ever increasing rate. Real-time airborne GPS techniques have already been used in aerial surveys of the Oak Ridge Reservation to collect high-resolution aerial photography with accurate positioning information. Computerized stereo techniques are being used with special goggles to help generate orthographic images (digital images corrected for camera, terrain, and other distortions) from stereo photography. Also, 3-D subsurface modeling and visualization are being done for hazardous waste studies.

To provide intelligent and efficient access to large amounts of geospatial data, work is under way to prepare and load this information on Internet and World Wide Web servers, which can be accessed by data browsing tools such as Mosaic and Netscape. These capabilities are important to the Oak Ridge user community and to the success of the National Spatial Data Infrastructure (NSDI) during the 1990s.

## Regional Modeling and GIS Development (1969–1976)

In 1969, the U.S. Congress passed the National Environmental Policy Act (NEPA), the National Science Foundation (NSF) initiated the Research Applied to National Needs (RANN) program, Ian McHarg published *Design with Nature*, and ORNL delved headlong into regional modeling and GIS. Clearly, NEPA was a major impetus to the other three events.

Before NEPA, research and development, infrastructural development, and resource management decisions had been based almost exclusively on engineering and cost-benefit considerations. Suddenly, NEPA thrust all large enterprises, including the federal government itself, into a new legal and ethical milieu in which comprehensive, interdisciplinary analyses were absolutely essential. Alvin Weinberg, director of the Laboratory from 1955 to 1973, immediately recognized the need and sought to diversify the Laboratory's missions.

For GIS, the most important development in the early days at ORNL was the Oak Ridge Regional Modeling Information System and associated tools that supported spatial data input and display. The primary purpose was "to provide the data management capability for analysis models which forecast the spatial distribution and ecological effects of activities within a geographical region." The land-use modeling efforts became the principal impetus to remote sensing development as well as to the GIS expansions.

Initial GIS software techniques were based on hierarchical grid cell systems. It became apparent that additional capabilities were needed for accurate cartographic representation and analysis of vector-based map data. By the mid-1970s, development of sophisticated polygonal-based GIS systems at ORNL were well under way. Our development of efficient storage and computational techniques for integrating raster-based grid cell and vector-based systems opened the door to addressing larger and more complex problems with a national scope. Incorporation of new algorithms designed by Phil Coleman and Bob Edwards provided a capability for analyzing and displaying large national data bases.

## Integrated Assessments (1977–1985)

In the mid-1970s, a shift in federal policy greatly reduced NSF funding for the DOE national laboratories. From then on, hardly another penny was received to support basic research, development, or operation of GIS systems at ORNL. The GCM group and the Energy Division shifted to applications-driven research, the funding for which allowed continued development and operations.

*ORNL systems were used. . . to support conflict resolution in power plant siting.*

We never had the luxury of focusing on a particular technology (remote sensing or computer cartography, for instance) to the exclusion of other technologies. We were then, and are still, comprehensive integrators with analytical purposes paramount in everything we do. In many respects, this approach has been advantageous because (1) the integrated GIS technologies were then applicable to a wide range of spatial problems, and (2) the applications-driven development minimized "ivory-tower" research looking for a problem to solve.

The first seeds of the new order were sown in 1975 when Richard Durfee and Bob Honea used ORRMIS tools for predictive modeling of coal strip mining and associated environmental problems. Results of this work were presented to Robert Seamans, head of the Energy Research and Development Administration (ERDA), predecessor to DOE. Soon afterward, we became heavily involved in siting analysis. In 1975 and 1976, ORNL systems were used, along with data from the Maryland Automated Geographic Information System, to support conflict resolution in power plant siting. By the late 1970s, these systems were heavily involved in decision support for federal energy policy and resource management. ORNL employed GIS extensively to evaluate the environmental impacts of various proposed National Energy Plans. Later, we predicted the amount of coal that could be produced from federally leased lands and evaluated the impacts on energy supply of designating certain lands as wilderness areas, thus protecting them from exploration for and extraction of oil, gas, and uranium.

During the mid-to-late 1970s, the Laboratory played a major role in the National Uranium Resource Evaluation (NURE) Program. ORNL's Computer Sciences Division (now the Computational Physics and Engineering Division), in cooperation with DOE's Grand Junction Office, was the national repository for all data collected and analyzed to assess the availability and location of potential uranium resources for future commercial nuclear power, research reactors, and other uses. ORNL staff were responsible for overall data management, GIS processing, spatial analysis, and mapping. Al Brooks was director of the Oak Ridge effort to support DOE in surveying the country for potential uranium resources and estimating possible reserves. Through a multitude of subcontractors, DOE conducted both aerial radiometric and geomagnetic surveys and hydrogeologic ground sampling on a quadrangle-by-quadrangle basis across the United States.

The aircraft had special sensors to detect radioactive isotopes of elements such as bismuth, thallium, and potassium as well as magnetic fields.

One example of highly specialized GIS work at ORNL was Ed Tinnel's development, in cooperation with Bill Hinze of Purdue University, of spatial filtering, interpolation, and contouring techniques to convert one-dimensional flight line data into meaningful maps of regional magnetic data. The purpose was to use these data to help study geologic features and identify magnetic anomalies that might indicate the presence of mineral deposits. These maps were also provided to the U.S. Geological Survey for publication. This was one of the earliest projects that required the handling of massive amounts of spatial, tabular, and textual information of many different types. During this time specialized GIS hardware systems were implemented to provide new ways of digitizing and displaying large amounts of geographic data.

Multiple energy assessments were early examples of policy analysis using GIS. A flurry of activity began each time President Jimmy Carter proposed a new National Energy Plan. Econometric models were run by the Energy Information Administration to project, as far as the year 2000, energy demand and fuel use by type in each major region of the country. These regional projections were passed to ORNL, where energy demand was disaggregated by Dave Vogt to Bureau of Economic Analysis Regions and supply was allocated to counties. Around 1980 Ed Hillsman and others of the Energy Division projected electrical generation from each existing plant and simulated construction or retirement of different plants by fuel type to determine if the president's goals would be met. Dobson and Alf Shepherd projected the amount of water needed for energy production and compared it with the amount of water available in each basin in the United States. ORNL's projections of electrical generation for different areas were passed to other national laboratories (Argonne, Brookhaven, Los Alamos, and the Solar Energy Research Institute), which used the information to evaluate effects on air quality, water quality, and labor supply. All results were reported to DOE, which conducted policy analysis of the feasibility of each proposed plan. Results of one GIS assessment of the projected water consumption by energy facilities in the Ohio River Basin were shown to President Carter in a live presentation using a graphics station when he visited ORNL in 1978.

In short, as early as the 1970s the nation's energy system and many pertinent physical and cultural features were simulated through GIS in linkage with econometric models, location-allocation models, environmental assessment models, and spatial data bases. The principal output was by county, but many of the data bases and computations covered details finer than the county level. For example, the data bases included population at the Enumeration District level, all power plants over 10 megawatts in generating capacity and all U.S. Geological Survey stream gauging station records. The models were as sophisticated as any in use at that time with or without GIS.

Another major multiyear effort involving ORNL researchers in the early 1980s was the development of a national abandoned mine lands inventory for the Office of Surface Mining (OSM) of the Department of the Interior. This effort, headed by Bob Honea, was based on federal legislation mandating that abandoned mine lands be reclaimed to protect human health, safety, welfare, and the environment, using funds collected as taxes on mining operations. A national inventory of abandoned mine lands was necessary to determine the affected areas in urgent need of reclamation and to establish priorities for reclamation of other sites. The effort was initially viewed as a technology-based project involving heavy use of remote sensing, GIS, record-based information systems, and statistical tools.

It was anticipated that analysis of Landsat satellite imagery would be a key ingredient for identifying detailed impacts from the disturbed, abandoned lands. However, an interesting turn of events made the project much more difficult than expected. When attempts were made to use results from satellite analyses to meet the mandates in the legislation, we found that the worst threats to human health and safety (e.g., open mine shafts, acid drainage, polluted water supplies) could not be determined from satellite data. Major environmental impacts could be addressed by analyzing satellite images, but health and safety impacts and reclamation cost estimates required field data collection and field assessment efforts. Thus, a major field collection effort, which included on-site interviews with affected populations, was carried out in conjunction with the state agencies of all the coal-mining states. Unique information handling techniques were devised to standardize and computerize textual, tabular, temporal, and spatial data from forms and maps that could then be linked with GIS for spatial aggregation, statistics, and mapping. Don Wilson was responsible for overseeing the computerization of all this information and development of a consolidated data base. These results could then support assessments at the state, regional, and national levels to aid OSM in allocating reclamation funds and overseeing mitigation of the severest problems.

Methodologies developed at ORNL for one application were readily adapted and applied to other problems. For example, our initial demographic work of the late 1970s was extended to compute detailed population distributions for any place or region in the United States.

The technique was used by Phil Coleman and Durfee to compute population distributions around all nuclear power plants in the United States. Our results, including the calculation of population exclusion zones, enabled the Nuclear Regulatory Commission to assess these exclusion areas—regions where additional nuclear power plants should not be built because too many people live or work there—to help make planning and licensing decisions.

## Issue-Oriented Research and Analysis (1986–1995)

Starting in the mid-1980s, the emphasis shifted again, this time in a very positive direction, as GIS became an important tool in topical research on scientific issues of national interest, as illustrated in these four examples.

**Lake Acidification and Acid Precipitation.** Acid precipitation can cause water in lakes to acidify, potentially reducing fish populations. Lake acidification and other environmental issues that may be related to acid precipitation were major themes of GIS work at ORNL in the late 1980s. The Environmental Sciences Division (ESD) was involved prominently in the National Acid Precipitation Assessment Program (NAPAP), especially the National Surface Water Survey. Through extensive collaboration with U.S. Environmental Protection Agency (EPA) laboratories and numerous universities and private firms, Dick Olson, Carolyn Hunsaker, and other ESD personnel collected, managed, and analyzed massive geographic data bases for lakes and watersheds throughout the United States. The goal was to characterize contemporary chemistry, temporal variability, and key biological resources of lakes and streams in regions potentially sensitive to acid precipitation.

*ORNL researchers concluded that forest blowdown facilitated the acidification of some lakes.*

Simultaneously, the Energy Division approached the same problem from a different perspective. While NAPAP focused on impacts of acid precipitation, this project focused on watersheds and investigated possible causes of lake acidification.

In 1950, a huge storm with heavy rain and 105-mile-per-hour winds blew down numerous trees in 171,000 hectares of forest in the Adirondack Mountains of New York. In the 1980s it was observed that several lakes in the area were acidified, so one hypothesis was that the blowdown of the forest might be a cause. To determine if a relationship existed between the forest blowdown and lake acidification, Dobson and Dick Rush of ORNL and Bob Peplies of East Tennessee State University used an approach that combined GIS and digital remote sensing with the traditional field methods of geography. The methods of analysis consisted of direct observation, interpretation of satellite images and aerial photographs, and statistical comparison of two geographical distributions—one representing forest blowdown and another representing lake chemistry.

Associations in time and space between surface water acidity levels (pH) and landscape disturbance were found to be strong and consistent in the Adirondacks. Evidence of a temporal association was found at Big Moose Lake and Jerseyfield Lake in New York and at the Lygners Vider Plateau of Sweden. The ORNL researchers concluded that forest blowdown facilitated the acidification of some lakes by altering pathways for water transport. They suggested that waters previously acidified by acid deposition or other sources were not neutralized by contact with subsurface soils and bedrock, as is normally the case. Increased water flow through "pipes"—small tunnels formed as roots decayed—was proposed as the mechanism that may link biogeochemical impacts of forest blowdown to lake chemistry.

*ORNL has led the technical effort to improve methods for analyses of changes in uplands and wetlands.*

Both efforts illustrate an ORNL strength—the ability to assemble multidisciplinary teams and multiple organizations to attack complex problems. GIS, in itself, is an integrating technology because it draws together different sciences that have a common need for spatial data, visualization, and analysis capabilities. Such was the case in the acidification studies just described. Although primary responsibility for these two efforts rested separately in the Energy and Environmental Sciences divisions, the GCM group was heavily involved in both efforts. Thus, considerable interaction took place between the two projects. Since then, ESD, in cooperation with GCM and other groups, has continued to expand its GIS capabilities and resources. ESD scientists now have hands-on access to GIS systems and data bases to support a multitude of research efforts.

**Coastal Change Analysis.** For decades, the National Marine Fisheries Service (NMFS) of the National Oceanic and Atmospheric Administration (NOAA) has been concerned about declining fish populations in U.S. coastal waters. Suspecting that these declines might be caused by losses of habitat, such as saltmarshes and seagrasses, and increases in pollution resulting from expanding urban and rural development, as well as agriculture, NMFS initiated a research effort to solve the technical, institutional, and methodological problems of large-area change analysis—methods for determining the time, location, and degree of changes in large areas to better understand changes in ecosystems and ecological processes. ORNL has led the technical effort to improve methods for analyses of changes in uplands and wetlands, detected by satellite sensors, and to perform prototype satellite change analysis of the Chesapeake Bay. Integration of these remote sensing and GIS methodologies in a laboratory environment, in field investigations, in workshop settings, and for presentations and briefings in policy and management arenas shows how much this evolving technology is becoming ingrained in all phases of earth-sciences work.

The Coastal Change Analysis Program (C-CAP) is developing a nationally standardized data base of land cover and land-cover change in the coastal regions of the United States. As part of the Coastal Ocean Program (COP), C-CAP inventories coastal and submerged wetland habitats and adjacent uplands and monitors changes in these habitats over one to five years. This type of information and frequency of detection are required to improve scientific understanding of the linkages of coastal and submerged wetland habitats with adjacent uplands and with the distribution, abundance, and health of living marine resources. Satellite imagery (primarily Landsat Thematic Mapper), aerial photographs, and field data are interpreted, classified, analyzed, and integrated with other digital data in a GIS. The resulting land-cover change data bases are disseminated in digital form for use by anyone wishing to conduct geographic analysis in the completed regions.

Land cover change analysis has been completed for the Chesapeake Bay based on Landsat Thematic Mapper (TM) data. The resulting data base consists of land cover by class for 1984, land cover by class for 1988 and 1989, and a matrix of changes by class from 1984 to 1988–89. We found that, contrary to popular opinion, marshland in the Chesapeake Bay region increased slightly during the period. However, both forested wetlands and upland forests declined significantly, while land development expanded rapidly. At greater detail, we observed the formation of a new barrier island and recorded lateral movement of portions of its tip by almost a kilometer.

Although the Chesapeake Bay prototype focused on a single region, its purpose was to provide a technical and methodological foundation for change analysis throughout the entire U.S. coast. Four regional workshops (Southeast, Northeast, Great Lakes, and Pacific) addressed a full range of generic issues and identified the issues of special interest in each major coastal division of the United States. Ultimately, the protocol development effort involved more than 250 technical specialists, regional experts, and agency representatives.

During the summer of 1994, field work was conducted in the Gulf of Maine, along the Oregon and Washington coast, and in Alaska. The Alaskan study is especially interesting.

In 1986, the Hubbard Glacier moved, closing the narrow opening between the glacier and Russell Fiord's Gilbert Point on the coastline of Alaska. The ice dam later burst as the fiord's water rose, and the narrow opening was restored. The event was worrisome to salmon fishermen because the fiord's alternative outlet to the sea could destroy the unique stock of sockeye salmon that spawn in the Situk River. The glacier is poised to move again, and the new, more permanent ice dam that is expected could cause the fiord to empty through the Situk watershed, drastically altering its ecosystem.

Using satellite images of the Alaskan coastline from various years, we are identifying changes in the Alaskan coastline that will help predict the impacts on fisheries when the glacier closes the gap again. If the Situk River salmon are threatened, it may be necessary to transplant some of them to less vulnerable streams.

In studying satellite images, we have looked for changes in land cover from 1986 on and tried to quantify these changes on a regional basis. For example, we have looked at changes in the size and shape of woodlands, wetlands, grasslands, and bare ground over a period of years to characterize coastal changes. We are trying to model the direct relationship between land-cover changes and ecological processes.

To verify the accuracy of our interpretations of the satellite data, we visit the imaged sites. In 1993 and 1994, Ed Bright and Dobson went to Alaska to conduct field verification of a 1986 land-cover classification in the Yakutat Foreland and Russell Fiord. Now, when we do field work, we use a hand-held GPS device linked directly to a color laptop computer. Commercial software integrates the live GPS location coordinates with raster images representing land cover and with vector images representing other features such as roads. The device has more than doubled productivity in the field. We are currently designing a modeling approach that will link GIS, transport models, and process models to address the linkage between land-cover change and fisheries.

**Environmental Restoration.** To clean up a legacy of environmental contamination and to comply with environmental regulations, U.S. government facilities must locate, characterize, remove or treat, and properly dispose of hazardous waste. In the 1980s, ORNL researchers helped develop geographic workstations, spatial algorithms, 3-D subsurface modeling techniques, and data base systems for handling hazardous waste problems at Air Force installations. Later, this work provided a foundation for supporting environmental restoration activities at DOE facilities. Since the late 1980s, environmental restoration has become a major theme for GIS activities at ORNL. The integration of GIS with other technologies provides an important resource to support hazardous waste assessment and management, remediation, and policy formulation for environmental cleanup at DOE facilities. The locations of waste areas (i.e., surface operable units) across the DOE Oak Ridge Reservation (ORR) are represented by the bold polygons shown on the following map.

*In studying satellite images, we have looked for changes in land cover... and tried to quantify these changes on a regional basis.*

In conducting successful cleanup efforts and meeting regulatory requirements at these facilities, GIS can assist in many ways. Key aspects include investigation of the types and characteristics of contaminants; the location of possible pollutant sources; previous waste disposal techniques; the spatial extent of contamination; relationships among nearby waste sites; current and past environmental conditions, including surface, subsurface, and groundwater characteristics; possible pollutant

# Article 32. ORNL and the Geographic Information Systems Revolution

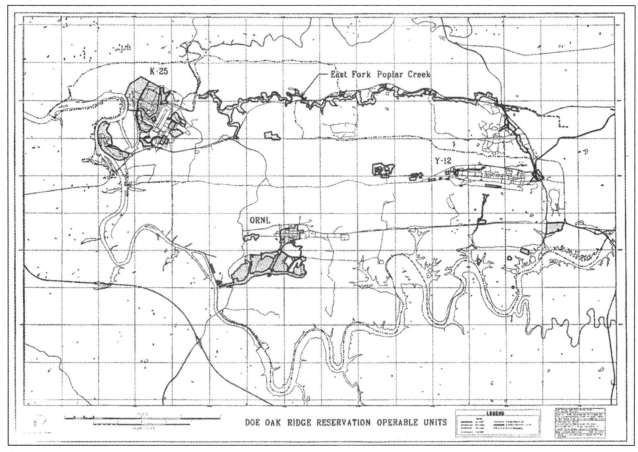

Waste areas designated as surface operable units on the DOE Oak Ridge Reservation.

transport mechanisms; efficient methods for analyzing and managing the information; effective cleanup strategies; and mechanisms for long-term monitoring to verify compliance.

## *GIS technologies support environmental monitoring and cleanup.*

Three programs that involve significant GIS activities in support of environmental restoration (ER) in Oak Ridge include the Oak Ridge Environmental Information System (OREIS), the Remote Sensing and Special Surveys (RSSS) Program, and the GIS and Spatial Technologies (GISST) Program. The OREIS effort is designed to meet environmental data management, analysis, storage, and dissemination needs in compliance with federal and state regulatory agreements for all five DOE facilities operated by Lockheed Martin Energy Systems. The primary focus of this effort has been to develop a consolidated data base, an environmental information system, and data management procedures that will ensure the integrity and legal defensibility of environmental and geographic data throughout the facilities. The information system is composed of an integrated suite of GIS, relational data base management, and statistical tools under the control of a user-friendly interface. The OREIS effort, previously led by Larry Voorhees and Raymond McCord, is now being directed by David Herr.

OREIS's data have been combined with other site-specific information to study a variety of environmental problems. ORNL has modeled contaminant leakage from underground waste lines and used historical aerial photos to assess potential pollutant migration.

## *Results from such analyses are useful in locating potentially contaminated... areas.*

The RSSS Program under Amy King supports ER site characterization, problem identification, and remediation efforts through the collection and analysis of data from aircraft and other remote sensors. One example has been helicopter radiometric surveys to determine gamma radiation levels across mapped areas of DOE facilities. GIS and remote sensing techniques also aid in the interpretation and visualization of airborne multispectral scanner data, thermal imagery, infrared and natural color photography, and electromagnetic and magnetic survey analyses. The following map shows examples of these types of processed information. Integrated results from such analyses are useful in locating potentially contaminated and affected areas, as well as possible underground structures that may be pertinent to hazardous waste burial and migration. Another example has been the delineation of waste trenches in burial ground areas that may be a source of waterborne contam-

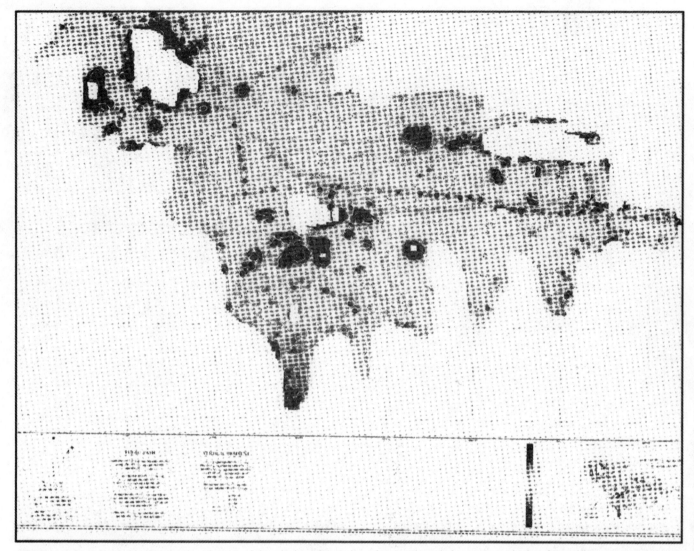

ORNL geophysicists led by Bill Doll perform spatial analysis and mapping of magnetic and electromagnetic data from airborne surveys to provide information that may aid in locating potentially contaminated areas on the Oak Ridge Reservation. Integration of this information with other types of remotely sensed data can help identify waste areas in burial grounds that may be a source of waterborne contaminants.

inants requiring remediation. The RSSS Program is also responsible for surveys of environmentally sensitive areas on the ORR.

## *GIS is used increasingly to plan, develop, and manage transportation infrastructures.*

The GISST effort, under Durfee, promotes the development, maintenance, and application of GIS technology, data bases, and standards throughout the ER Program. The largest activity currently under way is the development of base map data, digital orthophotos, and elevation models for all Energy Systems facilities using advanced stereo photogrammetric techniques based on real-time airborne GPS. When completed, these terrain data will be the most comprehensive GIS and orthoimage coverages of any DOE reservation. This project, under the technical direction of Mark Tuttle, is being carried out in cooperation with the Tennessee Valley Authority. Desktop mapping systems are being integrated into the daily operations of many Oak Ridge staff devoted to monitoring and cleaning up the ORR. To support these activities, a repository of the resulting data from this project is being made available to users networked into a local file server, which will soon be accessible as a World Wide Web server. These GIS data provide a consistent, current, and accurate base map that can be integrated with all other types of environmental and pollutant data for analysis and reporting.

The fusion of all types of spatial data is an important tool for any environmental activity on the ORR. Through these and other ER programs, facility data and environmental data bases have been developed to improve understanding of relationships among pollutant sources, surface and subsurface pathways, and receptors of environmental contaminants. Three-dimensional modeling, data management, and contaminant analysis have

## Article 32. ORNL and the Geographic Information Systems Revolution

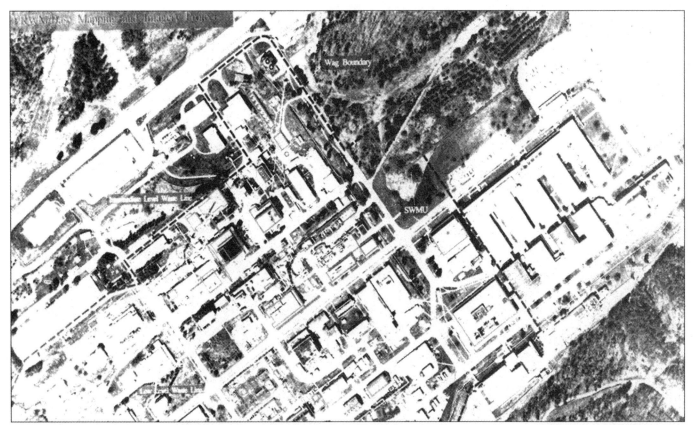

ORNL's fusion of orthoimagery with all types of spatial data on the Oak Ridge Reservation (note waste lines, operable unit boundaries, and solid waste management units in this figure) can help improve the identification and understanding of relationships among contaminant sources, support environmental monitoring activities, and assist in various types of facility management and land-use planning.

been enhanced through integration of computer tools and geospatial data. All these resources are becoming an integral part of the remediation planning and cleanup process, supported through communication networks linking scientists, engineers, and decision makers with analytical software and data bases.

**Transportation Modeling and Analysis.** Transportation systems and networks are crucial to the U. S. economy and way of life. GIS is used increasingly to plan, develop, and manage transportation infrastructures (e.g., highway, railway, waterway, and air transport networks) with the goal of improving efficiency in construction and operation.

Three main centers heavily involved in transportation modeling and geographic networks are the Energy Division (ED), the Chemical Technology Division (CTD), and the Computational Physics and Engineering Division (CPED). CTD has been primarily supporting DOE transportation needs in collaboration with CPED; ED has been supporting the Department of Transportation; and both ED and CPED have been supporting the Department of Defense. Collectively, the three groups have developed detailed representations of highway, railway, and waterway networks for the United States and military air transport networks for the entire world. ED, for example, is the developer and proprietor of the National Highway Planning System and the initial INTERLINE railway routing model. CTD has had a major responsibility for routing and assessing hazardous materials on the nation's highway and rail systems for many years (see figure next page). They have enhanced and adapted the INTERLINE and HIGHWAY routing models to assist in t his work. CPED has been a major developer of the Joint Flow and Analysis System for Transportation (JFAST), which is a multimodal transportation analysis model designed for the U.S. Transportation Command (USTRANSCOM) and the Joint Planning Community.

Operations Desert Shield and Desert Storm (1990–1991) involved the largest airlift of personnel and equipment from region to region ever accomplished. The U. S. Air Force's Military Airlift Command, now the Air Mobility Command (AMC), was responsible for this movement from the United States and Europe to the Persian Gulf region. Prior to that event, ORNL had worked with AMC to develop the Airlift Deployment Analysis System (ADANS), a series of scheduling algorithms and tools that enabled AMC to schedule missions to and from the Persian Gulf more rapidly and efficiently than ever before. ADANS is currently being used 24 hours per day by AMC to schedule peacetime, exercise, and contingency missions, as well as peacekeeping relief and humanitarian operations. Some of the key members of the ADANS team have included Glen Harrison, Mike Hilliard, Ron Kraemer, Cheng Liu, Steve Margle, and Irene Robbins.

The ADANS architecture is based on a relational data base management system, which operates on a network of powerful, UNIX-based workstations stretching across the United States with current installations at ORNL; Scott Air Force Base, Illinois; and

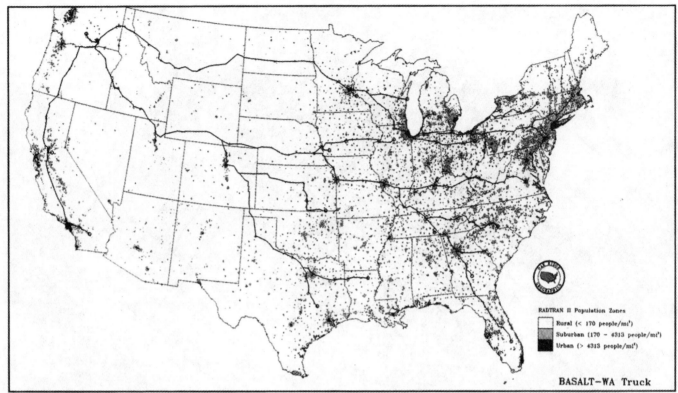

Results from Paul Johnson's and Dave Joy's work depict the computation display of possible truck routes for transporting radioactive waste material from groups of nuclear reactors to a candidate repository site. These routes are superimposed on zones of computed population densities across the United States.

Travis Air Force Base, California. PCs are used to perform some functions. The configuration includes a data base management system, a form generation tool, graphical display tools, a report generation system, communication software, a windowing system, and more than 500,000 lines of ADANS-unique code. All modules exchange data and run asynchronously. Thus, schedule planners can use the windowing system to keep track of and to modify multiple pieces of information. The three main components of the user interface are movement requirement and airlift resources data management, schedule analysis, and algorithm interaction.

All data and algorithms are geographically explicit. The user inputs data on a station-by-station basis with the textual network editor; the graphical network editor allows the user to establish a network and to enter or to edit information directly on a world map. With this system, it is easy to determine how cargo and passengers were moved, how many were moved as required, and to what aircraft they were assigned.

## *ORNL has advanced the use of GIS within our national infrastructure.*

The JFAST effort, initiated by Bob Hunter, is designed to determine transportation requirements, perform course of action analysis, and project delivery profiles of troops and equipment by air, land, and sea. JFAST was used in Desert Shield to analyze the airlift and sealift transportation requirements for deploying U.S. forces to the Middle East and predict their arrival dates in-theater. These deployment estimates provided input for establishing concepts and timing for military operations. Under Brian Jones' direction during Desert Storm, JFAST was also used to track ships, provide delivery forecasts, and analyze what-if scenarios such as canal closings and maintenance delays. In addition to analyzing support for humanitarian efforts such as those in Rwanda and Somalia, USTRANSCOM and the Joint Planning Community use JFAST to determine the transportation feasibility of deployment plans.

JFAST incorporates a graphical user interface that makes significant use of geographic display of transportation data as well as other graphic displays to aid the planner in understanding the output from the flow models. To assist in preparing briefings, all JFAST screens and graphic displays can be captured and inserted directly into presentation software while JFAST is running. Data from JFAST can also be sent directly to other Windows™-compliant applications, such as spreadsheets and word processing packages.

## ORNL's Role in the GIS Revolution

After a quarter of a century, how have GIS developments and applications at ORNL advanced science and served the national interest? ORNL has played an instrumental role in the GIS revolution by establishing and implementing a coherent vision that has been welcomed by scientific, policy, and management communities. ORNL has advanced the use of GIS within our national infrastructure. Today, commercial GIS products address many of the technical needs that required so much of our effort in the past, and the research frontiers have moved on to more complex methodological issues. However, no single commercial product today will handle all the current needs for GIS and related spatial technologies. One of our ongoing roles will be integration of multiple products with in-house technologies to best meet real-world needs that arise.

For the future, we envision linkages of GIS with environmental transport models and process models traditionally used by biologists, ecologists, and economists; implementation of GIS and digital remote-sensing techniques on supercomputers; 3-D GIS visualization and analysis; temporal analysis in a spatial context, and improved statistical analysis capabilities for geographic and other spatial data. The use of supercomputers will become even more important as new data collections—for example, the next generation of high-resolution satellite imagery—inundate the scientific community with terabytes of information. Justification for collecting and using these data will depend on the ability to extract meaningful information using supercomputer technology. We are currently addressing these and other technological issues, such as GIS animation, telecommunications, and real-time GPS and video linkage with GIS.

We hope to maintain a leadership position through continued advancement of hardware and graphics systems, GIS software, and data bases that will more effectively solve complex spatial problems. We think that knowledge-based expert systems will play a role in advancing future development and use of GIS technologies. We intend to assist the GIS community in improving standards and quality assurance procedures, and we look forward to assisting in enhancement of the National Spatial Data Infrastructure.

Ultimately, we view GIS as an integrating technology with the potential to improve all branches of science that involve location, place, or movement. Consider, for example, that most of the advances in medical imaging have been based on visual analysis. Imagine how much greater the potential would be if the images were enhanced by data structures, models, and analytical tools similar to those employed in analysis of the three-dimensional earth. We envision that certain technological thresholds will open the door to entire fields. For example, true 3-D analysis (more than visualization) and temporal GIS should provide new insights to geophysicists studying plate tectonics and the dynamic forces operating beneath the earth's surface. The single advancement of linking GIS with environmental transport models and process models will suddenly enable scientists and professionals in numerous other disciplines to incorporate spatial logic and geographical analysis alongside their traditional approaches. As these and other developments take place, a truly revolutionary new form of science should emerge.

## BIOGRAPHICAL SKETCHES

Jerome E. Dobson is a senior research staff member in ORNL's Computational Physics and Engineering Division. He currently serves as chairman of the Interim Research Committee of the University Consortium for Geographic Information Science, scientific editor of the *International GIS Sourcebook*, and a contributing editor and member of the editorial advisory board of *GIS World*. He holds a Ph.D. degree in geography from the University of Tennessee. He joined the ORNL staff in 1975. He is a former chairman of the Geographic Information Systems Specialty Group, Association of American Geographers, and member of the Steering Committee of the National Committee for Digital Cartographic Data Standards. He previously served as leader of the Resource Analysis Group in ORNL's Energy Division, as visiting associate professor with the Department of Geography at Arizona State University, as a member of the editorial board of *The Professional Geographer*, and as a member of the Steering Committee of the Applied Geography Conferences. Dobson was co-founder and first chairman of the Energy Specialty Group of the Association of American Geographers. He proposed the paradigm of automated geography, and he was instrumental in originating The National Center for Geographic Information and Analysis and in establishing the Coastal Change Analysis Program (C-CAP) of the National Oceanic and Atmospheric Administration. Employing geographic information systems (GIS) and automated geographic methods, he has proposed new evidence and theory regarding the mechanisms responsible for lake acidification and regarding continental drift and plate tectonics.

---

Richard Durfee is head of the Geographic Information Systems and Computer Modeling Group (GCM) in ORNL's Computational Physics and Engineering Division. He is also program manager of the GIS and Spatial Technologies Program for Environmental Restoration supporting DOE's Lockheed Martin Energy Systems facilities. He is responsible for an advanced GIS Computing and Technology Center at ORNL with special facilities for analyzing and displaying all types of geospatial information. Previously, he was head of the Geographic Data Systems Group in the former Computer Applications Division and section head in the former Computing and Telecommunications Division. He joined ORNL in 1965 as a member of the Mathematics Division. He has an M.S. degree in physics from the University of Tennessee. An expert in GIS and remote sensing technologies, he served on the initial steering committee for the DOE Environmental Restoration GIS Information Exchange conferences and for the early DOE Interlaboratory Working Group on Data Exchange. He was also an early member of the Federal Interagency Coordinating Committee on Digital Cartography. He is coauthor of many publications and presentations, including a GIS-related presentation to President Carter during his visit to ORNL in 1978. For more than 25 years, he has researched and directed a wide range of GIS technologies supporting hundreds of applications for more than 15 different federal agencies.

---

From *ORNL Review*, September 1, 2002. © 2002 by ORNL Review.

# Mapping the Outcrop

Teaching field geology with laptop computers and geographic information systems brings digital mapping to the outcrop.

By J. Douglas Walker and Ross A. Black

Over the last 20 years, geologists have evolved from die-hard computer-phobes to people using computers in virtually every aspect of their work. Now computers are making inroads into that last bastion of the "old-school" geologist: basic field mapping on the outcrop.

A field mapper can still pull on a pair of boots and grab a backpack, hammer and base map and conduct field work without a computer. Even with the advent of the personal computer, most outcrop mappers remain independent, self-reliant and proud of the low level of technology necessary to make original geological contributions to the scientific community.

**With the advent of the Internet, we expect new scientific information to be digital.**

But computers have become ubiquitous within our society. This is especially true in higher education. Today's student has been exposed to computers since an early age, and many students now actually see computers as a primary learning tool. Who better, then, to test digital mapping technologies than students? Over the last two years, we have been teaching our geology students to use laptop computers and geographic information system (GIS) software in the field.

Mapping itself is becoming a digital process (*Geotimes*, June 2000). With the advent of the Internet, we expect new scientific information to be digital. Large corporations, government agencies and, to some extent, universities began major efforts to put new information into a digital form many years ago. They also had to come to grips with the fact that it is tedious and expensive to convert older data into a usable digital form.

**Making mapping information digital from the earliest point of the mapping process—in the field—could save the mapper many time-consuming steps.**

Entering historic data into a system where it can be integrated with other information is a chore. Entering mapping data into the computer is one of the geologist's more onerous tasks. The endearing term "digi-slave" is common in laboratories and offices converting maps into digital data sets. Making mapping information digital from the earliest point of the mapping process—in the field—could save the mapper many time-consuming steps.

Geological data are location dependent: the elevation and location at which an observation is made is just as important as the observation itself. This is why most geological data are recorded in a map-based format. Geological outcrop observations are thus well suited to being recorded and manipulated with existing GIS software packages.

We and our students use GIS software to compile, compose and view geologic maps. We also use such software to merge maps with satellite imagery, aerial photography and other data sets, and for visualizing data in 3-D for modeling geological processes.

Our GIS lab has been integrating large geological and geophysical databases for the last six years. We have put a tremendous amount of time and effort into converting paper-based geological data into a digital form.

Existing geological maps are by far the hardest sources of information to enter into a database. The information on the map consists of on-the-fly interpretations of observations the field geologist made on the outcrop. The person digitizing the map must also interpret the map symbols on-the-fly.

Why not enter the data into the GIS package on the outcrop, eliminating the need for another step in the lab that is technical, time consuming, costly and prone to error?

After asking ourselves this same question, we started putting together a computer-based field-mapping program. Several private companies, government agencies and academic groups were pursuing the same goal, and we met them at meetings of the Geological Society of America, American Association of Petroleum Geologists, American Geophysical Union and the Environmental Research Institute Users Group. Some groups, most notably the Canadian Geological Survey, Bowling Green State University and the University of California at Berkeley, had digital mapping programs in place. But almost everyone was at about the same stage we were.

We investigated the technologies we could pursue and then sought funding for the project. We received funding from the University of Kansas (KU) Technology Fund, the Geothermal Program Office of the U.S. Navy and the KU Department of Geology.

## Students in the field

We first used the computers and software in a graduate level mapping course offered in January 1999. Four graduate students signed up. These students had previous mapping experience, and three of the four had used ArcView software.

The students first had to figure out how to carry their laptops into the field. They dug into a box of straps, buckles, clips and tape we had purchased from a camping store, and, after an hour of fiddling, were ready to go into the field. Most of the first day was spent getting used to the computers and trying to enter geologic data on the outcrop.

In a word, the students were very unhappy with the whole operation at the end of the day (mutiny might be a more accurate term). Carrying the nine-pound computers was not fun, digitizing was a pain, the screens were hard to read and the batteries were weak and heavy. We were not very encouraged.

To avoid a total loss of field time, the students printed maps and took them into the field to map on the next day. This went fine until it came time to enter the data into the computer. Predictably, this was a tedious operation. Sensing that the computers could actually save some time and effort, we went out with computers and paper maps the third day. By the end of the day the students were mapping pretty well on the computers and were in better spirits.

We spent another seven days in the field mapping and working out problems with the computers and software. By the end of two weeks, three of the four students were happy with the mapping setup. The fourth remained unconvinced that we had come up with anything useful.

After this session, it was with great uneasiness that we introduced computer mapping into our undergraduate field camp in June of 1999. We taught eight undergraduate students in the last week of a six-week field course. None of them had ever used a GIS program, let alone ArcView. We gave them the same short, three-hour introductory session we'd given the graduate students and then sent them into the field.

The undergraduate experience could not have been more different from the graduate one. The students were very excited about mapping with computers and welcomed the change from paper maps, photos and mylar overlays. Most students were comfortable with the computers by the end of the first day; by mid-morning of the second day they were asking questions about the geology and not about how to use the programs.

We attribute these different reactions to two factors. First, the undergraduates are not set in their ways about mapping. They did not have the background baggage the graduate students carried. Second, the undergraduates were more used to computers. The age difference of a couple of years is just enough that the younger students expect to use computers in all aspects of their education and most aspects of their lives.

## The future

We will continue using computers in University of Kansas field courses. We plan to expand the undergraduate component to about half of the six-week course. We still consider conventional mapping skills important. Introducing the computer adds a level of excitement for the students.

Many of our students are now taking GIS courses. This change is a grassroots effort among the undergraduates and not an idea the faculty pushed onto them. They will know more about the basic software components than will some of the field-camp faculty members.

---

**The widespread availability of inexpensive (or free) 7.5-minute topographic maps in various digital formats has been one of the important factors in making digital mapping systems easier.**

---

Some of the problems with field computers are being remedied. Laptops are getting lighter—four pounds instead of nine pounds. Battery life is steadily increasing. We can map all day on a single lithium-ion battery. Touchscreens are now readable in sunlight. Personal digital assistants should soon be powerful enough to run the software and handle the large image files needed to perform efficient mapping.

The widespread availability of inexpensive (or free) 7.5-minute topographic maps in various digital formats has been one of the important factors in making digital mapping systems easier to use. Now we need inexpensive, large scale, aerial photos at digital resolutions useable for geological field studies.

The final component for digital mapping is GPS receivers. Using the Global Positioning System (GPS) promises to give the field geologist real-time, accurate location information. Although we have a method for connecting the GPS unit directly to the laptop and downloading the location, the GIS packages do not automatically update the map with the proper map symbols. Thus it is easier to read the location from the GPS unit and then manually move the laptop cursor to that point on the map and record the field observation. We hope that GPS receivers soon become standard options for ruggedized laptops and that GIS packages include easy-to-use interfaces with standard GPS data. The demise of selective availability in GPS signals (Geotimes, June 2000) has made GPS coordinates almost 10 times more accurate.

Improved, publicly available GPS signals may be what pushes GPS/GIS-driven geological field mapping onto every student's field belt in a one-piece unit that will fit in a Gfeller field case.

## Additional Reading

"Bedrock geologic mapping using ArcInfo" by T.E. Waht, J.D. Miller and E.J. Bauer. *Proceedings* of ESRI Users Conference, 1995. p. 167.

"The Bedrock of Geologic Mapping" by P. Chirico. *Geo Info Systems*, 1997. v. 7, n. 10, p. 26–31.

"Development of Geographic Information Systems Oriented Databases for Integrated Geological and Geophysical Applications" by J.D. Walker, R.A. Black, J.K. Linn, A.J. Thomas, R. Wiseman and M.G. D'Attillio. *GSA Today*, 1996 v. 6, n. 3, p.1–7.

*Getting to know ArcView GIS* by ESRI Press. 1998.

---

**Walker and Black teach in the Department of Geology at the University of Kansas in Lawrence, Kan.**

# Gaining Perspective

The proliferation of satellite technology, from spy-quality photos to low-resolution radar images, is giving us new, more meaningful ways to envision complex information about the Earth. But whether we will act on the picture of ecological destruction this technology is cobbling together—from global climate change to wholesale clearing of forests—remains to be seen.

by Molly O'Meara Sheehan

During the last few decades of the 20th century it became evident that tropical rainforests were endangered not only by road-building, timber-cutting, and other incursions of the bulldozer and saw, but by thousands of wildfires. Historically, relatively small fires have been set by slash-and-burn farmers trying to clear patches of jungle for farm land. But starting in 1997, fires in the world's tropical forests from Brazil to Papua New Guinea raged on a scale never recorded before. The causes of these huge conflagrations raised questions, because tropical rainforests rarely burn naturally.

In the wake of several haze-induced accidents and public health warnings in smoke-covered Indonesia, the need for a clear answer to these questions was given legal significance when President Suharto, under pressure from neighboring countries, passed a decree making it illegal to set forest fires. The politically connected timber industry had managed to direct most of the blame for forest fires on the small-scale, slash-and-burn farmers, but Indonesia's rogue environment minister Sarwono Kusumaatmadja employed a relatively new intelligence-gathering technology to get a clear picture of the situation: he downloaded satellite images of burning Indonesian rainforests from a U.S. government website and compared them to timber concession maps. The satellite images confirmed that many of the blazes were being set in areas the timber companies wanted to clear for plantations. With the satellite evidence in hand, Kusnmaatmadja got his government to revoke the licenses of 29 timber companies.

## High Speed Intelligence

The environment minister's quick work on the rainforest issue is just one of many recent cases involving environmental questions in which satellite surveillance has been used to provide answers that might otherwise not have been known for years, if ever. The images captured by cameras circling high above the planet are proving effective not only because they scan far more extensively than ground observers can, but because they can be far faster than traditional information-gathering methods.

Pre-satellite studies of the oceans, for example, had to be done from boats, which can only reach a tiny fraction of the oceanic surface in any given month or year. And even after centuries of nautical exploration, most of the information scientists have gathered about winds, currents, and temperatures comes from the commercial trade routes of the North Atlantic between the United States and Europe. Satellites don't replace on-the-water research, as they can't collect water samples, but for some kinds of data collection they can do in minutes what might take boats centuries to do. Satellites can, in principle, watch the whole of the world's oceans, providing almost immediate assessments of environmental conditions everywhere.

Similarly quick surveillance is available for many parts of the biosphere that are otherwise difficult to reach—the polar ice, dense forest interiors, and atmosphere. As a result, says Claire Parkinson of the U.S. National Aeronautics and Space Administration (NASA), "theory and explanations no longer have a database restricted to areas and times where humans have physically [gone] and made observations or left instruments to record the measurements." The speed of environmental research has taken a quantum leap.

Speed isn't only a matter of technical capability, however. In practice, it's also a matter of access. Spy satellites began circling the globe soon after Russia's Sputnik went into orbit in 1957. But the information they relayed to Soviet and U.S. intelligence agencies was kept sequestered.

Article 34. GAINING PERSPECTIVE

The difference now is that satellites are increasingly being used for purposes other than espionage or military intelligence. The great majority are for telecommunications. However, more than 45—many owned by governments, but a growing number of them privately owned—are being used for monitoring various phenomena on the ground, on the water, or in the atmosphere. In addition, more than 70 launches are planned during the next 15 years by civil space agencies and private companies. How these instruments are used, and by whom, will have enormous consequences for the world.

THE RACE AGAINST TIME

If incidents like the Indonesian forest fire intervention are any indication, environmental monitoring by orbiting cameras could play a critical role in reversing the global trends of deforestation and ecological collapse that now threaten the long-term viability of civilization. Denis Hayes, the former Worldwatch Institute researcher who is the chairman of Earth Day 2000 asked a few years ago, in a speech, "How can we have won so many environmental battles, yet be so close to losing the war?" Since then, we have edged still closer. Clearly, the number of battles being won is too small, and the time it takes to win them is too long.

**FLOODING IN BANGLADESH**

SPOT SATELLITE IMAGERY: © CNES 2000. COURTESY SPOT IMAGE CORPORATION, WWW.SPOT.COM.

This medium resolution SPOT image of the confluence of the Meghna and Ganges Rivers and nearby Dhaka, the capital of Bangladesh, depicts a kind of collision between population growth and the ebbs and flows of the powerful rivers that feed the Ganges Delta. The land here is some of the most fertile--and heavily cultivated--in the world, replenished with rich silt washed down from the Himalayas by annual floods. The sprawling city of Dhaka and a couple of roads are perceptible in the northwest. And a closer view shows considerable development and cultivation throughout the image. (Even though the area is densely populated, this is difficult to see as most of the people live in small towns or villages.)

Another way of posing Hayes's now famous question might be to ask whether the processes of information-gathering and dissemination essential to changing human behavior can be speeded up enough to accelerate the environmental movement. Telecommunications satellites began providing part of the answer several decades ago by facilitating the formation of an active international environmental community that can mobilize quickly—whether to protest a dam on the Narmada River of India or to stop the use of genetically modified organisms in food production in Europe.

But while activism gained momentum, field work remained ominously slow—biologists slogging about in boots and rowboats, while the forces they were trying to understand raced over the Earth on the wings of global commerce, or ripped into it with the blades of industrial agriculture and resource extraction. However, satellite monitoring has begun to help researchers to more quickly assemble the data needed to bring decisive change. Remotely sensed images are contributing to critical areas of environmental research and management, including:

• **Weather**: The first meteorological satellites were launched in the early 1960s, and quickly became a key part of the U.N. World Meteorological Organization's World Weather Watch. In addition to the satellite data, virtually all nations contribute surface measurements of temperature, precipitation, and wind to this program to aid weather prediction, which has enormous social and economic benefits. In recent years, optical sensors that collect data on sea surface temperature and radar sensors that estimate ocean height have proven useful in understanding and predicting El Niño events, which can damage fisheries and agriculture by bringing warmth and wetness to much of the west coasts of South and North America, and drought to Southeast Asia, Australia, and parts of Africa.

• **Climate**: In the 1990s, researchers began to delve into satellite archives to study longer-term climate patterns. For instance, satellite images have helped reveal a decrease in snow in the Northern Hemisphere, a lengthening growing season in northern latitudes, and the breakup of major ice sheets. Radar sensors have been used to construct topographical maps of the ocean bottom, which in turn provide better understanding of the ocean currents, tides, and temperatures that affect climate. However, it was not until recently that space agencies began to design satellite systems dedicated specifically to climate research. In 1999, the United States launched Terra, which carries five different sensors for recording climatic variables such as radiative energy fluxes, clouds, water vapor, snow cover, land use, and the biological productivity of oceans. It is to be the first in a series of satellites that will create a consistent data set for at least 18 years.

• **Coastal boundary changes**: Whether as a result of warmer temperatures contributing to sea-level rise or irrigation projects shrinking lakes, coastal configurations change over time—sometimes dramatically. Scientists have compared declassified images from covert U.S. military satellites pointed at Antarctica in 1963 to recent images of the same regions, for example, to reveal changes in the continent's ice cover. Other comparisons show the extent to which central Asia's Aral Sea and Africa's Lake Chad have diminished in size.

• **Habitat Protection**: The destruction of habitats as a result of human expansion has been identified as the largest single cause of biodiversity loss. Satellite images have proved quite effective in revealing large-scale forest destruction, whether by fire or clearcutting, not only in Indonesia but in the Amazon and other biological hotspots. New, more detailed imagery may reveal small-scale habitat niches. In Australia, for example, the Australia Koala Foundation plans to use detailed satellite images to identify individual eucalyptus trees. Researchers will compare these images to field data to determine what this species of tree looks like from above, then use the information to more quickly map individual trees or groves than would be possible from the ground. In the oceans, the same satellite-generated maps of undersea topography and sea surface temperature used to study weather and climate can be used to track the upwellings of nutrient-rich water that help to sustain fisheries.

• **Environmental law enforcement**: International organizations and national governments can use remote imaging to put more teeth in environmental laws and treaties. One of the leading fishing nations, Peru, is monitoring its coastal waters to prevent the kind of heavy overfishing that has so often caused fisheries to collapse. In Italy, the city of Ancona plans to buy satellite images to detect illegal waste dumps. Within the next decade, large-scale use of this technology could give urgently needed new effectiveness to such agreements as the Kyoto Protocol to the Climate Convention, the Biodiversity Convention, the Convention on Illegal Trade in Endangered Species (CITES), or the Law of the Sea.

## MAKING SENSE OF NONSENSE

Look closely at a small detail of a newspaper or magazine photo—put it under a magnifying glass—and it may make no sense. The dots don't form any recognizable image. But stand back and see the photo as a whole, and it snaps into focus. Satellite images do the same thing, only on a vastly larger scale. Bits of information that might make no meaningful pattern when seen from the ground may, when seen from many kilometers above, resolve themselves into startling pictures.

The first pictures from space were photographs made from film, by astronauts aboard the first manned flights to the moon in the 1960s. These photos, of a fragile blue planet suspended in the vast blackness of space, helped to inspire the nascent environmental movement—one of them becoming the emblem of the first Earth Day in 1970.

In later surveillance from satellites, the imaging was digitized so that the data could be sent down in continuous streams and in much larger quantities than would be possible with film. Although one Russian satellite still uses regular camera film that is dropped to Earth in a canister and retrieved from the North Sea, most remote sensing satellites now use digital electronic sensors. The binary data they send down can be reconstructed into visual images by ground-based computers.

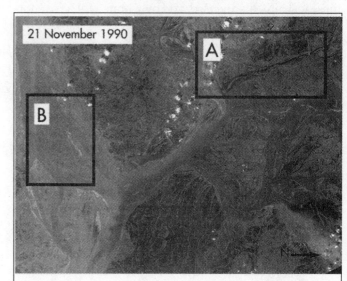

Bangladesh is one of the most densely populated countries in the world, and one of the most low-lying--the majority of the country is barely above sea level--so even the smallest of the annual floods can cause considerable damage. This image, in which darker areas are water-covered, was recorded during a devastating period of flooding, which took thousands of lives. Dhaka (A) is waterlogged and the once-cultivated islands in the Ganges, in the south of the image (B), are completely inundated.

The amount of detail varies with the type of sensor (see table, "Selected Satellite Systems"). For instance, an image of a 1,000 square kilometer tract of land obtained by a fairly low-resolution satellite such as AVHRR, which is used in continental and global studies of land and ocean, might contain 1,000 picture elements—or "pixels" (one piece of data per square kilometer). In contrast, an image of the same tract from the new, high-resolution Ikonos satellite would have 1 *billion* pixels (one per square meter). Between the broad perspective of satellites like AVHRR and the telescopic imaging of those like the new spy-quality satellite Ikonos, are medium-resolution sensors, such as those aboard the Landsat and SPOT satellites, in which one pixel represents a piece of land that is between 30 and 10 meters across. However, the level of detail is also limited by the size of the medium on which an image is displayed. For instance, if an Ikonos image with 1 billion pixels were reproduced in a magazine image one-quarter the size of this page, it would be reduced to 350,000 pixels.

Different tasks require different levels of detail. The value of high resolution lies in its enabling the viewer to hone in on a much smaller piece of the ground and see it in a kind of detail that the lower resolution camera could not capture. For a larger area, a lower resolution would suffice to give the human eye and brain as clear a pattern as it can recognize. Whereas the wide coverage provided by lower-resolution satellites has proved useful in understanding large-scale natural features such as geologic formations and ocean circulation, very detailed imagery may be best able to reveal niche habitats—such as the individual treetops that are home to the koala—and manmade structures, such as buildings, tanks, weapons, and refugee camps.

But satellite surveillance can do much more than provide huge volumes of sharp visual detail of the kind recorded by conventional optical cameras. The orbital industry also deploys a range of sensors that pick up information outside the range of the human eye, which can then be translated into visual form:

- Near-infrared emissions from the ground can be used to assess the health of plant growth, either in agriculture or in natural ecosystems, because healthy green vegetation reflects most of the near-infrared radiation it receives;
- Thermal radiation can reveal fires that would otherwise be obscured by smoke;
- Microwave emissions can provide information about soil moisture, wind speed, and rainfall over the oceans;
- Radar—short bursts of microwaves transmitted from the satellite—can penetrate the atmosphere in all conditions, and thus can "see" in the dark and through haze, clouds, or smoke. Radar sensors launched by European, Japanese, and Canadian agencies in the 1990s have been used mainly to detect changes in the freezing of sea ice in dark, northern latitudes, helping ships to navigate ice fields and steer clear of icebergs. Radar is what enabled satellites to map the ocean bottom, which would otherwise be obscured.

While satellites have the technical capability to monitor the Earth's entire surface—day or night, cloud-covered or clear, on the ground or underwater—this doesn't mean we now have updated global maps of whatever we want. Aside from the World Weather Watch, there is no process for coordinating a worldwide, long-term time series of comparable data from Earth observations. Rather, individual scientists collect data to answer specific questions for their own projects. In recent years, national space administrations have teamed up with research funding agencies and two international research programs to support an Integrated Global Observing Strategy that would create a framework for uniting environmental observations. Researchers are now trying to demonstrate the viability of this approach with a suite of projects, including one on forests and another on oceans.

In addition, to make sense of remotely sensed data requires comparison with field observations. An important element of the weather program's success, for instance, is the multitude of observations from both sky and land. And satellite estimations of sea-surface temperature can't generate El Niño forecasts automatically, but must be combined with other data sources, including readings from a network of ocean buoys that monitor wind speed and a satellite altimeter that measures water height. Similarly, the radar scans used to make maps of the ocean floor are calibrated and augmented by sounding surveys conducted by ships. Even a task as straightforward as the location of eucalyptus trees for the Australia koala project requires initial field observations to confirm that the typical visual pattern being searched out from above is indeed that of the eucalyptus, and not of some other kind of tree.

SPOT SATELLITE IMAGERY: © CNES 2000. COURTESY SPOT IMAGE CORPORATION, WWW.SPOT.COM.

This medium-resolution SPOT image of the Nile Delta shows the steady march of irrigation (indicated by the dark shades and crop circles) out into the desert near Cairo. While the Nile Delta and its floodplain have been farmed for thousands of years, only within the past 30 years have crops been planted intensively out in the desert. And with good reason: Cairo's population has grown from 5 million in 1970 to more than 11 million today.

Satellite imagery has become even more useful with the advent of geographic information systems (GIS), which allow users to combine satellite images with other data in a computer to create maps and model changes over time. In much the same way that old medical encyclopedias depict human anatomy, with transparencies of the skeleton, circulatory system, nervous system, and organs that can be laid over a picture of the body, a GIS stores multiple layers of geographically referenced information. The data layers might include satellite images, topography, political boundaries, rivers, highways, utility lines, sources of pollution, and wildlife habitat.

Maps that are stored in a GIS allow people to exploit the data storage capacity and calculating power of com-

## SELECTED SATELLITE SYSTEMS PRODUCING COMMERCIALLY AVAILABLE IMAGERY

| Satellite | Launch Date | Owner | Spatial Resolution | Spectral Range | Price per Square Mile |
|---|---|---|---|---|---|
| Landsat series | 1972 | NASA, (U.S. space agency) | 30-120m, | visible light (red, green blue); near-, short-wave, and thermal infrared | $.02-.03 |
| Landsat-7 | 1999 | | 15-60m | | |
| Terra | 1999 | NASA | 15m-22km | visible, near-, short-wave, mid-, and thermal infrared | N/A |
| SPOT series | 1986 | CNES (French space agency) | 10-30m | visible, near-and short-wave infrared | $1-3 |
| SPOT-4 | 1997 | | | | |
| AVHRR | 1979 | NOAA (U.S. agency) | 1.1km | visible, thermal infrared | $.08-80 per 10,000 square miles |
| IRS-1D | 1997 | Indian remote sensing agency | 6m | visible, near- and short-wave infrared | $1.30-6.20 |
| Ikonos | 1999 | Space Imaging Corp. | 1-4m | visible; near-infrared | $75-250 |
| Orb View | | Orbimage Corp. | | | |
| OrbView-1 | 1995 | | 10 km | visible; near-infrared | |
| OrbView-2 | 1997 | | 1 km | visible; near-infrared | $.0003 |
| OrbView-3 | 2000 | | 1-8m | visible; near-infrared | N/A |
| OrbView-4 | 2000-01 | | 1-8m | visible; near-infrared | N/A |
| QuickBird | 2000 | Earthwatch Corp. | 1-4m | visible; near-infrared | N/A |
| ERS SAR | | | | | |
| ERS-1 | 1991 | European space agencies | 25m | radar; C-band | |
| ERS-2 | 1995 | | | | |
| IERS-1 | 1992 | Japanese space agency | 18m | radar; L-band | |
| Radarsat | | | | | |
| Radarsat-1 | 1995 | Canadian space agency | 8-100m | radar; C-band | $.04-5.40 |

puters. Thus when geographically referenced data are entered into a GIS, the computer can be harnessed to look at changes over time, to identify relationships between different data layers, to change variables in order to ask "what if" questions, and to explore various alternatives for future action.

Because human perception can often identify patterns more easily on maps than in written text or numbers, maps can help people understand and analyze problems in ways that other types of information cannot. The Washington, D.C.-based World Resources Institute (WRI), for example, has used GIS to analyze threats to natural resources on a global scale. Researchers have combined ground and satellite data on forests with information about wilderness areas and roads to map the world's remaining large, intact "frontier" forests and identify hot-spots of deforestation. A similar WRI study, investigating threats to coral reefs, assembled information from 14 global data sets, local studies of 800 sites, and scientific expertise to conclude that 58 percent of the world's reefs are at risk from development.

With advances in computing power, some GIS software packages can now be run on desktop computers, allowing more people to take advantage of them. In fact, the number of people using GIS is swelling by roughly 20 percent each year, and the leading software company,

ESRI, grew from fewer than 50,000 clients in 1990 to more than 220,000 in 1999.

A related technology spurring the market for geographic information is the Global Positioning System—a network of 24 navigation satellites operated by the U.S. Department of Defense. A GPS receiver on the ground uses signals from different satellites to triangulate position. (For security reasons, the Defense Department purposefully introduces a distortion into the signal so that the location is correct only to within 100 meters.) As GPS receivers have become miniaturized, their cost has come down. The technology is now built into some farm machines, cars, and laptop computers. Researchers can take air or water samples and feed the data directly into a GIS, with latitude and longitude coordinates supplied by the GPS receiver in their computers.

The relatively new field of "precision agriculture" demonstrates how satellite imagery, GIS, and GPS systems can all be used to show farmers precisely how their crops are growing. Conditions in every crop row can be monitored when a farmer walks into the field—or when a satellite flies over—and recorded for analysis in a GIS. During the growing season, satellite monitoring can track crop conditions, such as the amount of pest damage or water stress, and allow farmers to attend to affected areas. The central component of a precision agriculture operation is a yield monitor, which is a sensor in a harvesting combine that receives GPS coordinates. As the combine harvests a crop such as corn, the sensor records the quality and quantity of the harvest from each section of the field. This detailed information provides indicators about the soil quality and irrigation needs of different parts of the field, and allows the farmer to apply water, fertilizer, and pesticides more accurately the following season.

Meanwhile, the growth of the Internet is allowing satellite images and GIS data to be more easily distributed. In late 1997, Microsoft Corporation teamed up with the Russian space agency Sovinformsputnik and image providers such as Aerial Images, Inc. and the U.S. Geological Survey to create TerraServer, the first website to allow people to view, download, and purchase satellite images. Some satellite operators have begun to offer catalogs of their images on the Internet.

## Whose Picture Is It?

Throughout history, people have fought for possession of pieces of the Earth's land and water. It was not until the advent of Earth observation satellites that ownership of the images of those places was seriously debated. To legitimize its satellite program, the United States argued strongly that the light reflected off the oceans or mountains, like the air, should be in the public domain—a claim now accepted by many other countries. But for the last two decades, the United States has also promoted the involvement of private U.S. companies in earth observation. Until the end of the Cold War, companies were reluctant to enter the satellite remote sensing business for fear of restrictions related to national security concerns. But in September 1999, a U.S.-based firm called Space Imaging launched the first high-resolution commercial satellite; this year, two other U.S. enterprises, Orbimage and EarthWatch, plan to launch similar instruments.

This trend raises questions about the tension between public and private interests in exploiting space. On the one hand, the widespread availability of detailed images means greater openness in human conduct. With satellite imagery, it is impossible to hide (or not find out about) such harmful or threatening activities as Chernobyl-scale nuclear accidents, or widespread forest clearing, or major troop movements. On the other hand, whether the information will be used to its full potential is up to governments and citizens.

There are obvious benefits to commercially available high-resolution images. Governments wary of revealing secrets have traditionally restricted the circulation of detailed satellite information. Now, images of the Earth that were once available to a select few intelligence agencies are accessible to anyone with a credit card. The information may be valuable to many non-military enterprises—public utilities, transportation planners, telecommunications firms, foresters, and others who already rely on up-to-date maps for routine operations.

In the world of NGOs, the impact of high-resolution imagery may be most dramatic for groups that keep an eye on arms control agreements and government military activities. "When one-meter black-and-white pictures hit the market, a well-endowed non-governmental organization will be able to have pictures better than [those] the U.S. spy satellites took in 1972 at the time of the first strategic arms accord," writes Peter Zimmerman, a remote-sensing and arms control expert, in a 1999 *Scientific American* article.

Indeed, when Ikonos released imagery of a top-secret North Korean missile base last January, the Federation of American Scientists (FAS), a U.S.-based nonprofit group, published a controversial analysis of the images contradicting U.S. military claims that the site is one of the most serious missile threats facing the United States. (The potential threat from this base is a key argument for proponents of a multi-billion dollar missile shield, the construction of which would violate the anti-ballistic missile treaty.) Noting the absence of transportation links, paved roads, propellant storage, and staff housing, the FAS report found the facility "incapable of supporting the extensive test program that would be needed to fully develop a reliable missile system."

In addition, the private entrants in the satellite remote sensing business may spur the whole industry to become more accessible. Already, the new companies are beginning to seek partnerships with government imagery providers, so that customers are able to go to one place to buy a number of different types of images. For instance,

Orbimage, which is scheduled to launch its first high resolution satellite this year, has made agreements to sell medium resolution images from the French SPOT series and radar images from the Canadian Radarsat. Such arrangements may make it easier for people to find and use imagery.

## FISHBONE FORESTS

### Rondonia, Brazil, 1986

LANDSAT IMAGES COURTESY OF THE TROPICAL RAINFOREST INFORMATION CENTER OF THE BASIC SCIENCE AND REMOTE SENSING INITIATIVE AT MICHIGAN STATE UNIVERSITY, A NASA EARTH SCIENCE INFORMATION PARTNER. WWW.BSRSI.MSU.EDU

These medium-resolution Landsat images show the legacy of the Brazilian government's Polonoreste project, which held out the offer of cheap land to bring farmers to this remote area of the Amazon. The centerpiece of this project, the Trans-Amazon Highway, can be seen following the Jiparaná River, running from the southeast to the northwest through the booming rural cities of Pimenta Bueno (center) and then Cacoal (top left). A tiny outpost in the early 1970s, Cacoal is now home to more than 50,000 people.

### Rondonia, Brazil, 1999

LANDSAT IMAGES COURTESY OF THE TROPICAL RAINFOREST INFORMATION CENTER OF THE BASIC SCIENCE AND REMOTE SENSING INITIATIVE AT MICHIGAN STATE UNIVERSITY, A NASA EARTH SCIENCE INFORMATION PARTNER. WWW.BSRSI.MSU.EDU

Highway BR 364, as it is called, has unzipped this part of Amazon, opening the vast forests to the tell-tale "fishbone pattern" of forest clearing. The roads that radiate out from the highway are flanked with once-forested land cleared for agriculture. Heavy rainfall leaches the soil of nutrients within a few seasons and crop yields quickly decline. The cycle of clearing starts anew when the degraded land is sold to cattle ranchers for pasture lands, which can be seen here as large rectangular clearings. The untouched tract of forest to the northwest is an indigenous preserve.

But private ownership of satellite data—making it available only at a price that not all beneficiaries can pay, or protecting it with copyright agreements—could also cause serious impediments to reversing ecological decline:

• One of the most important potential applications of remote sensing could be its use by non-governmental public interest groups, which provide a critical counterweight to the government and corporate sectors. NGOs that monitor arms control or humanitarian emergencies, for example, can use satellite data to pressure governments to live up to international nonproliferation agreements and foreign aid commitments. But a group that buys images from a private satellite company in order to publicize a humanitarian disaster or environmental threat could—depending on copyright laws and restrictions—find itself prohibited from posting the images on its website or distributing them to the media. While low-resolution imagery from government sources is easily shared, citizen groups and governments should quickly set a precedent for sharing information from high-resolution imagery.

• Only a few, very large agricultural operations can afford the high-tech equipment required to practice precision agriculture. Because farmers with small plots of land can simply walk into their fields for an assessment, the technology will remain most useful to larger operations. So although this tool could improve the way large-scale agriculture is practiced in the short term, it could also delay the long-term transition to more sustainable agricultural practices, which tend to require smaller-scale farms.

• Large companies looking for places to extract oil, minerals, or biological resources could purchase detailed satellite data and gain unfair advantage over cash-strapped developing nations in which the resources are located, whose governments cannot afford satellite imagery, GIS software, and technical support staff and systems needed to make and maintain such maps.

Finally, along with the question of who will own the technology, there is the related question of whether there is significant risk of its being badly misused. A few decades ago, the advent of commercially available high-resolution images would likely have been met with great alarm, as a manifestation of the "Big-Brother-is-watching-you" fear that pervaded the Cold War years. That fear may have receded, but what remains is a conundrum that has haunted every powerful new technology, from steel blades to genetic engineering. Could access to detailed images of their enemies cause belligerent nations to become more dangerous than they already are?

In any case, there's no turning back now. Many of the remote sensing satellites now scanning the Earth and scheduled for launch this year will be in orbit until long after the basic decisions affecting the planet's long-term environmental future—and perhaps the future of civilization—have been made. Ultimately, an educated global citizenry will be needed to make use of the flood of data being unleashed from both publicly and privately owned satellites. Policy analyst Ann Florini of the Carnegie Endowment writes: "With states, international organizations, and corporations all prodding one another to release ever more information, civil society can take that information, analyze and compile it, and disseminate it to networks of citizen groups and consumer organizations."

If the Internet has given the world's technological infrastructure a new nervous system, the Earth-observation satellites are giving it a new set of eyes. In precarious times, that could be useful. U.S. Vice President Al Gore, who understands the potential of remote sensing, has called for the completion of a "Digital Earth," a 1-meter resolution map of the world that would be widely accessible. According to Brian Soliday of Space Imaging, the Ikonos instrument alone might be able to assemble a cloud-free map of the world at 1-meter resolution within four to five years. That's about as long as it has taken to do the vaunted Human Genome map, and this map would be much bigger. Arguably, because it covers not only us humans but also the biological and climatic systems in which our genome evolved and must forever continue to depend, it could also be at least as valuable.

---

Molly O'Meara Sheehan is a research associate at the Worldwatch Institute.

---

From *World Watch*, March/April 2000, pp. 14–24. © 2000 by the WorldWatch Institute. Reprinted by permission.

# Article 35

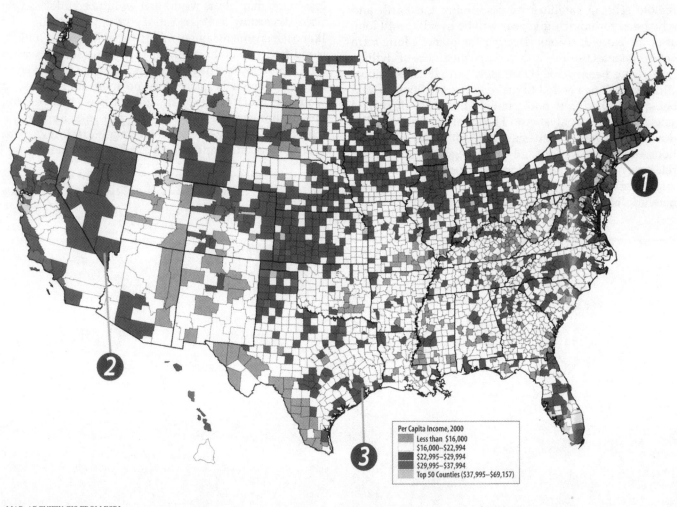

MAP: ARCVIEW GIS FROM ESRI

# Counties With Cash

*When you're hot, you're hot. When you're not—there's probably a reason.*

**BY JOHN FETTO**

As April ushers in the tax season, it seemed like a good time to investigate where America's wealth is concentrated these days.

There are numerous ways to gauge wealth, from household income to median home value. But only one measure, per-capita income, indicates the amount of money available for each person to spend. Data provided by Washington, D.C.-based market research firm Woods & Poole Economics, Inc. was used to create our map that illustrates the distribution of per-capita income by county.

The 50 counties with the highest per-capita income... range from $37,995 per person in Union County, New Jersey, to $69,157 in New York, New York (1). The national average is $28,309.

The heaviest concentrations of wealth are found in and around New York City,

## ROLLING IN IT

Counties with the highest per-capita income in 2000 and projected income for 2010, in 1999 dollars.

| COUNTY | PER CAPITA INCOME | |
|---|---|---|
| | 2000 | 2010 |
| New York, NY | $69,157 | $99,088 |
| Marin, CA | $54,608 | $85,243 |
| Pitkin, CO | $54,076 | $79,455 |
| Fairfield, CT | $53,474 | $80,234 |
| Somerset, NJ | $51,605 | $81,320 |
| Alexandria (independent city), VA | $50,752 | $78,255 |
| Westchester, NY | $50,402 | $74,206 |
| Morris, NJ | $49,640 | $76,613 |
| Bergen, NJ | $48,137 | $72,053 |
| Arlington, VA | $47,252 | $71,679 |
| Montgomery, MD | $46,911 | $69,648 |
| Teton, WY | $45,758 | $65,623 |
| San Francisco, CA | $45,694 | $68,942 |
| Montgomery, PA | $45,553 | $67,748 |
| Fairfax + Fairfax City + Falls Church, VA | $45,493 | $69,378 |
| Lake, IL | $45,218 | $69,154 |
| Nassau, NY | $45,176 | $68,399 |
| Oakland, MI | $44,767 | $69,393 |
| Nantucket, MA | $44,534 | $66,055 |
| San Mateo, CA | $43,884 | $65,378 |

Source: Woods & Poole Economics Inc.

San Francisco, and Washington, D.C. In fact, 20 of the top 50 counties are clustered around one of these three metro areas—ten in New York City, five in San Francisco, and five in D.C. As our map shows, there is plenty of wealth scattered throughout the country, mostly in major metropolitan areas.

But not all major metros have a county in the money: Las Vegas, Miami, Houston, and Salt Lake City are four that don't make the cut.

What gives? Factors such as the percent of the population employed by the service sector, percent of retirees, household size, and urban landscape all play a role, says Martin Holdrich, senior economist at Woods & Poole. The economy of Las Vegas (2), for example, is driven by the service sector, which pays less than any other industry. "Unless, there is growth in the other sectors, Las Vegas counties will not be in the top 50," says Holdrich. Other cities affected by a high percentage of service workers are Reno, Honolulu, and Orlando.

Retirement counties also tend to rank lower on the wealth scale because typically, retiree incomes are low. "Retired persons do have income, but the flow of dollars going through their households annually is much lower than it was when they were working," says Holdrich. Phoenix, Tucson, Miami, and their surrounding counties are all affected in part by the retirement factor.

Another thing to consider in places like Phoenix, Tucson, and Miami is immigration. Immigrants tend to be younger and work in the service sector—a double whammy where income is concerned.

Immigrant households are also typically larger than those of nonimmigrants. "Income per capita is biased against places with large families," says Holdrich. That's also why you won't find Salt Lake City or many other Utah counties with large numbers of Mormon households near the top of the list, simply because they have larger-than-average households. In fact, we found that nine of the 20 lowest-ranking counties for per-capita income are among the top 20 counties with the most persons per household.

County size is important to consider when examining a map of income as well. Huge geographic areas, like Houston (3), incorporate vast disparities in income: from inner city to affluent suburbs, says Holdrich. Which is why you're less likely to see a county with high per-capita income in areas of the country with geographically large counties. Western counties' per-capita income tends to skew low because of their large size. Where counties are smaller, and wealth and poverty can be easily separated by a county line, per-capita income is up.

Just as there are metro areas that don't appear to have much wealth, there are non-metro areas that have loads of it, including seasonal escapes for the wealthy. Non-metro areas that score in the top 20 for per-capita income include Pitkin County, Colorado ($54,076), home to Aspen; Teton County, Wyoming ($45,758), Jackson Hole; and the island county of Nantucket, Massachusetts ($44,534).

Several upscale Florida communities appear near the top of the list as well. They are Sarasota County, Martin County, Collier County, and Indian River County.

Remember the old adage, it takes money to make money? Well, the richest counties are doing their best not to disprove that. In fact, very little will change with respect to wealth distribution in the next ten years.

In 2010, the only changes to our list of the ten wealthiest counties will be a little place-swapping here and there. Still, nobody will be packing up and moving off the list. The average per-capita income of the top ten is projected to increase by 51 percent between now and 2010. Now *that's* a raise.

---

Reprinted from *American Demographics*, April 2000, pp. 42–43. © 2000 by Intertec Publishing Corp., Stamford, CT. All rights reserved. Reprinted by permission.

# A City of 2 Million Without a Map

SOMEWHERE IN this lakeside Central American town, there's a woman who lives beside a yellow car. But it's not her car. It's her address. If you were to write to her, this is where you would send the letter: "From where the Chinese restaurant used to be, two blocks down, half a block toward the lake, next door to the house where the yellow car is parked, Managua, Nicaragua."

Try squeezing that onto the back of a postcard. Come to that, try putting yourself in the place of the letter carriers who have to deliver such unruly epistles. How, for example, would they know where the Chinese restaurant used to be if it isn't there anymore? How would they know which way is "down," considering that "down," as employed by people in these parts, could as easily mean "up"?

How would they know which way the lake lies, when most of the time—in this topsy-turvy capital, punctured by the tall green craters of half a dozen ancient volcanoes—they cannot even see the lake? Finally, how would they know where the yellow car is parked, if its owner happens to be out for a spin?

Somehow, the people who live here have figured these things out. Granted, they've had practice. After all, most Managua street addresses take this cumbersome and inscrutable form. "We don't have a real street map," concedes Manuel Estrada Borge, vice president of the Nicaragua Chamber of Commerce, "so we have an amusing little system that no one from anywhere else can understand."

Welcome to Managua, quite possibly the only place on Earth where upward of 2 million people manage to live, work, and play—not to mention find their way around—in a city where the streets have no names.

No numbers, either. Well, that isn't quite true. A few Managua streets do indeed have conventional names. Some houses even have numbers. But no one hereabouts ever uses them. Why bother? Managuans have their own amusing little system to sort these matters out, a system that has the amusing little side-effect of driving most visitors crazy.

"For people who've just come here," says a long-time Canadian resident of the city, "there's no way on God's Earth that they'd know what you're talking about."

What Managuans are talking about, when all is said and done, is an earthquake that shattered this city three decades ago. Before that time, Managua was an urban conglomeration much like any other, at least in the sense that it had a recognizable center. It also had streets that ran east and west or north and south, and those streets not infrequently bore names. And numbers.

But then, on Dec. 23, 1972, the seismological fault lines that zigzag beneath Managua shifted and buckled, with horrific results. Upward of 20,000 people were killed in the quake, and the city was pretty much reduced to rubble. The catastrophe thoroughly disrupted the old grid pattern of Managua's streets, so the city's surviving residents were obliged to devise a new way of locating things. They started with a landmark—a certain tree, for example, or a pharmacy or a plaza or a soft-drink bottling plant—and they went from there.

Nowadays, for example, if you wished to visit the small Canadian Consulate in Managua, you would present yourself at the following address: *De Los Pipitos, dos cuadras abajo*. In English, this means: From Los Pipitos, two blocks down.

Any self-respecting inhabitant of Managua knows that "Los Pipitos" refers to a child-welfare agency whose headquarters are located a little south of the Tiscapa Lagoon. Managuans also know that *abajo*, in this context, does not mean "down" in a topographical sense. It means "west," because the sun goes down in the west. (By the same token, in Managua street talk, "*arriba*," or "up," means "east." *Al lago*, which literally means "to the lake," is how Managuans say "to the north." For some inexplicable reason, when they want to say "to the south," Managuans say "*al sur*," which means "to the south.")

Just to make a complicated process even more perplexing, Managuans, who normally use the metric system, will often give directions by employing an ancient Spanish unit of measurement called the *vara*. They will say, "From the little tree, two blocks to the south, 50 *varas* to the east." Visitors will therefore need to know how long a *vara* is (0.847 meters). They will also need to know that the "little tree" is no longer little. It is actually quite tall.

A few years ago, the Nicaraguan postal agency considered scrapping the jerry-rigged system of street addresses. But nothing came of the project. Besides, the scheme actually does seem to work. Nedelka Aguilar, for example, has learned that you merely have to have a little faith. Born in Nicaragua, she left as a young girl and spent most of her youth in southern Ontario. Now she lives in Managua once more. Shortly after her return four years ago, she arranged to visit a woman who dwelled at that outlandish address—"From where the Chinese restaurant used to be, two blocks down, half a block toward the lake, next door to the house where the yellow car is parked." By this time, Aguilar spoke the Managua dialect of street addresses well enough to take in the gist of this information. But what about that yellow car?

"I said to the woman, 'How will I find you if the yellow car isn't there?'" Aguilar smiles and shakes her head at the memory. "The woman laughed. She said, 'The yellow car is always there.'"

—OAKLAND ROSS, *The Toronto Star* (liberal), Toronto, Canada, April 21, 2002

# China Journal I

Henry Petroski

The Yangtze is the third longest river in the world. Originating from 5,800-meter-high Mount Tanggula on the Tibet Plateau, the Yangtze follows a sinuous west-to-east route for more than 6,000 kilometers before emptying into the East China Sea at Shanghai. The river has 3,600 tributaries and drains almost 2 million square kilometers, almost 19 percent of China's land area.

During flood season, the water level in the river can rise as much as 15 meters, affecting 15 million people and threatening 1.5 million hectares of cultivated land. Historic floods have been devastating. The flood of 1870 is still talked about along the middle reaches of the river, and the flood of 1954 inundated 3 million hectares of arable land and claimed 30,000 lives. Altogether in the 20th century, as many as half a million people may have died in the Yangtze's flood waters.

The Yangtze also has some of the most beautiful scenery in the world in the region known as the Three Gorges, with spectacular cliffs and steep mountains rising as high as 1,500 meters. Interspersed with gently rolling hills and long sloping riverbanks, the gorges have been compared in majesty to the Grand Canyon. Cruising the river through the Three Gorges is considered a classic travel experience, as each bend in the river reveals a new perspective on the marvels that geological change has wrought.

## Making Choices

Balancing the desire to preserve the river in all its natural glory against that to tame it to control flooding, generate power and provide more reliable shipping conditions presents a classical dilemma involving engineering and society. When nationalist leader Sun Yat-Sen proposed a Three Gorges Dam in 1919, the ecological costs were overshadowed by the economic benefits for China. In the mid-1940s, a preliminary survey, along with planning and design efforts, was carried out by the U.S. Bureau of Reclamation under the direction of John Lucian Savage, designer of the Hoover and Grand Coulee dams. In his exploratory role, Savage became the first non-Chinese engineer to visit the Three Gorges with the thought of locating an appropriate dam site. Savage's work is the likely inspiration for John Hersey's novel, *A Single Pebble*, whose opening sentence is, "I became an engineer." In the story, the unnamed engineer travels up the Yangtze in a junk pulled by trackers in the ancient and, once, the only way to make the river journey upstream.

Chairman Mao Zedong was a staunch supporter of a Three Gorges Dam, which he felt would provide a forceful symbol of China's self-sufficiency and ability to develop its resources without Western aid. As early as 1953, Mao expressed his preference for a single large dam rather than a series of smaller ones, and he suggested that he would resign the chairmanship of the Communist Party in China to assist in the design of the project. Mao's poem about being at ease swimming across the turbulent Yangtze reflects on how all things change, like the swift river and the gorges through which it flows. He knows that Goddess, a prominent peak in the middle reaches of the Three Gorges, will marvel at the accomplishment of a dam.

In 1992, the Chinese government announced officially its determination to tame the Yangtze with what would be the world's largest hydroelectric dam, ultimately to be fitted with 26 generators rated at 700 megawatts each. The total of 18,200 megawatts is equivalent to the output of approximately 15 of the largest nuclear-power plants operating in the world today. Since one of China's most pressing environmental problems is pollution from burning fossil fuels, the prospect of a clean hydroelectric dam generating about 10 percent of the country's power is very appealing to the Chinese leadership. In addition to providing flood control and power generation, the dam will open up the Yangtze as far upriver as Chongqing to 10,000-metric-ton ships, providing an opportunity for China to develop container ports almost 2,000 kilometers inland. This purpose goes hand in hand with China's plan to soon become a full partner in world trade operations.

For all its practical benefits to China, the Three Gorges Dam project has been opposed by numerous groups both domestic and international. Especially vocal have been human-rights advocates, environmentalists and historians. Among the most persistent opposing voices has been Dai Qing, who was educated as an engineer but became disillusioned during the Cultural Revolution and finally turned to investigative journalism. Her 1989 book, *Yangtze! Yangtze!*, was highly critical of the idea of a Three

BOB SACHA/NATIONAL GEOGRAPHIC SOCIETY

Figure 1. Three Gorges Dam will contain the world's largest hydroelectric plant when completed, with generators expected to provide 10 percent of China's electricity. The impoundment will prevent flooding, yet the project is not without detractors on environmental grounds.

Gorges Dam and led to her temporary imprisonment. That book and her subsequent one, *The River Dragon Has Come!*, published in 1998, state the fundamental case against the project but seem to have had little, if any, effect on the progress of the dam.

When complete, Three Gorges Dam, which will stretch about 2 kilometers across the Yangtze at Sandouping, will be 185 meters high and will create a reservoir 600 kilometers long, reaching all the way to Chongqing. The filling of the reservoir will displace on the order of a million people, inundate almost 50,000 hectares of prime farmland, submerge archaeological treasures and forever alter the appearance of the Three Gorges.

The project has been described as "perhaps the largest, most expensive, and perhaps most hazardous hydroelectric project ever attempted." Vocal protesters and international politics have no doubt influenced the World Bank's refusal to finance the project. Bowing to pressure from environmental groups, the Clinton Administration opposed competitive financing through the Export-Import Bank, effectively discouraging American companies from participating. The Chinese government has nonetheless been resolute.

As in all large dam projects, choosing the site was of fundamental importance. Of 15 locations seriously con-

sidered, the final choice was made on the basis of geological foundation conditions and accessibility to construction equipment and materials. The chosen dam site is 28 kilometers upriver from Yichang, where the Yangtze runs wide between gently sloping banks that provide staging areas for the construction project. The geology in the area is ideal, in that it is underlain with solid granite for some 10 kilometers surrounding the dam site, providing a stable construction base.

The project was planned to be completed in three stages. Phase I, stretching from 1993 to 1997, consisted of building coffer dams within which the river bottom could be excavated and the foundation of the dam begun. A temporary ship lock was also constructed during this phase in order to allow shipping to pass the construction site throughout the project. Phase II, extending from 1998 to 2003, involves the construction of a good part of the dam proper. Concrete is being poured 24 hours a day to complete the spillway of the dam, the intake portions of the dam designed for the power generation and the initial stage of the power plant itself. At the end of Phase II, the reservoir will be filled and the dam will begin to generate power, which will produce revenue to fund the final phase of the project. Phase III, stretching from 2004 to 2009, will involve the completion of the dam across the river, including additional powerhouse units. According to the China Yangtze Three Gorges Project Development Corporation (CTGPC), the government-authorized entity created to own the project, construction is on schedule for the completion of the present phase in 2003.

[PHOTOGRAPH COURTESY OF ALAN R. MILLER, NEW MEXICO TECH.]

Figure 2. Yangtze River, although offering spectacular scenery, may have claimed a half-million lives during the 1900s in devastating floods.

## A First-hand Perspective

I was invited recently to lead a civil engineering delegation from the United States to visit the Three Gorges Dam construction site and talk with Chinese engineers about the project. The delegation would have the opportunity to see first-hand the scale and technical nature of this gargantuan engineering project and to visit the areas along the Yangtze that will be permanently altered by the creation of the reservoir. The delegation would see the towns that will be submerged and from which so many people will be displaced. Our group would also have an opportunity to experience China in this time of rapid emergence as a full player on the world economic scene.

[PHOTOGRAPH COURTESY OF ALAN R. MILLER.]

Figure 3. Lock at Gezhouba Dam, completed in 1981, provides shipping access above Yichang.

Our delegation consisted of 40 engineers. They were mostly civil engineers, some with extensive experience in dam construction and power generation, but there were also a number of electrical and mechanical engineers, among others, as well as a geologist, reflecting the inherent interdisciplinary nature of large engineering projects. About 25 guests traveled with the delegation. Most were spouses of the delegates, but there were also a half-dozen professional sociologists who were interested in the relocation problems associated with the Three Gorges Project.

The conventional wisdom in the United States about the project is that it is technologically risky, environmentally unsound, sociologically devastating and economically unwise for China at this time. Thus, the overall view of the project that is held by most Americans is that it is ill-advised at best and a disaster in the making at worst. Different members of the delegation took different preconceptions with them to China, and some brought home altered perceptions.

The delegation assembled in Los Angeles in mid-November for a predeparture briefing. The 15-hour flight to Hong Kong was pleasant and uneventful, with most delegates getting a good night's sleep. A five-hour layover in Hong Kong gave some of us an opportunity to ride the new Airport Express into the city that is now one part of but still apart from China.

The world of difference between Hong Kong and the interior of China was emphasized by the fact that our flight from Hong Kong to Wuhan was classified as an international one. Wuhan created in 1950 out of the merger of three cities separated only by the Yangtze and Han rivers, a region rich in Chinese history. The consolidated city is located midway on a north-south line between Beijing and Canton and an east-west line between Shanghai and Chongqing. In contrast to the bright stainless-steel expansiveness of the new airport at Hong Kong, the dated one at Wuhan was tiny, dingy and drab. Wuhan, one of China's industrial cities, has a population of about 4 million. It was our point of entry into China's interior because it is on the Yangtze River and conveniently connected by a modern toll highway to Yichang, headquarters of the CTGPC and only 30 kilometers downstream from the construction site.

## A Land of Contrasts

Although a straight shot on a new superhighway, the bus ride to Yichang took about four-and-a-half hours, so we spent the night in Wuhan. This important river port evokes 19th-century technology as much as Hong Kong does 21st. That is not to say that Wuhan is without its buses and cars, the increase of which throughout China is creating enormous traffic and pollution problems, or its better hotels complete with CNN on television. Rather, the enormous reliance of the people on muscle power harks back to a previous century. Myriad bicycles have their dedicated lanes, and they carry goods that in America would be found in pickup trucks and delivery vans. It is common to see bicyclists struggling up the slightest incline under a load of reinforcing steel or plastic pipe that extends several meters ahead of and behind the bike. Smaller loads, though not always smaller by much, are carried in bundles hung from the ends of bamboo poles balanced on the shoulders of bearers, who trek along among the bicycles.

The few hours we had in Wuhan were spent riding a bus to and from the Yellow Crane Tower, a restored ancient hilltop pagoda, perhaps the city's most famous tourist attraction. On the ride to and from the Yellow Crane Tower, which was apparently named after a mythic bird—there are no cranes colored yellow in China, our tour guide informed us—we crossed and recrossed a road-and-railroad bridge spanning the Yangtze. Through the haze with which we would become quite familiar, we could glimpse a newer cable-stayed crossing in the distance, one of the many newly constructed modern bridges we would encounter as we crisscrossed the country.

The next day, the bus ride to Yichang was through primitive farmland. In sharp contrast to the new cars and buses traveling the highway on which we rode, the farms showed no sign of mechanization. Those farmers who did not walk behind a water buffalo worked bent over in their fields. During harvest, farmers stay in the tents and tiny shacks that abound beside the fields. Clusters of small run-down farmhouses marked simple villages, with virtually all buildings oriented with their entrance facing south, in the tradition of much-grander Chinese houses. Although in the days of collective farming there was some machinery, our guide told us, that has not survived into the present era when smaller plots of land are worked by individual farmers. That is not to say that they own the land, however, for we were also told that the state owns all the land in China.

Since it was late fall, there were few crops in evidence. The clearly irregular fields followed the contours of irrigation ditches, and some worked-out fields were excavated deeper than their neighbors to allow for fish farming and lotus cultivation. Hubei province's great Jingbei Plain west of Wuhan is extraordinarily flat, and after several hours' riding we had become so accustomed to the flatness of the land that the sudden appearance of hills with terraced fields worked many of us out of a torpor.

The presence of hills soon yielded to mountains, which signaled our land approach to Xiling Gorge, the most downriver of the Three Gorges and thus often referred to as the third gorge. The twists and turns of the highway through the mountains caused the city of Yichang to appear as suddenly as new stretches of river would when we would sail through the gorges a few days hence. The most prominent building to first come into view in the city was the modern China Telecom Building, the city's tallest. It and the headquarters building of the CTGPC dominate the skyline of hilly Yichang, which has come to be known as "dam city" and "electricity city," in recognition of the many hydroelectric power plants in the area.

Our first technical visit was to Three Gorges University, a consortium of several institutions in the area that is the national center for teaching hydraulic and electrical engineering. At our meeting, we received an academic background briefing on the Three Gorges Dam project, which prepared us for our visit to the site the next day.

Before leaving the Yichang area, we visited the Gezhouba Dam, completed in 1981 and a prototype of sorts for the Three Gorges Dam. Located about 30 kilometers downstream from Sandouping, this hydroelectric dam has all the features of the larger structure. In particular, it has sediment-control gates, which by design when opened scour out accumulated sand and silt from behind the dam and distribute it in a controlled manner downstream. The issue of accumulating material behind the Three Gorges Dam is an objection raised by opponents, who argue that in time the reservoir will fill with silt and become unnavigable. Impounding silt behind the dam would also deprive the agricultural land downstream of natural replenishment. The reportedly successful operation of Gezhouba Dam, however, appears to have allayed immediate concerns about silt, at least among the engineers.

[PHOTOGRAPH COURTESY OF CATHERINE PETROSKI.]

**Figure 4. Xiling Bridge crosses the Yangtze just downstream from the Three Gorges Dam site.**

It was dark when we left Gezhouba Dam and boarded the buses for the ride to Sandouping, the base town for the Three Gorges Dam project. Since it was dark, we could not see the terrain through which we were riding, but the grades of the hills and the rock slopes visible in the bus's headlights made it clear that we were traveling through rough territory. The road was new, constructed in the past few years to serve the project site, and it led through a heavily guarded check station. At one point we passed through a tunnel estimated to be about two kilometers long, suggesting that the mountains above us were too high or steep to put a road over. After maybe 45 minutes of riding without seeing any significant number

Figure 5. Ship locks for Three Gorges Dam have been carved out of solid granite.

of lights, we came upon Sandouping, a small town by Chinese standards but a bustling center beside the Yangtze. Our hotel was a relatively new high rise. From its windows we could see the outline of lights on the cables of a major suspension bridge, suggesting that we were beside the river.

In the daylight, we would learn that the graceful structure was the Xiling Bridge. On the way to the dam-construction site, we passed numerous warehouses and dormitories for workers. The latter, we were told, will be converted into tourist accommodations, since the recreational lake to be formed behind Three Gorges Dam is expected to bring large numbers of vacationers to this region. Dominating the route to the construction site was a large pit where granite is being crushed into pieces of aggregate for the concrete. A system of conveyor belts carries the stone over and along the road to the concrete plant.

## The Meaning of Big

The construction site is so large, extending well over a kilometer out from the river bank and into the riverbed, that it is hard to encompass it in a single view. Perhaps the dominant first impression is the countless number of tall construction cranes, literally countless because they blend into each other and disappear behind each other. Our guide joked that we were now seeing a real yellow tower crane, as opposed to the mythical source of the name of the Yellow Crane Tower that we puzzled over in Wuhan.

The first stop on the site was at the location of the locks. Twin pairs of five locks are being carved out of solid granite and lined with concrete. They will carry 10,000-metric-ton ships and barges in stages through the difference in water level behind and in front of the dam. Viewed from near their bottom, the scale of this one aspect of the Three Gorges Project is enormous. I certainly have never seen anything like it, and I imagine that it rivals even the construction of the individually larger locks of the Panama Canal. Workers at the bottom of the manmade granite box canyons looked minuscule, and it seemed impossible that these locks were blasted out of the granite in only a few years's time. Our guide told us that the Chinese calligraphy atop a nearby promontory motivated the workers to keep at the task with "first-class management, high-quality workmanship, first-rate construction."

For American engineers, one notable feature of this Chinese construction site was the freedom with which we visitors were allowed to move among the piles of construction materials and debris. Such traipsing around is unheard of at construction sites in the U.S. Only a simple railing with wide openings separated us from a 30-meter-or-so fall into one of the ship-lock excavations, but neither the Chinese nor the visitors seemed to be bothered by their proximity to the precipice.

After spending some time at the locks, we reboarded our buses and were driven over to the dam proper, which we viewed head-on from its downriver side. The scale of this part of the project was even grander than that of the locks, for it rose higher into the air and stretched over a kilometer wide before us. Under construction to our left was the spillway, with one section of it raised to the dam's final height, giving a sense of how the completed structure will loom over this part of the river. To our right was the power-plants section, which will hold 14 hydraulic turbine generator units capable of generating 9,800 megawatts of power when the dam is completed. (The remaining 8,400-megawatt capacity of the dam's power plant will not be realized until the third phase of the project is completed and all potential generating capacity is in place.)

Behind us stood the batch plant, where concrete was being mixed constantly for the 24-hour-a-day work schedule. One of the major considerations in placing concrete in such a massive structure is how to dissipate the heat of hydration that is generated in the concrete. If the concrete experiences too much thermal expansion as it sets, cracks will develop when it cools and contracts. Taking the heat away in a controlled and timely manner obviates this unwanted behavior. At the Three Gorges Dam, the thermal problem is being handled in several ways. As with Hoover Dam, cooling pipes are being imbedded into the concrete to carry away some of the heat. The amount of undesirable heat is itself being eliminated at the source by mixing and placing the concrete at the lowest temperature possible. This is accomplished through cooling the aggregate by blowing cold air over it, by using ice water in the mixing process and by ensuring that no concrete comes out of the batch plant at over seven degrees Celsius. The measures appear to be working. So far only one significant crack has appeared in the part of the dam in place, and the Chinese engineers seem confident that it has been satisfactorily repaired.

After the dam itself and the tower cranes—red, white and yellow—the next most prominent feature of the construction site is the conveyor-belt system that rises up to great heights on temporary concrete columns. The conveyor system to deliver the concrete is a crucial component of the job, for the rate at which concrete is placed largely determines if the project can be kept on schedule. The unique Rotec conveyor system is one of the rare American presences in the project (Caterpillar and General Electric being others). Unfortunately, shortly before our visit to the site, there was an accident with one of the conveyors, killing some workers. Before that accident, we were told, the safety record of the project had been excellent. At the time of our visit, the conveyor system was not operating at the desired capacity, which irritated the Chinese, and local papers were carrying stories of a law suit against the conveyor company for breach of contract.

While at the construction site of the Three Gorges Dam, it is hard not be be awed by the enormity of the project and the confidence of engineers working to hold back the legendary Yangtze, building on their experience with Gezhouba Dam and the many other flood-control and hydroelectric projects completed throughout their country in recent decades. (A recent survey by the World Commission on Dams found that 46 percent of the world's 45,000 large dams are located in China. It also reported that, although they have contributed significantly to human development, dams have been the cause of considerable social and environmental damage.)

The convincing official arguments that the Chinese put forth about the multifarious good that the Three Gorges Dam will bring to their emerging economy impress visitors from a country that is what it is today in part because its engineers also tamed great and scenic rivers like the Colorado and the Snake. The preconceived opposition to the Chinese project as being merely irresponsible and antienvironmental that some members of our delegation brought with them from America was allayed as we stood before this monument-in-progress. For the time being, at least, we saw the Three Gorges Dam through Chinese eyes on Chinese ground. The next day we would board a riverboat to cruise past the dam site, up through the storied gorges and past their Goddess Peak, reflecting on the changes the world-class engineering project is bringing to the great Yangtze.

---

Henry Petroski is A. S. Vesic Professor of Civil Engineering and a professor of history at Duke University. Address: Box 90287, Durham, NC 27708-0287

---

From *American Scientist,* May/June 2001, pp. 198-203 by Henry Petroski. © 2001 by American Scientist. Reprinted by permission.

# UNIT 5
# Population, Resources, and Socioeconomic Development

## Unit Selections

38. **Before the Next Doubling**, Jennifer D. Mitchell
39. **A Turning-Point for AIDS?** *The Economist*
40. **AIDS Scourge in Rural China Leaves a Generation of Orphans**, Elisabeth Rosenthal
41. **The Next Oil Frontier**, *Business Week*
42. **Gray Dawn: The Global Aging Crisis**, Peter G. Peterson
43. **A Rare and Precious Resource**, Houria Tazi Sedeq
44. **In Race to Tap the Euphrates, the Upper Hand Is Upstream**, Douglas Jehl
45. **The Coming Water Crisis**, Marianne Lavelle and Joshua Kurlantzick

## Key Points to Consider

- How are you personally affected by the population explosion?
- Give examples of how economic development adversely affects the environment. How can such adverse effects be prevented?
- How do you feel about the occurrence of starvation in developing world regions?
- What might it be like to be a refugee?
- In what forms is colonialism present today?
- For how long are world systems sustainable?

 **Links: www.dushkin.com/online/**
These sites are annotated in the World Wide Web pages.

**African Studies WWW (U.Penn)**
http://www.sas.upenn.edu/African_Studies/AS.html

**Geography and Socioeconomic Development**
http://www.ksg.harvard.edu/cid/andes/Documents/Background%20Papers/Geography&Socioeconomic%20Development.pdf

**Human Rights and Humanitarian Assistance**
http://www.etown.edu/vl/humrts.html

**Hypertext and Ethnography**
http://www.umanitoba.ca/faculties/arts/anthropology/tutor/aaa_presentation.new.html

**Research and Reference (Library of Congress)**
http://lcweb.loc.gov/rr/

**Space Research Institute**
http://arc.iki.rssi.ru/eng/

**World Population and Demographic Data**
http://geography.about.com/cs/worldpopulation/

The final unit of this anthology includes discussions of several important problems facing humankind. Geographers are keenly aware of regional and global difficulties. It is hoped that their work with researchers from other academic disciplines and representatives of business and government will help bring about solutions to these serious problems.

Probably no single phenomenon has received as much attention in recent years as the so-called population explosion. World population continues to increase at unacceptably high rates. The problem is most severe in the less developed countries where, in some cases, populations are doubling in less than 20 years.

The human population of the world passed the 6 billion mark in 1999. It is anticipated that population increase will continue well into the twenty-first century, despite a slowing in the rate of population growth globally since the 1960s. The first article in this section deals with issues of population growth. The next two articles deal with the devastation of AIDS in Africa and China. They are followed by a focus on the rise of U.S. oil interests in the Central Asian Region. "Gray Dawn: The Global Aging Crisis" reviews the global increase in the number of elderly. The availability of fresh water is considered in "A Rare and Precious Resource," followed by two more articles on the issue of fresh water quality and quantity.

# Before the Next Doubling

*Nearly 6 billion people now inhabit the Earth—almost twice as many as in 1960. At some point over the course of the next century, the world's population could double again. But we don't have anything like a century to prevent that next doubling; we probably have less than a decade.*

by Jennifer D. Mitchell

In 1971, when Bangladesh won independence from Pakistan, the two countries embarked on a kind of unintentional demographic experiment. The separation had produced two very similar populations: both contained some 66 million people and both were growing at about 3 percent a year. Both were overwhelmingly poor, rural, and Muslim. Both populations had similar views on the "ideal" family size (around four children); in both cases, that ideal was roughly two children smaller than the actual average family. And in keeping with the Islamic tendency to encourage large families, both generally disapproved of family planning.

But there was one critical difference. The Pakistani government, distracted by leadership crises and committed to conventional ideals of economic growth, wavered over the importance of family planning. The Bangladeshi government did not: as early as 1976, population growth had been declared the country's number one problem, and a national network was established to educate people about family planning and supply them with contraceptives. As a result, the proportion of couples using contraceptives rose from around 6 percent in 1976 to about 50 percent today, and fertility rates have dropped from well over six children per woman to just over three. Today, some 120 million people live in Bangladesh, while 140 million live in Pakistan—a difference of 20 million.

Bangladesh still faces enormous population pressures—by 2050, its population will probably have increased by nearly 100 million. But even so, that 20 million person "savings" is a colossal achievement, especially given local conditions. Bangladeshi officials had no hope of producing the classic "demographic transition," in which improvements in education, health care, and general living standards tend to push down the birth rate. Bangladesh was—and is—one of the poorest and most densely populated countries on earth. About the size of England and Wales, Bangladesh has twice as many people. Its per capita GDP is barely over $200. It has one doctor for every 12,500 people and nearly three-quarters of its adult population are illiterate. The national diet would be considered inadequate in any industrial country, and even at current levels of population growth, Bangladesh may be forced to rely increasingly on food imports.

All of these burdens would be substantially heavier than they already are, had it not been for the family planning program. To appreciate the Bangladeshi achievement, it's only necessary to look at Pakistan: those "additional" 20 million Pakistanis require at least 2.5 million more houses, about 4 million more tons of grain each year, millions more jobs, and significantly greater investments in health care—or a significantly greater burden of disease. Of the two nations, Pakistan has the more robust economy—its per capita GDP is twice that of Bangladesh. But the Pakistani economy is still primarily agricultural, and the size of the average farm is shrinking, in part because of the expanding population. Already, one fourth of the country's farms are under 1 hectare, the standard minimum size for economic viability, and Pakistan is looking increasingly towards the international grain markets to feed its people. In 1997, despite its third consecutive year of near-record harvests, Pakistan attempted to double its wheat imports but was not able to do so because it had exhausted its line of credit.

And Pakistan's extra burden will be compounded in the next generation. Pakistani women still bear an average of well over five children, so at the current birth rate, the 10 million or so extra couples would produce at least 50 million children. And these in turn could bear nearly 125 million children of their own. At its current fertility rate, Pakistan's population will double in just 24 years—that's more than twice as fast as Bangladesh's population is growing. H. E. Syeda Abida Hussain, Pakistan's Minister of Population Welfare, explains the problem bluntly: "If we achieve success in lowering our population growth substantially, Pakistan has a future. But if, God forbid, we should not—no future."

## The Three Dimensions of the Population Explosion

Some version of Mrs. Abida's statement might apply to the world as a whole. About 5.9 billion people currently inhabit the Earth. By the middle of the next century, according to U.N. projections, the population will probably reach 9.4 billion—and all of the net increase is likely to occur in the developing world. (The total population of the industrial countries is expected to decline slightly over the next 50 years.) Nearly 60 percent of the increase will occur in Asia, which will grow from 3.4 billion people in 1995 to more than 5.4 billion in 2050. China's population will swell from 1.2 billion to 1.5 billion, while India's is projected to soar from 930 million to 1.53 billion. In the Middle East and North Africa, the population will probably more than double, and in sub-Saharan Africa, it will triple. By 2050, Nigeria alone is expected to have 339 million people—more than the entire continent of Africa had 35 years ago.

Despite the different demographic projections, no country will be immune to the effects of population growth. Of course, the countries with the highest growth rates are likely to feel the greatest immediate burdens—on their educational and public health systems, for instance, and on their forests, soils, and water as the struggle to grow more food intensifies. Already some 100 countries must rely on grain imports to some degree, and 1.3 billion of the world's people are living on the equivalent of $1 a day or less.

But the effects will ripple out from these "front-line" countries to encompass the world as a whole. Take the water predicament in the Middle East as an example. According to Tony Allan, a water expert at the University of London, the Middle East "ran out of water" in 1972, when its population stood at 122 million. At that point, Allan argues, the region had begun to draw more water out of its aquifers and rivers than the rains were replenishing. Yet today, the region's population is twice what it was in 1972 and still growing. To some degree, water management now determines political destiny. In Egypt, for example, President Hosni Mubarak has announced a $2 billion diversion project designed to pump water from the Nile River into an area that is now desert. The project—Mubarak calls it a "necessity imposed by population"—is designed to resettle some 3 million people outside the Nile flood plain, which is home to more than 90 percent of the country's population.

Elsewhere in the region, water demands are exacerbating international tensions; Jordan, Israel, and Syria, for instance, engage in uneasy competition for the waters of the Jordan River basin. Jordan's King Hussein once said that water was the only issue that could lead him to declare war on Israel. Of course, the United States and the western European countries are deeply involved in the region's antagonisms and have invested heavily in its fragile states. The western nations have no realistic hope of escaping involvement in future conflicts.

Yet the future need not be so grim. The experiences of countries like Bangladesh suggest that it is possible to build population policies that are a match for the threat. The first step is to understand the causes of population growth. John Bongaarts, vice president of the Population Council, a non-profit research group in New York City, has identified three basic factors. (See figure on the next page.)

*Unmet demand for family planning.* In the developing world, at least 120 million married women—and a large but undefined number of unmarried women—want more control over their pregnancies, but cannot get family planning services. This unmet demand will cause about one-third of the projected population growth in developing countries over the next 50 years, or an increase of about 1.2 billion people.

*Desire for the large families.* Another 20 percent of the projected growth over the next 50 years, or an increase of about 660 million people, will be caused by couples who may have access to family planning services, but who choose to have more than two children. (Roughly two children per family is the "replacement rate," at which a population could be expected to stabilize over the long term.)

*Population momentum.* By far the largest component of population growth is the least commonly understood. Nearly one-half of the increase projected for the next 50 years will occur simply because the next reproductive generation—the group of people currently entering puberty or younger—is so much larger than the current reproductive generation. Over the next 25 years, some 3 billion people—a number equal to the entire world population in 1960—will enter their reproductive years, but only about 1.8 billion will leave that phase of life. Assuming that the couples in this reproductive bulge begin to have children at a fairly early age, which is the global norm, the global population would still expand by 1.7 billion, even if all of those couples had only two children—the longterm replacement rate.

## Meeting the Demand

Over the past three decades, the global percentage of couples using some form of family planning has increased dramatically—from less than 10 to more than 50 percent. But due to the growing population, the absolute number of women not using family planning is greater today than it was 30 years ago. Many of these women fall into that first category above—they want the services but for one reason or another, they cannot get them.

Sometimes the obstacle is a matter of policy: many governments ban or restrict valuable methods of contraception. In Japan, for instance, regulations discourage the use of birth control pills in favor of condoms, as a public health measure against sexually transmitted diseases. A study conducted in 1989 found that some 60 countries required a husband's permission before a woman can be sterilized; several required a husband's consent for all forms of birth control.

Elsewhere, the problems may be more logistical than legal. Many developing countries lack clinics and pharmacies in rural areas. In some rural areas of sub-Saharan Africa, it takes an average of two hours to reach the nearest contraceptive provider. And often contraceptives are too expensive for most people. Sometimes the products or services are of such poor quality that they are not simply ineffective, but dangerous. A woman who has been injured by a badly made or poorly inserted IUD may well be put off by contraception entirely.

In many countries, the best methods are simply unavailable. Sterilization is often the only available nontraditional option, or the only one that has gained wide acceptance. Globally, the procedure accounts for about 40 percent of contraceptive use and in some countries the fraction is much higher: in the Dominican Republic and India, for example, it stands at 69 percent. But women don't generally resort to sterilization until well into their childbearing years, and in some countries, the procedure isn't permitted until a woman reaches a certain age or bears a certain number of children. Sterilization is therefore no substitute for effective temporary methods like condoms, the pill, or IUDs.

There are often obstacles in the home as well. Women may be prevented from seeking family planning services by disapproving husbands or in-laws. In Pakistan, for example, 43 percent of husbands object to family planning. Frequently, such objections reflect a general social disapproval inculcated by religious or other deeply-rooted cultural values. And in many places, there is a crippling burden of ignorance: women simply may not know what family planning services are available or how to obtain them.

Yet there are many proven opportunities for progress, even in conditions that would appear to offer little room for it. In Bangladesh, for instance, contraception was never explicitly illegal, but many households follow the Muslim custom of *purdah*, which largely secludes women in their communities.

### Population of Developing Countries, 1950--95, with Projected Growth to 2050

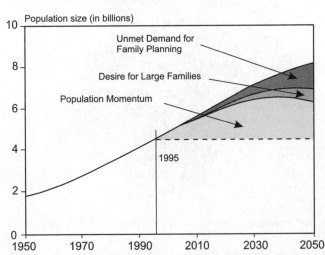

Source: U.N., World Population Prospects: The 1996 Revision (New York: October 1998); and John Bongaarts, "Population Policy Options in the Developing World," Science, 11 February 1994.

Since it's very difficult for such women to get to family planning clinics, the government brought family planning to them: some 30,000 female field workers go door-to-door to explain contraceptive methods and distribute supplies. Several other countries have adopted Bangladesh's approach. Ghana, for instance, has a similar system, in which field workers fan out from community centers. And even Pakistan now deploys 12,000 village-based workers, in an attempt to reform its family planning program, which still reaches only a quarter of the population.

Reducing the price of contraceptives can also trigger a substantial increase in use. In poor countries, contraceptives can be an extremely price-sensitive commodity even when they are very cheap. Bangladesh found this out the hard way in 1990, when officials increased contraceptive prices an average of 60 percent. (Under the increases, for example, the cheapest condoms cost about 1.25 U.S. cents per dozen.) Despite regular annual sales increases up to that point, the market slumped immediately: in 1991, condom sales fell by 29 percent and sales of the pill by 12 percent. The next year, prices were rolled back; sales rebounded and have grown steadily since then.

Additional research and development can help broaden the range of contraceptive options. Not all methods work for all couples, and the lack of a suitable method may block a substantial amount of demand. Some women, for instance, have side effects to the pill; others may not be able to use IUDs because of reproductive tract infections. The wider the range of available methods, the better the chance that a couple will use one of them.

## Planning the Small Family

Simply providing family planning services to people who already want them won't be enough to arrest the population juggernaut. In many countries, large families are still the ideal. In Senegal, Cameroon, and Niger, for example, the average woman still wants six or seven children. A few countries have tried to legislate such desires away. In India, for example, the Ministry of Health and Family Welfare is interested in promoting a policy that would bar people who have more than two children from political careers, or deny them promotion if they work within the civil service bureaucracy. And China's well-known policy allows only one child per family.

But coercion is not only morally questionable—it's likely to be ineffective because of the backlash it invites. A better starting point for policy would be to try to understand why couples want large families in the first place. In many developing countries, having lots of children still seems perfectly rational: children are a source of security in old age and may be a vital part of the family economy. Even when they're very young, children's labor can make them an asset rather than a drain on family income. And in countries with high child mortality rates, many births may be viewed as necessary to compensate for the possible deaths (of course, the cumulative statistical effect of such a reaction is to *over*-compensate).

Religious or other cultural values may contribute to the big family ideal. In Pakistan, for instance, where 97 percent of the population is Muslim, a recent survey of married women found that almost 60 percent of them believed that the number of children they have is "up to God." Preference for sons is another widespread factor in the big family psychology: many large families have come about from a perceived need to bear at least one son. In India, for instance, many Hindus believe that they need a son to perform their last rites, or their souls will not be released from the cycle of births and rebirths. Lack of a son can mean abandonment in this life too. Many husbands desert wives who do not bear sons. Or if a husband dies, a son is often the key to a woman's security: 60 percent of Indian women over 60 are widows, and widows tend to rely on their sons for support. In some castes, a widow has no other option since social mores forbid her from returning to her birth village or joining a daughter's family. Understandably, the fear of abandonment prompts many Indian women to continue having children until they have a son. It is estimated that if son preference were eliminated in India, the fertility rate would decline by 8 percent from its current level of 3.5 children per woman.

Yet even deeply rooted beliefs are subject to reinterpretation. In Iran, another Muslim society, fertility rates have dropped from seven children per family to just over four in less than three decades. The trend is due in some measure to a change of heart among the government's religious authorities, who had become increasingly concerned about the likely effects of a population that was growing at more than 3 percent per year. In 1994, at the International Conference on Population and Development (ICPD) held in Cairo, the Iranian delegation released a "National Report on Population" which argued that according to the "quotations from prophet Mohammad… and verses of [the] holy Quran, what is standing at the top priority for the Muslims' community is the social welfare of Muslims." Family planning, therefore, "not only is not prohibited but is emphasized by religion."

Promotional campaigns can also change people's assumptions and behavior, if the campaigns fit into the local social context. Perhaps the most successful effort of this kind is in Thailand, where Mechai Viravidaiya, the founder of the Thai Population and Community Development Association, started a program that uses witty songs, demonstrations, and ads to encourage the use of contraceptives. The program has helped foster widespread awareness of family planning throughout Thai society. Teachers use population-related examples in their math classes; cab drivers even pass out condoms. Such efforts have paid off: in less than three decades, contraceptive use among married couples has risen from 8 to 75 percent and population growth has slowed from over 3 percent to about 1 percent—the same rate as in the United States.

Better media coverage may be another option. In Bangladesh, a recent study found that while local journalists recognize the importance of family planning, they do not understand population issues well enough to cover them effectively and objectively. The study, a collaboration between the University Research Corporation of Bangladesh and Johns Hopkins University in the United States, recommended five ways to improve coverage: develop easy-to-use information for journalists (press releases, wall charts, research summaries), offer training and workshops, present awards for population journalism, create a forum for communication between journalists and family planning professionals, and establish a population resource center or data bank.

Often, however, the demand for large families is so tightly linked to social conditions that the conditions themselves must be viewed as part of the problem. Of course, those conditions vary greatly from one society to the next, but there are some common points of leverage:

*Reducing child mortality* helps give parents more confidence in the future of the children they already have. Among the most effective ways of reducing mortality are child immunization programs, and the promotion of "birth spacing"—lengthening the time between births. (Children born less than a year and a half apart are twice as likely to die as those born two or more years apart.)

*Improving the economic situation of women* provides them with alternatives to child-bearing. In some countries, officials could reconsider policies or customs that limit women's job opportunities or other economic rights, such as the right to inherit property. Encouraging "micro-leaders" such as Bangladesh's Grameen Bank can also be an effective tactic. In Bangladesh, the Bank has made loans to well over a million villagers—mostly impoverished women—to help them start or expand small businesses.

*Improving education* tends to delay the average age of marriage and to further the two goals just mentioned. Compulsory school attendance for children undercuts the economic incentive for larger families by reducing the opportunities for child labor. And in just about every society, higher levels of education correlate strongly with small families.

## Momentum: The Biggest Threat of All

The most important factor in population growth is the hardest to counter—and to understand. Population momentum can be easy to overlook because it isn't directly captured by the statistics that attract the most attention. The global growth rate, after all, is dropping: in the mid-1960s, it amounted to about a 2.2 percent annual increase; today the figure is 1.4 percent. The fertility rate is dropping too: in 1950, women bore an average of five children each; now they bear roughly three. But despite these continued declines, the absolute number of births won't taper off any time soon. According to U.S. Census Bureau estimates, some 130 million births will still occur annually for the next 25 years, because of the sheer number of women coming into their child-bearing years.

The effects of momentum can be seen readily in a country like Bangladesh, where more than 42 percent of the population is under 15 years old—a typical proportion for many poor countries. Some 82 percent of the population growth projected for Bangladesh over the next half century will be caused by momentum. In other words, even if from now on, every Bangladeshi couple were to have only two children, the country's population would still grow by 80 million by 2050 simply because the next reproductive generation is so enormous.

The key to reducing momentum is to delay as many births as possible. To understand why delay works, it's helpful to think of momentum as a kind of human accounting problem in which a large number of births in the near term won't be balanced by a corresponding number of deaths over the same period of time. One side of the population ledger will contain those 130 million annual births (not all of which are due to momentum, of course), while the other side will contain only about 50 million annual deaths. So to put the matter in a morbid light, the longer a substantial number of those births can be delayed, the longer the death side of the balance sheet will be when the births eventually occur. In developing countries, according to the Population Council's Bongaarts, an average 2.5-year delay in the age when a woman bears her first child would reduce population growth by over 10 percent.

One way to delay childbearing is to postpone the age of marriage. In Bangladesh, for instance, the median age of first marriage among women rose from 14.4 in 1951 to 18 in 1989, and the age at first birth followed suit. Simply raising the legal age of marriage may be a useful tactic in countries that permit marriage among the very young. Educational improvements, as already mentioned, tend to do the same thing. A survey of 23 developing countries found that the median age of marriage for women with secondary education exceeded that of women with no formal education by four years.

Another fundamental strategy for encouraging later childbirth is to help women break out of the "sterilization syndrome" by providing and promoting high-quality temporary contraceptives. Sterilization might appear to be the ideal form of contraception because it's permanent. But precisely because it is permanent, women considering sterilization tend to have their children early, and then resort to it. A family planning program that relies heavily on sterilization may therefore be working at cross purposes with itself: when offered as a primary form of contraception, sterilization tends to promote early childbirth.

## What Happened to the Cairo Pledges?

At the 1994 Cairo Conference, some 180 nations agreed on a 20-year reproductive health package to slow population growth. The agreement called for a progressive rise in annual funding over the life of the package; according to U.N. estimates, the annual price tag would come to about $17 billion by 2000 and $21.7 billion by 2015. Developing countries agreed to pay for two thirds of the program, while the developed countries were to pay for the rest. On a global scale, the package was fairly modest: the annual funding amounts to less than two weeks' worth of global military expenditures.

Today, developing country spending is largely on track with the Cairo agreement, but the developed countries are not keeping their part of the bargain. According to a recent study by the U.N. Population Fund (UNFPA), all forms of developed country assistance (direct foreign aid, loans from multilateral agencies, foundation grants, and so on) amounted to only $2 billion in 1995. That was a 24 percent increase over the previous year, but preliminary estimates indicate that support declined some 18 percent in 1996 and last year's funding levels were probably even lower than that.

The United States, the largest international donor to population programs, is not only failing to meet its Cairo commitments, but is toying with a policy that would undermine international family planning efforts as a whole. Many members of the U.S. Congress are seeking reimposition of the "Mexico City Policy" first enunciated by President Ronald Reagan at the 1984 U.N. population conference in Mexico City, and repealed by the Clinton administration in 1993. Essentially, a resurrected Mexico City Policy would extend the current U.S. ban on funding abortion services to a ban on funding any organization that:

- funds abortions directly, or
- has a partnership arrangement with an organization that funds abortions, or
- provides legal services that may facilitate abortions, or
- engages in any advocacy for the provision of abortions, or
- participates in any policy discussions about abortion, either in a domestic or international forum.

The ban would be triggered even if the relevant activities were paid for entirely with non-U.S. funds. Because of its draconian limits even on speech, the policy has been dubbed the "Global Gag Rule" by its critics, who fear that it could stifle, not just abortion services, but many family planning operations involved only incidentally with abortion. Although Mexico City proponents have not managed to enlist enough support to reinstate the policy, they have succeeded in reducing U.S. family planning aid from $547 million in 1995 to $385 million in 1997. They have also imposed an unprecedented set of restrictions that meter out the money at the rate of 8 percent of the annual budget per month—a tactic that *Washington Post* reporter Judy Mann calls "administrative strangulation."

If the current underfunding of the Cairo program persists, according to the UNFPA study, 96 million fewer couples will use modern contraceptives in 2000 than if commitments had been met. One-third to one-half of these couples will resort to less effective traditional birth control methods; the rest will not use any contraceptives at all. The result will be an additional 122 million unintended pregnancies. Over half of those pregnancies will end in births, and about 40 percent will end in abortions. (The funding shortfall is expected to produce 16 million more abortions in 2000 alone.) The unwanted pregnancies will kill about 65,000 women by 2000, and injure another 844,000.

Population funding is always vulnerable to the illusion that the falling growth rate means the problem is going away. Worldwide, the annual population increase had dropped from a high of 87 million in 1988 to 80 million today. But dismissing the problem with that statistic is like comforting someone stuck on a railway crossing with the news that an oncoming train has slowed from 87 to 80 kilometers an hour, while its weight has increased. It will now take 12.5 years instead of 11.5 years to add the next billion people to the world. But that billion will surely arrive—and so will at least one more billion. Will still more billions follow? That, in large measure, depends on what policymakers do now. Funding alone will not ensure that population stabilizes, but lack of funding will ensure that it does not.

## The Next Doubling

In the wake of the Cairo conference, most population programs are broadening their focus to include improvements in education, women's health, and women's social status among their many goals. These goals are worthy in their own right and they will ultimately be necessary for bringing population under control. But global population growth has gathered so much momentum that it could simply overwhelm a development agenda. Many countries now have little choice but to tackle their population problem in as direct a fashion as possible—even if that means temporarily ignoring other social problems. Population growth is now a global social emergency. Even as officials in both developed and developing countries open up their program agendas, it is critical that they not neglect their single most effective tool for dealing with that emergency: direct expenditures on family planning.

The funding that is likely to be the most useful will be constant, rather than sporadic. A fluctuating level of commitment, like sporadic condom use, can end up missing its objective entirely. And wherever it's feasible, funding should be designed to develop self-sufficiency—as, for instance, with UNFPA's $1 million grant to Cuba, to build a factory for making birth control pills. The factory, which has the capacity to turn out 500 million tablets annually, might eventually even provide the country with a new export product. Self-sufficiency is likely to grow increasingly important as the fertility rate continues to decline. As Tom Merrick, senior population advisor at the World Bank explains, "while the need for contraceptives will not go away when the total fertility rate reaches two—the donors will."

Even in narrow, conventional economic terms, family planning offers one of the best development investments available. A study in Bangladesh showed that for each birth prevented, the government spends $62 and saves $615 on social services expenditures—nearly a tenfold return. The study estimated that the Bangladesh program prevents 890,000 births a year, for a net annual savings of $547 million. And that figure does not include savings resulting from lessened pressure on natural resources.

Over the past 40 years, the world's population has doubled. At some point in the latter half of the next century, today's population of 5.9 billion could double again. But because of the size of the next reproductive generation, we probably have only a relatively few years to stop that next doubling. To prevent all of the damage—ecological, economic, and social—that the next doubling is likely to cause, we must begin planning the global family with the same kind of urgency that we bring to matters of trade, say, or military security. Whether we realize it or not, our attempts to stabilize population—or our failure to act—will likely have consequences that far outweigh the implications of the military or commercial crisis of the moment. Slowing population growth is one of the greatest gifts we can offer future generations.

---

Jennifer D. Mitchell is a staff researcher at the Worldwatch Institute.

---

From *World Watch* magazine, January/February 1998, pp. 20–27. © 1998 by the Worldwatch Institute, Washington, DC. Reprinted by permission.

SCIENCE AND TECHNOLOGY

# A turning-point for AIDS?

The impact of the global AIDS epidemic has been catastrophic, but many of the remedies are obvious. It is now a question of actually doing something

DURBAN

WHEN Thabo Mbeki, South Africa's president, opened the world AIDS conference in Durban on July 9th, he was widely expected to admit that he had made a mistake. Mr Mbeki has been flirting with the ideas of a small but vociferous group of scientists who, flying in the face of all the evidence, maintain that AIDS is not caused by the human immunodeficiency virus (HIV). His speech at the opening ceremony would have been the ideal opportunity for a graceful climbdown. Instead, he blustered and prevaricated, pretending that there was a real division of opinion among scientists about the matter, and arguing that the commission that he has appointed to look into this non-existent division would resolve it.

AIDS is the most political disease around. People talk a lot about AIDS "exceptionalism", and in many ways it is exceptional. For a start, it is difficult to think of another disease that would have brought the host country's head of state out of his office to open a conference, confused though his ideas may be. It is also exceptional, in modern times, in the attitudes of the healthy towards the infected. Illness usually provokes sympathy. But in many parts of the world those who have HIV are treated rather as lepers were in biblical days. Indeed, Gugu Dlamini, a community activist in KwaZulu Natal, the South African province in which Durban lies, was stoned to death by her neighbours when she revealed that she had the virus. In many parts of the world, as Kevin De Cock, of America's Centres for Disease Control (CDC), pointed out to the conference, attitudes to AIDS can be summed up in four words: silence, stigma, discrimination and denial. Mr Mbeki himself is at least guilty of denial.

Nevertheless, the fact that this, the 13th such AIDS conference, was held in Africa shows that some progress is being made. The previous conference, in Geneva in 1998, claimed to be "bridging the gap" between the treatment of the disease in the rich and poor worlds. It did no such thing. This one set as its goal to "break the silence". It may have succeeded. What needed to be shouted from the rooftops was that, contrary to some popular views, AIDS is not primarily a disease of gay western men or of intravenous drug injectors. It is a disease of ordinary people leading ordinary lives, except that most of them happen to live in a continent, Africa, that the rich countries of the world find it easy to ignore.

In some places the problem is so bad that it is hard to know where to begin. According to United Nations estimates, 25m of the 34m infected people in the world live in Africa. In absolute terms, South Africa has the most cases (4m, or about 20% of the adult population), but several of its neighbours have even worse ratios. In Botswana, for example, 36% of the adult population is now infected with HIV. Barring some currently unimaginable treatment—unimaginable in both efficacy and cost—almost all of these people will die as a result.

## The hydra-headed monster

And mere numbers are not the only issue. People talk, rhetorically, of waging war on diseases. In the case of AIDS, the rhetoric could be inverted, for the effects of the illness on human populations are similar to those of war. Most infectious diseases tend to kill infants and the old. AIDS, like war, kills those in the prime of life. Indeed, in one way it is worse than war. When armies fight, it is predominantly young men who are killed. AIDS kills young women, too.

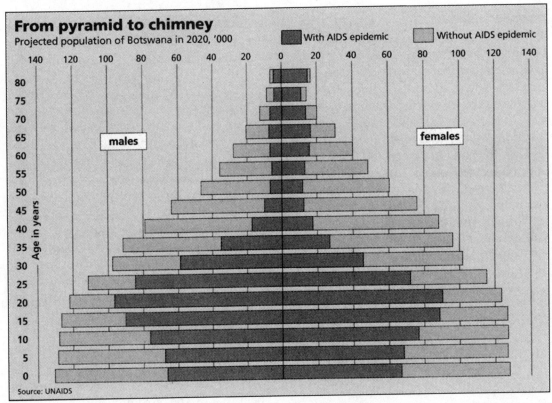

**From pyramid to chimney**
Projected population of Botswana in 2020, '000
■ With AIDS epidemic  ■ Without AIDS epidemic
Source: UNAIDS

The result is social dislocation on a grand scale. As the diagram on the next page shows, the age-distribution of Botswana's population will change from the "pyramid" that is typical of countries with rapidly growing populations, to a "chimney-shaped" graph from which the young have been lopped out. Ten years from now, according to figures released at the conference by USAID, the American government's agency for international development, the life expectancy of somebody born in Botswana will have fallen to 29. In 20 years' time, the old will outnumber the middle-aged. Nor are things much better in other countries in southern Africa. In Zimbabwe and Namibia, two of Botswana's neighbours, life expectancy in 2010 will be 33. In South Africa it will be 35.

The destruction of young adults means that AIDS is creating orphans on an unprecedented scale. There are 11.2m of them, of whom 10.7m live in Africa. On top of that, vast numbers of children are infected as they are born. These are the exception to the usual rule that infants do not get the disease. Children are rarely infected in the womb, but they may acquire the virus from their mothers' vaginal fluids when they are born, or from breast milk. More than 5m children are reckoned to have been infected in this way. Almost 4m of them are already dead.

It sounds hopeless. And yet it isn't. Two African countries, Uganda and Senegal, seem to have worked out how to cope with the disease. Their contrasting experiences serve both as a warning and as a lesson to other countries in the world, particularly those in Asia that now have low infection rates and may be feeling complacently smug about them. The warning: act early, or you will be sorry. The lesson: it is, even so, never too late to act. Senegal began its anti-AIDS programme in 1986, before the virus had got a proper grip. It has managed to keep its infection rate below 2%. Uganda began its programme in the early 1990s, when 14% of the adult population was already infected. Now that figure is down to 8% and falling. In these two countries, the epidemic seems to have been stopped in its tracks.

As Roy Anderson, a noted epidemiologist from Oxford University, pointed out to the conference, stopping an epidemic requires one thing: that the average number of people infected by somebody who already has the disease be less than one. For a sexually transmitted disease, this average has three components: the "transmissibility" of the disease, the average rate that an infected person acquires new and uninfected partners, and the average length of time for which somebody is infectious.

## Cutting off the hydra's heads

The easiest of these to tackle has been transmissibility. Surprisingly, perhaps, AIDS is not all that easily transmissible compared with other diseases. But there are three ways—one certain, one as yet a pious hope, and one the subject of some controversy—to reduce the rate of transmission between adults still further. The first is to use condoms. The second is to develop a microbicide that will kill the virus in the vagina. And the third is to treat other sexually transmitted diseases.

Both Senegal and Uganda have been strong on the use of condoms. In Senegal, for example, the annual number of condoms used rose from 800,000 in 1988 to 9m in 1997.

Nevertheless, it still takes a lot of encouragement to persuade people to use them. Partly, this is a question of discounting the future. For decades African lives have been shorter, on average, than those in the rest of the world. With AIDS, they are getting shorter still. A Botswanan who faces the prospect of death before his 30th birthday is likely to be more reckless than an American who can look forward to well over twice that lifespan; a short life might as well be a merry one.

There is also the question of who wears the condom. Until recently, there was no choice. Only male condoms were available. And women in many parts of Africa are in a weak negotiating position when it comes to insisting that a man put one on. The best way out of this is to alter the balance of power. That, in general, means more and better education, particularly for girls. This, too, has been an important component of the Senegalese and Ugandan anti-AIDS programmes. A stop-gap, though, is the female condom, a larger version of the device that fits inside the vagina, which is proving surprisingly popular among groups such as Nairobi prostitutes. But an even less intrusive—and to a man invisible—form of protection would be a vaginal microbicide that kills the virus before it can cross the vaginal wall.

Here, however, the news is bad. Much hope had been pinned on a substance called nonoxynol-9 (the spermicide used to coat condoms that are intended to prevent pregnancy rather than disease). Unfortunately, the results of a major United Nations trial announced at the conference have confirmed the suspicion that nonoxynol-9 does not work against HIV. So researchers have gone back to the drawing-board and are searching for suitable (and suitably cheap) substances among the cast-offs from anti-viral drugs used to treat AIDS in rich countries.

More equivocal is the value of treating other sexually transmitted diseases as a way of preventing the transmission of HIV. Clearly, such treatment is a good thing in its own right. But a study carried out a few years ago at Mwanza, Tanzania, suggested that it also stymies HIV. That would not be surprising, since the vaginal lesions that other venereal diseases produce should make excellent entry points for the virus. Yet a more recent study at Rakai, Uganda, suggests that other venereal diseases make no difference; the matter is now a subject of much debate.

Sex is not the only way that HIV is transmitted. Infected mothers can give it to their children. But here, too, transmissibility can be reduced dramatically.

A first way of doing this is to test pregnant women to see if they have the virus. If they do, they are unlikely to pass it on to the fetus in the womb, but they are quite likely to do so in the act of giving birth. According to figures presented to the conference by Ruth Nduati, of the University of Nairobi, up to 40% of children born to untreated infected women catch AIDS this way. But that number can be reduced drastically—to around 20%—by giving infected pregnant women a short course of an antiviral drug just before they give birth.

Until recently, the preferred drug was AZT. Many African governments balked at using this because, although it is cheap by western standards, it can stretch African health budgets to breaking point. However, recent studies carried out in Kenya and South Africa have shown that an even cheaper drug called nevirapine will do just as well. A course of this costs $4, still a fair whack for an impoverished country, but worth it both for the life of a child and for the cost-saving of not having to treat that child's subsequent illness.

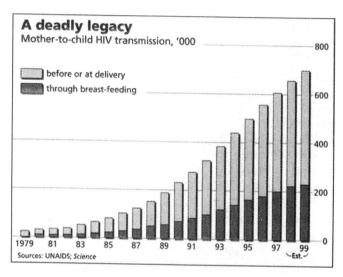

Once safely born, the child of an infected mother is still not out of the woods. This is because it can be infected via its mother's milk. Oddly, this is a more intractable problem than transmission at birth. Nobody knows (because nobody has tried to find out) if carrying on with AZT or nevirapine would keep a mother's milk virus-free. But the cost would be prohibitive anyway. The only alternative is not to breast-feed.

That may sound easy, but it is not. First, the formula milk that could substitute for breast milk costs money. Second, unless clean water is available to mix with it, the result is likely to be a diarrhoeal disease that may kill the child anyway. And third, by failing to breast-feed, a mother in many parts of Africa is in effect announcing that she has the virus, and thus exposing herself to both stigma and discrimination. Not breast-feeding is, nevertheless, an effective addition to pre-natal antiviral drugs. According to Dr Nduati, combining both methods can bring the infection rate below 8%.

The second of Dr Anderson's criteria, the rate of acquisition of new and uninfected partners, is critical to the speed with which AIDS spreads, but is also far harder to tackle. The reasons why AIDS has spread faster in some places than in others are extremely complicated. But one important factor is so-called disassortative mating.

As far as is known, all AIDS epidemics start with the spread of the disease in one or more small, high-risk groups. These groups include prostitutes and their clients, male homosexuals and injecting drug users. The rate at which an epidemic spreads to lower-risk groups depends a great deal on whether different groups mate mainly among themselves (assortative mating) or whether they mate a lot with other people (disassortative mating). The more disassortative mating there is, the faster the virus will spread.

Sub-Saharan Africa and the Caribbean (the second-worst affected part of the world) have particularly high levels of disassortative mating between young girls and older men. And in an area where AIDS is already highly prevalent, older men are a high-risk group; they are far more likely to have picked up the virus than younger ones. This helps to explain why the rate of infection is higher in young African women than it is in young African men.

Inter-generational churning may thus, according to Dr Anderson's models, go a long way towards explaining why Africa and the Caribbean have the highest levels of HIV infection in the world. And it suggests that, as with condom use, a critical part of any anti-AIDS campaign should be to give women more power. In many cases, young women are coerced or bribed into relationships with older men. This would diminish if girls were better educated—not least because they would then find it easier to earn a living.

The partial explanation for Africa's plight that disassortative mating provides should not, however, bring false comfort in other areas. Dr Anderson's models suggest that lower levels of disassortative mating cannot stop an epidemic, they merely postpone it. Those countries, such as Ukraine, where HIV is spreading rapidly through a high-risk group (in Ukraine's case, injecting drug users), need to act now, even if the necessary action, such as handing out clean needles, is politically distasteful. Countries such as India, where lower-risk groups are starting to show up in the statistics, and where the prevalence rates in some states are already above 2%, needed to act yesterday, and to aim their message more widely. The example of Senegal (and, indeed, the strongly worded, morally neutral advertising campaigns conducted in many western countries in the 1980s), shows the value of early action as surely as do Dr Anderson's models.

To tackle the third element of those models—the length of time that somebody is infectious—really requires a vaccine. Drugs can reduce it to some extent, by bringing people's viral load down to the point where they will not pass on the disease. But effective therapies are currently expensive and, despite the widespread demands at the conference for special arrangements that would lower their price in poor countries, are unlikely to become cheap enough for routine use there for some time. On top of that, if drugs are used carelessly, resistant strains of the virus can emerge, rendering the therapies useless. A study by the CDC, published to coincide with the conference, showed resistant strains in the blood of three-quarters of the participants in a United Nations AIDS drug-access programme in Uganda.

## The search for a vaccine

Vaccines are not immune to the emergence of resistant strains. But they are one-shot treatments and so are not subject to the whims of patient compliance with complex drug regimes. Non-compliance is the main cause of the emergence of resistant strains, since the erratic consumption of a particular drug allows populations of resistant viruses to evolve and build up.

In total, 21 clinical trials of vaccines are happening around the world, but only five are taking place in poor countries, and only two are so-called phase 3 trials that show whether a vaccine will work effectively in the real world. Preliminary results from these two trials, which are being conducted by an American company called VaxGen, are expected next year. They are eagerly awaited, for even a partially effective vaccine could have a significant impact on the virus's spread. A calculation by America's National Institutes of Health shows that, over the course of a decade, a 60%-effective vaccine introduced now would stop nearly twice as many infections as a 90%-effective one introduced five years hence.

Even then, there is the question of cost. This is being addressed by the International AIDS Vaccine Initiative (IAVI), a New York-based charity. IAVI, according to its boss Seth Berkley, acts like a venture-capital firm. At the moment, that capital amounts to about $100m, gathered from various governments and foundations. IAVI provides small firms with seed money to develop new products, but instead of demanding a share of the equity in return, it requires that the eventual product, if any, should be sold at a low profit margin—about 10%. If a sponsored firm breaks this arrangement, IAVI can give the relevant patents to anybody it chooses. At the moment, IAVI has four such partnerships, and it chose the conference to announce that one—a collaboration with the Universities of Oxford and Nairobi—has just received regulatory approval and will start trials in September.

None of these things alone will be enough to stop the epidemic in its tracks, but in combination they may succeed. And one last lesson from Dr Anderson's equations is not to give up just because a policy does not seem to be working. Those equations predict that applying a lot of effort to an established epidemic will have little initial effect. Then, suddenly, infection rates will drop fast. The message is: "hang in there". AIDS may be exceptional, but it is not that exceptional. Good science and sensible public policy can defeat it. There is at least a glimmer of hope.

---

Reprinted with permission from *The Economist*, July 15, 2000, pp. 77–79. © 2000 by The Economist, Ltd. Distributed by The New York Times Special Features.

# AIDS Scourge in Rural China Leaves a Generation of Orphans

By ELISABETH ROSENTHAL

DONGHU, China—Neighbors remember when young Dong Yangnan was a "xiao pangzi," or little fatty, the kind of husky, moon-cheeked child that Chinese grandmothers adore. Today, at 12, he is orphaned, stick thin and dressed in tattered clothes.

Last summer, his mother died of AIDS. His father, coughing and feverish, succumbed to the disease in May. Yangnan lives with an elderly grandfather, surviving on rice gruel and steamed buns.

"Before, I had a happy life, and my parents took good care of me," he said listlessly, his big eyes staring away to a lost past. "Now I have to look after myself and often have no money."

AIDS is creating an explosion of destitute orphans here in China's rural heartland and is driving large numbers of families into such dire poverty that they can no longer afford to feed or clothe, much less educate, their children.

At the start of last year, there were no orphans in this village in southern Henan Province. Today, because of AIDS, there are nearly 20, and hundreds more are likely to face a similar fate within a year or two. Residents estimate that 200 of the village's 600 families have one parent dead and the other ill, often too frail to work or even rise from bed. They receive little government help.

According to unpublished statistics from the United Nations Development Program, the number of families living below the official poverty line in Xincai, the county that includes Donghu, skyrocketed last year, to 270,000 from 40,000. Breadwinners fell ill, and families spent whatever they could scrape together for food and care.

> In some places, selling blood served as a source of emergency income, but in others, most adults sold blood at least occasionally, and many sold it every week.

Experts say the blow dealt by AIDS to villages like Donghu has been sharper and crueler than anywhere else in the world because of the unusual and efficient way the disease spread here.

Nearly the entire adult population of some villages was infected almost simultaneously in the 1990's as poor farmers flocked en masse to blood collection stations whose unsterile practices introduced hefty doses of H.I.V., the virus that causes AIDS, directly into their veins. Now, the victims—including many married couples—are falling ill and dying almost in unison.

In other countries suffering epidemics, grandparents or aunts and uncles have helped the sick or taken in children. But here those relatives are often themselves overwhelmed by AIDS. Also, because China's family planning policies have limited families to one or two children, there is rarely an older sibling to serve as a surrogate parent.

Ren Genqing, 16, dropped out of school three years ago because the money that would have gone for his school fees was needed to buy medicine for his parents. His father died of AIDS in 2000, his mother in 2001. One uncle has died of AIDS, and another is sick. He alone is responsible for his 12-year-old brother.

"I'm growing up, but my brother is still young," he said, a slightly cocky teenager, old before his time. "Before, the children here used to play soccer and other games, but you rarely see that these days. Lots of people are dying, and nobody's in the mood for that sort of thing."

Some Chinese experts estimate that selling blood was common in dozens of Henan province's counties before it was banned in the mid-90's, leaving at least a million people infected with H.I.V. In some places, selling blood served as a source of emergency income—fast cash to fix a roof or pay off a debt—but in others, like Donghu, most adults sold blood at least occasionally, and many sold it every week.

Like many of the most severely affected villages, Donghu was near a blood collection station, one with government ties. Commercials on local television assured villagers that selling their blood was safe.

Villagers here estimate that more than half of adults in Donghu were infected with H.I.V. in the early 1990's. A decade later, the death rate is gathering steam, with several people dying each week. The effects are largely hidden since local officials monitor access to the village and have warned residents not to speak with reporters.

> **There is really nowhere most families can turn for help. ...Their children leave school and go hungry.**

"The situation is worsening very rapidly because, once a spouse dies, the burden on the remaining one escalates and, of course, they are all infected too," one villager said.

Extreme poverty has quickly and predictably followed, as able-bodied adults can no longer work and families sell their possessions to pay for basic needs. They borrow to buy medicine for suffering loved ones, but the simple remedies they can afford are ineffective against AIDS.

Compounding the financial woes, grain, fruit and vegetables grown in these villages are almost impossible to sell in nearby cities, whose residents are afraid of contagion.

"It really brings you to tears," said a medical worker who has visited villages in the province. "You see these pretty decent houses, built with the money from selling blood, but inside there is nothing. They've sold the farm tools, the animals, even the furniture. People who are dying are lying on the floor."

For families like Ren Dahua's, it has been a vicious cycle: poverty begat AIDS, but AIDS has begotten previously unimaginable poverty.

Mr. Ren started selling blood to patch his mud and brick hut, to keep his children dry when it rained. He also used the money to repay debts incurred from the purchase of an ox, fertilizer and wheat seed.

When the blood stations opened in 1992, he and his wife rushed to sell their blood, for about $5 a bag. He regarded it as an opportunity and sold blood more than 30 times.

When two more blood stations opened nearby—one affiliated with the local Red Cross and another run out of a hospital less than 100 yards from his front door—he sometimes visited daily.

At the time, blood from several farmers was pooled and centrifuged to skim off the plasma, which the blood stations sold to companies to make medicines. The remaining red cells were pooled and transfused back into the sellers—providing a gruesomely efficient method for transmitting blood-borne diseases, including hepatitis and AIDS.

By 1993, both Mr. Ren and his wife, Diao Yuhuan, were disqualified from selling blood because they had obvious symptoms of hepatitis C: jaundice, swollen waists and almost constant nausea. They did not know that they had also contracted H.I.V., which often takes years to show symptoms.

Last year, Ms. Diao fell ill with tuberculosis, an infection that is often severe in people who have H.I.V. Selling his possessions, Mr. Ren scraped together 3,500 yuan, which covered a brief—but useless—hospital stay in Beijing. His wife died at home in January.

"Because I spent so much money when my wife was ill, my children cannot go to school," said Mr. Ren, who also has H.I.V. "My son passed the high school entrance exam, but there's no money for him to go."

In some families, like that of Wei Zhanjun, two generations of adults are dead or dying, leaving a single child carrying an unimaginable burden. Mr. Wei, whose wife died of AIDS in 2000, is so short of breath he can barely walk. His body is covered with painful sores. His parents, in their 50's, are bedridden with similar symptoms. Only his 8-year-old son, Wei Zhicheng, is healthy.

"He is a good boy, but ever since my wife fell ill, there has been no money in this home and not enough food," he wrote in a letter describing his plight. "Now, nobody farms our family's land, and we have heavy debts that we cannot repay." Money donated by neighbors to pay his son's school fees was quickly diverted to buy painkillers.

There is really nowhere most families can turn for help. Most people die in horrible pain with little care. Their children leave school and go hungry. Although a few villages have been given simple medicine and a bit of financial aid, some by private groups and some by the government, overwhelmed health officials have been slow to react.

> **Many poor farmers with AIDS have shifted their focus from securing treatment for themselves to ensuring a future for their children.**

In some villages, dozens of children have dropped out of school because their families can no longer afford the fees, and proposals to offer such children discounts have proved ineffective. Some children from homes where a family member has H.I.V. say they have been barred from school. Others say the discounts are often so small, about 20 percent, that school remains unaffordable.

Wang Beibei, 10, a star pupil from Suixian, a county in northern Henan, was expelled from third grade last year after school officials discovered that her father had died of AIDS.

"They were afraid to let me in, and my friends stopped playing with me," she said by phone, from the home of a sympathetic neighbor. About a third of the families in her village had sold blood—fewer than in Donghu—in large part because the village was farther from blood stations.

In June, Beibei's mother died of AIDS. School is out of the question. There is no one to work the family's land, and she and her brother struggle just to look out for each other. "My brother cooks for me, and we eat noodles," she said. "We have no money for eggs or meat."

In Donghu, the school still admits such children if they can pay but offers no significant tuition breaks.

Likewise, though government plans have called for families unable to farm because of AIDS to be exempt from grain taxes, families here and elsewhere say they are still required to pay in full. "The government doesn't do anything for me, and likewise it didn't do anything for my family," said Gao Li, 14, an orphan from Donghu, with cropped hair and a quiet, matter-of-fact voice.

"I'm responsible for my brother, who is 10," she said. "Nobody among my relatives can help. My dad had brothers but one is dead, and the others are sick, too. My biggest difficulty is, I have no future."

Indeed with so much death and so little reason to hope, many poor farmers with AIDS have shifted their focus from securing treatment for themselves to ensuring a future for their children.

Since late last year, Xie Yan, who is in her late 30's and is H.I.V. positive, has had an obsession: She wants to find someone to adopt her 4-year-old son, who is not infected, as well as someone to support her two daughters, 13 and 9. Her husband died of AIDS last year, and last winter she watched her best friend bleed to death on a hospital's doorstep while the friend's 4-year-old watched in terror.

"I try not to think about myself since I know I won't be cured," she said. "But at night I can't sleep—I have nightmares and wild thoughts—worrying about what will happen to the kids."

From the *New York Times.com,* August 25, 2002. © 2002 by The New York Times Company. Reprinted by permission.

# Article 41

**CENTRAL ASIA**

# THE NEXT OIL FRONTIER

## America carves out a sphere of influence on Russia's borders

It's Happy Hour at Fisherman's Wharf, an expatriate hangout in Baku, a port on the Caspian Sea in the former Soviet republic of Azerbaijan. The place is just around the corner from the town's only McDonald's, and on a Friday in April, a gaggle of Brits, Americans, and Aussies are gathered on bar stools to munch peanuts, quaff beer, and shoot the bull. Talking about Web access in this authoritarian, Muslim country, one guy, looking as if he had just returned from a long stint on an offshore oil rig, says to his buddy: "Yeah, but can you get hustler.com?"

The rugged oil worker is a type Americans can readily identify. Most Americans, though, couldn't find Azerbaijan on the map. And they probably wouldn't be able to find—or spell—Kyrgyzstan, Uzbekistan, Kazakhstan, or Tajikistan. But American soldiers, oilmen, and diplomats are rapidly getting to know this remote corner of the world, the old underbelly of the Soviet Union. The game the Americans are playing has some of the highest stakes going. What they are attempting is nothing less than the biggest carve-out of a new U.S. sphere of influence since the U.S. became engaged in the Mideast 50 years ago. The result could be a commitment of decades that exposes America to the threat of countless wars and dangers. But this huge venture—call it an Accidental Empire—could also stabilize the fault line between the West and the Muslim world and reap fabulous energy wealth for the companies rich enough and determined enough to get it.

The buildup has been breathtakingly fast. Consider:

- A year ago, not a single U.S. soldier was in the region. Today, roughly 4,000 servicemen and women are building bases, assisting the Afghan war, and training anti-insurgency troops along a rim of peril stretching 2,000 miles from Kyrgyzstan, on China's border, to Georgia, on the Black Sea. In early May, U.S. advisers started training antiguerrilla forces in the Pankisi Gorge in Georgia, where Muslim insurgents believed to be connected to al Qaeda are taking refuge from their struggle against Russian troops in Chechnya. A few days before that, Defense Secretary Donald H. Rumsfeld declared on a visit to Kyrgyzstan, where the U.S. Air Force has a base, that coalition troops would stay there "as long as necessary."

- From incidental sums fewer than five years ago, the amount of U.S. investment in the region has jumped to $20 billion.
- The energy giants have revved up their commitment to the Caspian region, one of the world's last undeveloped clusters of fields. Major investors include ChevronTexaco Corp., Exxon Mobil Corp., BP PLC, and Halliburton. BP alone plans to plow as much as $12 billion into the region over the next eight years.
- U.S. government aid is on track to jump 50% from pre-September 11 levels, to $809 million a year.

Every day, Americans are digging themselves in deeper into this part of the world, where 74 million people bring an exotic mix of Turkic, Mongol, Persian, and Slavic influence. What is fast evolving is a policy focused on guns and oil. The guns are to protect the local regimes from Islamic radicals and provide a staging area for attacks on Afghanistan. The goal is "to get rid of terrorism, not just get it out of Afghanistan," says A. Elizabeth Jones, Assistant Secretary of State for European and Eurasian Affairs. The guns, of course, will also protect the oil—oil that Washington hopes will lessen the West's dependence on the Persian Gulf and also lift the nations of the Caucasus and Central Asia out of their grinding poverty. "If you have prosperity, you have stability," Jones says.

Estimates of the Caspian oil pool vary greatly—from 200 billion barrels, on the level of a Saudi Arabia, to fewer than 100 billion barrels, still on a par with the reserves of the North Sea and at current oil prices worth $2.7 trillion. The Caspian could have a huge impact on the ability of OPEC to influence the oil market, says a U.S. government energy analyst. By 2010, the Caspian could account for 3% of global oil output, according to Moscow brokerage Renaissance Capital.

ChevronTexaco was the pioneer: In 1993, it bought into the huge Tengiz field in Kazakhstan. In October, 2001, almost $4 billion in investment later, a Chevron-led consortium opened its 980-mile pipeline from Tengiz to the Russian port of Novorossisk on the Black Sea. BP's Caspian project is one of its biggest anywhere. ExxonMobil has stakes in the Tengiz field and in offshore deposits belonging to Kazakhstan and Azerbaijan. All three majors are hungry to get in on future finds. "I don't think ChevronTexaco's appetite for investment in this part of the world is satisfied yet," says Dennis Fahy, general manager of Chevron in Kazakhstan.

Article 41. THE NEXT OIL FRONTIER

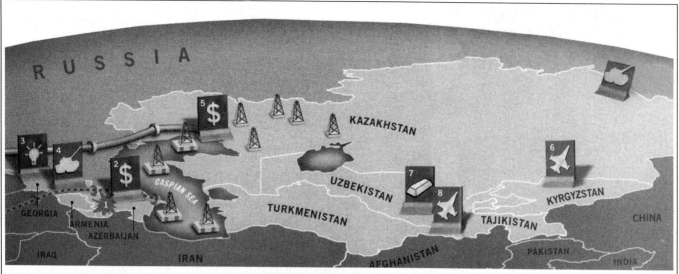

# America's Game in Russia

### Armenia

**Population** 3.8 million
**GDP** $2 billion
**Natural Resources** copper, gold
**U.S. Objective 1** Deal with Azerbaijan over the disputed region of Nagorno-Karabakh

### Azerbaijan

**Population** 8.1 million
**GDP** $5.7 billion
**Natural Resources** 0.9 billion tons of oil reserves
**Investment Picture 2** BP will invest up to $12 billion in off-shore Caspian oil fields
**U.S. Objective** Thwart Iranian-sponsored militancy and strengthen Azeri Navy to secure oil investments

### Georgia

**Population** 5.4 million
**GDP** $3.1 billion
**Investment Picture 3** AES has invested $250 million in power
**U.S. Objective 4** Use military to flush out militants

### Kazakhstan

**Population** 14.9 million
**GDP** $20 billion
**Natural Resources** 1.82 trillion cubic meters of gas; 1.1 billion tons of oil
**Investment Picture 5** ChevronTexaco, ExxonMobil, and others have invested $12.5 billion
**U.S. Objective** Oil security and access to ex-Soviet air bases

### Kyrgyzstan

**Population** 4.7 million
**GDP** $1.4 billion
**Natural Resources** gold
**Investment Picture:** Procter & Gamble's sales last year grew by 40%
**U.S. Objective 6** Troops stationed at ex-Soviet Manas air base for Afghanistan campaign

### Tajikistan

**Population** 6.2 million
**GDP** $1.1 billion
**U.S. Objective** Thwart opium smuggling from Afghanistan

### Turkmenistan

**Population** 5.4 million
**GDP** $3.3 billion
**Natural Resources** 2.83 trillion cubic meters of gas reserves

### Uzbekistan

**Population** 25 million
**GDP** $6.5 billion
**Natural Resources** 1.85 trillion cubic meters of gas reserves, plus gold and uranium
**Investment Picture 7** Newmont Mining plans more gold mines
**U.S. Objective 8** Troops at Karshi air base are set for an Afghanistan campaign

Data: European Bank for Reconstruction & Development, U.S. Government, *BusinessWeek*

Key to the game are the pipelines, where diplomacy and oilcraft meet. The U.S. wants a pipeline that will help its friends in the region and freeze out its enemies—especially the Iranians, also located on the Caspian. That's why Washington is strongly discouraging plans by some oil majors to lay a pipeline across Iran, lobbying instead for a proposed $3 billion, 1,090-mile pipeline to carry up to 1 million barrels of oil a day from Baku through Georgia to the Mediterranean port of Ceyhan in NATO ally Turkey. BP, which is seeking to recruit other investors for the Baku-Ceyhan pipeline, is expected to make a final decision by June about going ahead. "Construction is going to be approved," says Richard Pegge, a senoir manager in BP's Baku office.

## THE SOUTHERN RIM: OIL-RICH, BUT STILL DIRT-POOR

| | GDP PER CAPITA |
|---|---|
| SOUTHERN-RIM REPUBLICS | $586 |
| CHINA | 914 |
| RUSSIA | 1,790 |
| IRAN | 1,797 |
| TURKEY | 2,848 |
| HUNGARY | 4,757 |
| U.S. | 36,500 |

Data: *BusinessWeek*, Eurpean Bank for Reconstruction & Development, World Bank

Nothing is easy in this part of the world, however. Georgia has been wracked by civil war, organized crime, and terrorism. It's hardly a safe place for a pipeline. So the Pentagon is sending 150 military trainers to Georgia to help with anti-terrorism efforts and is helping Azerbaijan to bolster its Navy and modernize an air base for potential use by U.S. forces.

Not everyone is putting out the welcome mat. Russian hardliners see the southern rim thrust as U.S. encirclement. "Your foreign policy," a group of ex-military officers recently wrote President Vladimir V. Putin, is "the policy of licking the boots of the West."

Putin is trying to calm the hotheads. He may be calculating that his struggling country, barely able to supply its own armed forces, can benefit from the Pentagon's thrust. Putin and Bush plan to discuss U.S. military involvement in the Caucasus and Central Asia at their summit on May 24 in Moscow.

There's certainly plenty to talk about. On an April trip to the region, Defense Secretary Rumsfeld met with Kazakh President Nursultan A. Nazarbayev to discuss Pentagon access to local airfields. Some 1,000 troops of the U.S. Army's 10th Mountain Div. are already stationed at the ex-Soviet Khanabad Air Force Base in southern Uzbekistan. Fascinated by the female soldiers at the base, Uzbek guards offer to sell snapshots of women G.I.s riding motor scooters.

Russians are not the only ones nervous about U.S. troops in Central Asia. The State Dept.'s research shows that most people in Uzbekistan, Kazakhstan, and Kyrgyzstan oppose an extended U.S. military presence. "If the U.S. overstays its welcome in the region, it could alienate key allies in the war against terrorism," the department concluded in its Apr. 4 analysis. That risk also exists in oil-rich, BP-dominated Azerbaijan. "Bush sees us as the 51st state," scoffs Teymur Mamedov, a 32-year-old logistics manager for a Western oil-services company in Baku. "But it doesn't work that way. There's nothing to hold us together—only money, and that's not enough."

Then there's China, whose leaders suspect that the Pentagon's real goal is to keep an eye on, and if need be, contain China's activities in the region. Not to be outdone by the U.S., the Chinese are helping equip the Kazakh military.

The Chinese can play the power game, but in this chess match the U.S. has more pieces. Uzbek President Islam A. Karimov is grateful that the Pentagon-led campaign in Afghanistan dealt a blow to the local Islamic guerrilla group that fought alongside the Taliban. He's opening up the country's state-owned gold mines to $100 million in investment from Denver's Newmont Mining Corp.

For those expats who battle unyielding officials, impossible infrastructure and the sheer remoteness of it all, a stint in this part of the world can have its rewards. Even though most Baku residents lack properly filtered water, Western executives tied to the oil business are spending millions of dollars renovating 19th century townhouses with wrought-iron balconies as finely crafted as those in Paris. Most of the expat executives are middle-aged men, and with their fat wallets—let's face it, it's not their bulging waistlines—they are magnets for beautiful young local women. "Certainly, sexual harassment rules don't apply here," says one American male fortysomething businessman, recounting the perks of life in Baku.

Sensitive to the imperialism rap, the Bush Administration says its goal in the southern rim is to nurture prosperous, democratic societies. This is why the U.S. in mid-March inked an agreement with Uzbekistan. America pledged to protect the country from external threats in return for its pledge to liberalize its Soviet-style economy, improve its human-rights record, and ease government-imposed press censorship. In southern Azerbaijan, the State Dept. is funding a human-rights center in the town of Lenkoran, 25 miles from Iran. Still, even among center leaders, there's skepticism about America's purposes. "If there was no oil in Azerbaijan, I am sure America would not help us," says one of the staffers.

The Kremlin is sympathetic but not optimistic. "It was Russia's mission for so long to protect Western civilization from the Asians," says Vyacheslav A. Nikonov of the Polity Foundation, a Moscow political think tank. "If Americans are going to take over this job, God bless them."

Such sentiments aren't souring the American can-do spirit. James C. Cornell, president of RWE Nukem Inc. in Danbury, Conn., plans to double its uranium production in Uzbekistan. "When the U.S. is engaged militarily, it creates an umbrella for so many activities—not just business, but also education, culture," he says. "All things become possible." Trouble is, quagmires become possible, too.

*By Paul Starobin, with Catherine Belton in Moscow, Stan Crock in Washington, and bureau reports*

# Gray Dawn: The Global Aging Crisis

*Peter G. Peterson*

## DAUNTING DEMOGRAPHICS

THE LIST of major global hazards in the next century has grown long and familiar. It includes the proliferation of nuclear, biological, and chemical weapons, other types of high-tech terrorism, deadly super-viruses, extreme climate change, the financial, economic, and political aftershocks of globalization, and the violent ethnic explosions waiting to be detonated in today's unsteady new democracies. Yet there is a less-understood challenge—the graying of the developed world's population—that may actually do more to reshape our collective future than any of the above.

Over the next several decades, countries in the developed world will experience an unprecedented growth in the number of their elderly and an unprecedented decline in the number of their youth. The timing and magnitude of this demographic transformation have already been determined. Next century's elderly have already been born and can be counted—and their cost to retirement benefit systems can be projected.

Unlike with global warming, there can be little debate over whether or when global aging will manifest itself. And unlike with other challenges, even the struggle to preserve and strengthen unsteady new democracies, the costs of global aging will be far beyond the means of even the world's wealthiest nations—unless retirement benefit systems are radically reformed. Failure to do so, to prepare early and boldly enough, will spark economic crises that will dwarf the recent meltdowns in Asia and Russia.

How we confront global aging will have vast economic consequences costing quadrillions of dollars over the next century. Indeed, it will greatly influence how we manage, and can afford to manage, the other major challenges that will face us in the future.

For this and other reasons, global aging will become not just the transcendent economic issue of the 21st century, but the transcendent political issue as well. It will dominate and daunt the public-policy agendas of developed countries and force the renegotiation of their social contracts. It will also reshape foreign policy strategies and the geopolitical order.

The United States has a massive challenge ahead of it. The broad outlines can already be seen in the emerging debate over Social Security and Medicare reform. But ominous as the fiscal stakes are in the United States, they loom even larger in Japan and Europe, where populations are aging even faster, birthrates are lower, the influx of young immigrants from developing countries is smaller, public pension benefits are more generous, and private pension systems are weaker.

Aging has become a truly global challenge, and must therefore be given high priority on the global policy agenda. A gray dawn fast approaches. It is time to take an unflinching look at the shape of things to come.

*The Floridization of the developed world.* Been to Florida lately? You may not have realized it, but the vast concentration of seniors there—nearly 19 percent of the population—represents humanity's future. Today's Florida is a demographic benchmark that every developed nation will soon pass. Italy will hit the mark as early as 2003, followed by Japan in 2005 and Germany in 2006. France and Britain will pass present-day Florida around 2016; the United States and Canada in 2021 and 2023.

*Societies much older than any we have ever known.* Global life expectancy has grown more in the last fifty years than over the previous five thousand. Until the Industrial Revolution, people aged 65 and over never amounted to more than 2 or 3 percent of the population. In today's developed world, they amount to 14 percent. By the year 2030, they will reach 25 percent and be closing in on 30 in some countries.

*An unprecedented economic burden on working-age people.* Early in the next century, working-age populations in most developed countries will shrink. Between 2000 and 2010, Japan, for example, will suffer a 25 percent drop in the number of workers under age 30. Today the ratio of working taxpayers to nonworking pensioners in the developed world is around 3:1. By 2030, absent reform, this ratio will fall to 1.5:1, and in some countries, such as Germany and Italy, it will drop all the way down to 1:1 or even lower. While the longevity revolution represents a miraculous triumph of modern medicine and the extra years of life will surely be treasured by the elderly and their families, pension plans and other retirement benefit programs were not designed to provide these billions of extra years of payouts.

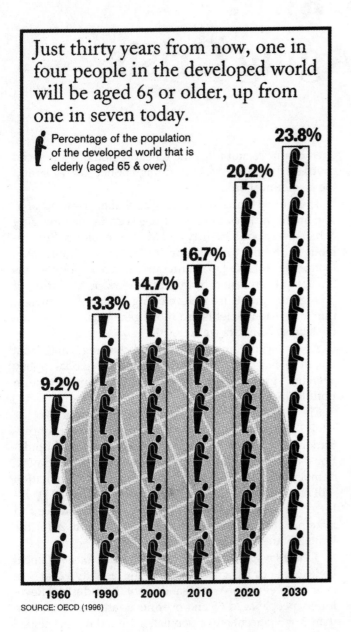

*The aging of the aged: the number of "old old" will grow much faster than the number of "young old."* The United Nations projects that by 2050, the number of people aged 65 to 84 worldwide will grow from 400 million to 1.3 billion (a threefold increase), while the number of people aged 85 and over will grow from 26 million to 175 million (a sixfold increase)—and the number aged 100 and over from 135,000 to 2.2 million (a sixteenfold increase). The "old old" consume far more health care than the "young old"—about two to three times as much. For nursing-home care, the ratio is roughly 20:1. Yet little of this cost is figured in the official projections of future public expenditures.

*Falling birthrates will intensify the global aging trend.* As life spans increase, fewer babies are being born. As recently as the late 1960s, the worldwide total fertility rate (that is, the average number of lifetime births per woman) stood at about 5.0, well within the historical range. Then came a behavioral revolution, driven by growing affluence, urbanization, feminism, rising female participation in the workforce, new birth control technologies, and legalized abortion. The result: an unprecedented and unexpected decline in the global fertility rate to about 2.7—a drop fast approaching the replacement rate of 2.1 (the rate required merely to maintain a constant population). In the developed world alone, the average fertility rate has plummeted to 1.6. Since 1995, Japan has had fewer births annually than in any year since 1899. In Germany, where the rate has fallen to 1.3, fewer babies are born each year than in Nepal, which has a population only one-quarter as large.

*A shrinking population in an aging developed world.* Unless their fertility rates rebound, the total populations of western Europe and Japan will shrink to about one-half of their current size before the end of the next century. In 1950, 7 of the 12 most populous nations were in the developed world: the United States, Russia, Japan, Germany, France, Italy, and the United Kingdom. The United Nations projects that by 2050, only the United States will remain on the list. Nigeria, Pakistan, Ethiopia, Congo, Mexico, and the Philippines will replace the others. But since developing countries are also experiencing a drop in fertility, many are now actually aging faster than the typical developed country. In France, for example, it took over a century for the elderly to grow from 7 to 14 percent of the population. South Korea, Taiwan, Singapore, and China are projected to traverse that distance in only 25 years.

*From worker shortage to rising immigration pressure.* Perhaps the most predictable consequence of the gap in fertility and population growth rates between developed and developing countries will be the rising demand for immigrant workers in older and wealthier societies facing labor shortages. Immigrants are typically young and tend to bring with them the family practices of their native culture—including higher fertility rates. In many European countries, non-European foreigners already make up roughly 10 percent of the population. This includes 10 million to 13 million Muslims, nearly all of whom are working-age or younger. In Germany, foreigners will make up 30 percent of the total population by 2030, and over half the population of major cities like Munich and Frankfurt. Global aging and attendant labor shortages will therefore ensure that immigration remains a major issue in developed countries for decades to come. Culture wars could erupt over the balkanization of language and religion; electorates could divide along ethnic lines; and émigré leaders could sway foreign policy.

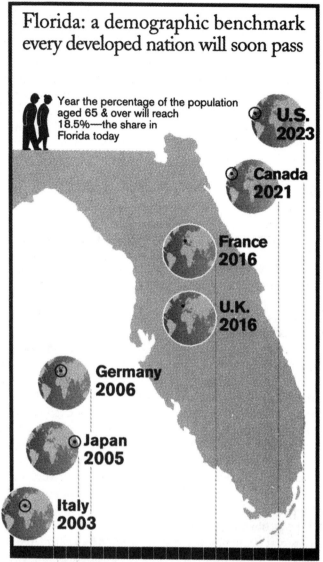

Florida: a demographic benchmark every developed nation will soon pass

Year the percentage of the population aged 65 & over will reach 18.5%—the share in Florida today

U.S. 2023
Canada 2021
France 2016
U.K. 2016
Germany 2006
Japan 2005
Italy 2003

SOURCE: OECD (1996); author's calculations

## GRAYING MEANS PAYING

OFFICIAL PROJECTIONS suggest that within 30 years, developed countries will have to spend at least an extra 9 to 16 percent of GDP simply to meet their old-age benefit promises. The unfunded liabilities for pensions (that is, benefits already earned by today's workers for which nothing has been saved) are already almost $35 trillion. Add in health care, and the total jumps to at least twice as much. At minimum, the global aging issue thus represents, to paraphrase the old quiz show, a $64 trillion question hanging over the developed world's future.

To pay for promised benefits through increased taxation is unfeasible. Doing so would raise the total tax burden by an unthinkable 25 to 40 percent of every worker's taxable wages—in countries where payroll tax rates sometimes already exceed 40 percent. To finance the costs of these benefits by borrowing would be just as disastrous. Governments would run unprecedented deficits that would quickly consume the savings of the developed world.

And the $64 trillion estimate is probably low. It likely underestimates future growth in longevity and health care costs and ignores the negative effects on the economy of more borrowing, higher interest rates, more taxes, less savings, and lower rates of productivity and wage growth.

There are only a handful of exceptions to these nightmarish forecasts. In Australia, total public retirement costs as a share of GDP are expected to rise only slightly, and they may even decline in Britain and Ireland. This fiscal good fortune is not due to any special demographic trend, but to timely policy reforms—including tight limits on public health spending, modest pension benefit formulas, and new personally owned savings programs that allow future public benefits to shrink as a share of average wages. This approach may yet be emulated elsewhere.

Failure to respond to the aging challenge will destabilize the global economy, straining financial and political institutions around the world. Consider Japan, which today runs a large current account surplus making up well over half the capital exports of all the surplus nations combined. Then imagine a scenario in which Japan leaves its retirement programs and fiscal policies on autopilot. Thirty years from now, under this scenario, Japan will be importing massive amounts of capital to prevent its domestic economy from collapsing under the weight of benefit outlays. This will require a huge reversal in global capital flows. To get some idea of the potential volatility, note that over the next decade, Japan's annual pension deficit is projected to grow to roughly 3 times the size of its recent and massive capital exports to the United States; by 2030, the annual deficit is expected to be 15 times as large. Such reversals will cause wildly fluctuating interest and exchange rates, which may in turn short-circuit financial institutions and trigger a serious market crash.

As they age, some nations will do little to change course, while others may succeed in boosting their national savings rate, at least temporarily, through a combination of fiscal restraint and household thrift. Yet this too could result in a volatile disequilibrium in supply and demand for global capital. Such imbalance could wreak havoc with international institutions such as the European Union.

In recent years, the EU has focused on monetary union, launched a single currency (the euro), promoted cross-border labor mobility, and struggled to harmonize fiscal, monetary, and trade policies. European leaders expect to have their hands full smoothing out differences between members of the Economic and Monetary Union (EMU)—from the timing of their business cycles to the diversity of their credit institutions and political cultures. For this reason, they established official public debt and deficit criteria (three percent of GDP for EMU membership) in order to discourage maverick nations from placing undue economic burdens on fellow members. But the EU has yet to face up to the biggest challenge to its future viability: the likelihood of varying national responses to the fiscal pressures of demographic aging. Indeed, the EU does not even include unfunded pension liabilities in the official EMU debt and deficit criteria—which is like measuring icebergs without looking beneath the water line.

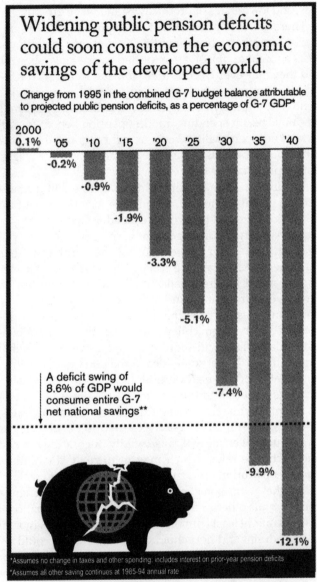

Widening public pension deficits could soon consume the economic savings of the developed world.

Change from 1995 in the combined G-7 budget balance attributable to projected public pension deficits, as a percentage of G-7 GDP*

2000: 0.1%
'05: -0.2%
'10: -0.9%
'15: -1.9%
'20: -3.3%
'25: -5.1%
'30: -7.4%
'35: -9.9%
'40: -12.1%

A deficit swing of 8.6% of GDP would consume entire G-7 net national savings**

*Assumes no change in taxes and other spending; includes interest on prior-year pension deficits
**Assumes all other saving continues at 1985-94 annual rate

SOURCE: OECD (1996); author's calculations

When these liabilities come due and move from "off the books" to "on the books," the EU will, under current constraints, be required to penalize EMU members that exceed the three percent deficit cap. As a recent IMF report concludes, "over time it will become increasingly difficult for most countries to meet the deficit ceiling without comprehensive social security reform." The EU could, of course, retain members by raising the deficit limit. But once the floodgates are opened, national differences in fiscal policy may mean that EMU members rack up deficits at different rates. The European Central Bank, the euro, and a half-century of progress toward European unity could be lost as a result.

The total projected cost of the age wave is so staggering that we might reasonably conclude it could never be paid. After all, these numbers are projections, not predictions. They tell us what is likely to happen if current policy remains unchanged, not whether it is likely or even possible for this condition to hold. In all probability, economies would implode and governments would collapse before the projections ever materialize. But this is exactly why we must focus on these projections, for they call attention to the paramount question: Will we change course sooner, when we still have time to control our destiny and reach a more sustainable path? Or later, after unsustainable economic damage and political and social trauma cause a wrenching upheaval?

## A GRAYING NEW WORLD ORDER

WHILE THE fiscal and economic consequences of global aging deserve serious discussion, other important consequences must also be examined. At the top of the list is the impact of the age wave on foreign policy and international security.

*Will the developed world be able to maintain its security commitments?* One need not be a Nobel laureate in economics to understand that a country's GDP growth is the product of workforce and productivity growth. If workforces shrink rapidly, GDP may drop as well, since labor productivity may not rise fast enough to compensate for the loss of workers. At least some developed countries are therefore likely to experience a long-term decline in total production of goods and services—that is, in real GDP.

Economists correctly focus on the developed world's GDP per capita, which can rise even as its workforce and total GDP shrink. But anything with a fixed cost becomes a national challenge when that cost has to be spread over a smaller population and funded out of shrinking revenues. National defense is the classic example. The West already faces grave threats from rogue states armed with biological and chemical arsenals, terrorists capable of hacking into vulnerable computer systems, and proliferating nuclear weapons. None of these external dangers will shrink to accommodate our declining workforce or GDP.

Leading developed countries will no doubt need to spend as much or more on defense and international investments as they do today. But the age wave will put immense pressure on governments to cut back. Falling birthrates, together with a rising demand for young workers, will also inevitably mean smaller armies. And how many parents will allow their only child to go off to war?

With fewer soldiers, total capability can be maintained only by large increases in technology and weaponry. But boosting military productivity creates a Catch-22. For how will governments get the budget resources to pay for high-tech weaponry if the senior-weighted electorate demands more money for high-tech medicine? Even if military capital is successfully substituted for military labor, the deployment options may be dangerously limited. Developed nations facing a threat may feel they have only two extreme (but relatively inexpensive) choices: a low-level response (antiterrorist strikes and cruise-missile diplomacy) or a high-level response (an all-out attack with strategic weapons).

*Will Young/Old become the next North/South fault line?* Historically, the richest industrial powers have been growing, capital-exporting, philanthropic giants that project their power and mores around the world. The richest industrial powers of the future may be none of these things. Instead, they may be demographically imploding, capital-importing, fiscally starving neutrals who twist and turn to avoid expensive international entanglements. A quarter-century from now, will the divide between today's "rich" and "poor" nations be better described as a divide between growth and decline, surplus and deficit, expansion and retreat, future and past? By the mid-2020s, will the contrast between North and South be better described as a contrast between Young and Old?

If today's largest low-income societies, especially China, set up fully funded retirement systems to prepare for their own future aging, they may well produce ever larger capital surpluses. As a result, today's great powers could someday depend on these surpluses to keep themselves financially afloat. But how should we expect these new suppliers of capital to use their newly acquired leverage? Will they turn the tables in international diplomacy? Will the Chinese, for example, someday demand that the United States shore up its Medicare system the way Americans once demanded that China reform its human rights policies as a condition for foreign assistance?

As Samuel Huntington recently put it, "the juxtaposition of a rapidly growing people of one culture and a slowly growing or stagnant people of another culture generates pressure for economic and/or political adjustments in both societies." Countries where populations are still exploding rank high on any list of potential trouble spots, whereas the countries most likely to lose population—and to see a weakening of their commitment to expensive defense and global security programs—are the staunchest friends of liberal democracy.

In many parts of the developing world, the total fertility rate remains very high (7.3 in the Gaza Strip versus 2.7 in Israel), most people are very young (49 percent under age 15 in Uganda), and the population is growing very rapidly (doubling every 26 years in Iran). These areas also tend to be the poorest, most rapidly urbanizing, most institutionally unstable—and most likely to fall under the sway of rogue leadership. They are the same societies that spawned most of the military strongmen and terrorists who have bedeviled the United States and Europe in recent decades. The Pentagon's long-term planners predict that outbreaks of regional anarchy will occur more frequently early in the next century. To pinpoint when and where, they track what they call "youth bulges" in the world's poorest urban centers.

Is demography destiny, after all? Is the rapidly aging developed world fated to decline? Must it cede leadership to younger and faster-growing societies? For the answer to be no, the developed world must redefine that role around a new mission. And what better way to do so than to show the younger, yet more tradition-bound, societies—which will soon age in their turn—how a world dominated by the old can still accommodate the young.

## WHOSE WATCH IS IT, ANYWAY?

FROM PRIVATE discussions with leaders of major economies, I can attest that they are well briefed on the stunning demographic trends that lie ahead. But so far they have responded with paralysis rather than action. Hardly any country is doing what it should to prepare. Margaret Thatcher confesses that she repeatedly tried to raise the aging issue at G-7 summit meetings. Yet her fellow leaders stalled. "Of course aging is a profound challenge," they replied, "but it doesn't hit until early in the next century—after my watch."

Americans often fault their leaders for not acknowledging long-term problems and for not facing up to silent and slow-motion challenges. But denial is not a peculiarly American syndrome. In 1995, Silvio Berlusconi's *Forza Italia* government was buffeted by a number of political storms, all of which it weathered—except for pension reform, which shattered the coalition. That same year, the Dutch parliament was forced to repeal a recent cut in retirement benefits after a strong Pension Party, backed by the elderly, emerged from nowhere to punish the reformers. In 1996, the French government's modest proposal to trim pensions triggered strikes and even riots. A year later the Socialists overturned the ruling government at the polls.

Each country's response, or nonresponse, is colored by its political and cultural institutions. In Europe, where the welfare state is more expansive, voters can hardly imagine that the promises made by previous generations of politicians can no longer be kept. They therefore support leaders, unions, and party coalitions that make generous unfunded pensions the very cornerstone of social democracy. In the United States, the problem has less to do with welfare-state dependence than the uniquely American notion that every citizen has personally earned and is therefore entitled to whatever benefits government happens to have promised.

How governments ultimately prepare for global aging will also depend on how global aging itself reshapes politics. Already some of the largest and most strident interest groups in the United States are those that claim to speak for senior citizens, such as the American Association of Retired Persons, with its 33 million members, 1,700 paid employees, ten times that many trained volunteers, and an annual budget of $5.5 billion.

Senior power is rising in Europe, where it manifests itself less through independent senior organizations than in labor unions and (often union-affiliated) political parties that formally adopt pro-retiree platforms. Could age-based political parties be the wave of the future? In Russia, although the Communist resurgence is usually ascribed to nationalism and nostalgia, a demographic bias is at work as well. The Communists have repositioned themselves as the party of retirees, who are aggrieved by how runaway inflation has slashed the real value of their pensions. In the 1995 Duma elections, over half of those aged 55 and older voted Communist, versus only ten percent of those under age 40.

Commenting on how the old seem to trump the young at every turn, Lee Kuan Yew once proposed that each tax-paying worker be given two votes to balance the lobbying clout of each retired elder. No nation, not even Singapore, is likely to enact Lee's suggestion. But the question must be asked: With ever more electoral power flowing into the hands of elders, what can motivate political leaders to act on behalf of the long-term future of the young?

A handful of basic strategies, all of them difficult, might enable countries to overcome the economic and political challenges of an aging society: extending work lives and postponing retirement; enlarging the workforce through immigration and increased labor force participation; encouraging higher fertility and investing more in the education and productivity of future workers; strengthening intergenerational bonds of responsibility within families; and targeting government-paid benefits to those most in need while encouraging and even requiring workers to save for their own retirements. All of these strategies unfortunately touch raw nerves—by amending existing social contracts, by violating cultural expectations, or by offending entrenched ideologies.

## TOWARD A SUMMIT ON GLOBAL AGING

ALL COUNTRIES would be well served by collective deliberation over the choices that lie ahead. For that reason I propose a Summit on Global Aging. Few venues are as well covered by the media as a global summit. Leaders have been willing to convene summits to discuss global warming. Why not global aging, which will hit us sooner and with greater certainty? By calling attention to what is at stake, a global aging summit could shift the public discussion into fast forward. That alone would be a major contribution. The summit process would also help provide an international framework for voter education, collective burden-sharing, and global leadership. Once national constituencies begin to grasp the magnitude of the global aging challenge, they will be more inclined to take reform seriously. Once governments get into the habit of cooperating on what in fact is a global challenge, individual leaders will not need to incur the economic and political risks of acting alone.

This summit should launch a new multilateral initiative to lend the global aging agenda a visible institutional presence: an Agency on Global Aging. Such an agency would examine how developed countries should reform their retirement systems and how developing countries should properly set them up in the first place. Perhaps the most basic question is how to weigh the interests and well-being of one generation against the next. Then there is the issue of defining the safety-net standard of social adequacy. Is there a minimum level of retirement income that should be the right of every citizen? To what extent should retirement security be left to people's own resources? When should government pick up the pieces, and how can it do so without discouraging responsible behavior? Should government compel people in advance to make better life choices, say, by enacting a mandatory savings program?

Another critical task is to integrate research about the age wave's timing, magnitude, and location. Fiscal projections should be based on assumptions that are both globally consistent and—when it comes to longevity, fertility, and health care costs—more realistic than those now in use. Still to be determined: Which countries will be hit earliest and hardest? What might happen to interest rates, exchange rates, and cross-border capital flows under various political and fiscal scenarios?

But this is not all the proposed agency could do. It could continue to build global awareness, publish a high-visibility annual report that would update these calculations, and ensure that the various regular multilateral summits (from the G-7 to ASEAN and APEC) keep global aging high on their discussion agendas. It could give coherent voice to the need for timely policy reform around the world, hold up as models whatever major steps have been taken to reduce unfunded liabilities, help design funded benefit programs, and promote generational equity. On these and many other issues, nations have much to learn from each other, just as those who favor mandatory funded pension plans are already benefiting from the examples of Chile, Britain, Austria, and Singapore.

Global aging could trigger a crisis that engulfs the world economy. This crisis may even threaten democracy itself. By making tough choices now world leaders would demonstrate that they genuinely care about the future, that they understand this unique opportunity for young and old nations to work together, and that they comprehend the price of freedom. The gray dawn approaches. We must establish new ways of thinking and new institutions to help us prepare for a much older world.

---

PETER G. PETERSON is the author of *Gray Dawn: How the Coming Age Wave Will Transform America—and the World*. He is Chairman of The Blackstone Group, a private investment bank, Chairman of The Institute for International Economics, Deputy Chairman of The Federal Reserve Bank of New York, Co-founder and President of The Concord Coalition, and Chairman of The Council on Foreign Relations.

# A rare and precious resource

**Fresh water is a scarce commodity. Since it's impossible to increase supply, demand and waste must be reduced. But how?**

Houria Tazi Sadeq*

Water is a bond between human beings and nature. It is ever-present in our daily lives and in our imaginations. Since the beginning of time, it has shaped extraordinary social institutions, and access to it has provoked many conflicts.

But most of the world's people, who have never gone short of water, take its availability for granted. Industrialists, farmers and ordinary consumers blithely go on wasting it. These days, though, supplies are diminishing while demand is soaring. Everyone knows that the time has come for attitudes to change.

Few people are aware of the true extent of fresh water scarcity. Many are fooled by the huge expanses of blue that feature on maps of the world. They do not know that 97.5 per cent of the planet's water is salty—and that most of the world's fresh water—the remaining 2.5 per cent—is unusable: 70 per cent of it is frozen in the icecaps of Antarctica and Greenland and almost all the rest exists in the form of soil humidity or in water tables which are too deep to be tapped. In all, barely one per cent of fresh water—0.007 per cent of all the water in the world, is easily accessible.

## Sharper vision

Desalinization, state of the art irrigation systems, techniques to harvest fog—technological solutions like these are widely hailed as the answer to water scarcity. But in searching for the "miracle" solution, hydrologists and policy-makers often lose sight of the question: how can we use and safeguard this vital resource? UNESCO's International Hydrological Programme (IHP) takes an interdisciplinary approach to this question. On the one hand, IHP brings together scientists from 150 countries to develop global and regional assessments of water supplies and, for example, inventories of groundwater contamination. At the same time, the programme focuses on the cultural and socio-economic factors involved in effective policy-making. For example, groundwater supplies in Gaza (Palestinian Authority) are coming under serious strain, partly because of new business investment in the area. IHP has a two-pronged approach. First, train and help local hydrologists accurately assess the supplies. Second, work with government officials to set up a licensing system for pumping groundwater.

By joining forces with the World Water Council, an international think-tank on hydrological issues, IHP is now hosting one of the most ambitious projects in the field: World Water Vision. Hundreds of thousands of hydrologists, policy-makers, farmers, business leaders and ordinary citizens will take part in public consultations to develop regional scenarios as to how key issues like contamination will evolve in the next 25 years.

**Global water withdrawals (1900-2000) in thousands of km³ per year**

- A person can survive for about a month without food, but only about a week without water.
- About 70 per cent of human skin consists of water.
- Women and children in most developing regions travel an average of 10 to 15 kilometres each day to get water.
- Some 34,000 people die a day from water-related diseases like diarrhoea and parasitic worms. This is the equivalent to casualties from 100 jumbo jets crashing every day!
- A person needs five litres of water a day for drinking and cooking and another 25 litres for personal hygiene.
- The average Canadian family uses 350 litres of water a day. In Africa, the average is 20 liters and in Europe, 165 litres.
- A dairy cow needs to drink about four litres of water a day to produce one litre of milk.
- A tomato is about 95 per cent water.
- About 9,400 litres of water are used to make four car tires.
- About 1.4 billion litres of water are needed to produce a day's supply of the world's newsprint.

Sources: International Development Initiative of McGill University, Canada; Saint Paul Water Utility, Minnesota, USA

### Lack of access to safe water and basic sanitation, by region, 1990–1996 (percent)

| Region | People without access to safe water | People without access to basic sanitation |
|---|---|---|
| Arab States | 21 | 30 |
| Sub-Saharan Africa | 48 | 55 |
| South-East Asia and the Pacific | 35 | 45 |
| Latin America and the Caribbean | 23 | 29 |
| East Asia | 32 | 73 |
| East Asia (excluding China) | 13 | – |
| South Asia | 18 | 64 |
| Developing countries | 29 | 58 |
| Least developed countries | 43 | 64 |

Source: *Human Development Report 1998*, New York, UNDP

### Periods of complete renewal of the earth's water resources

| Kinds of water | Period of renewal |
|---|---|
| Biological water | several hours |
| Atmospheric water | 8 days |
| Water in river channels | 16 days |
| Soil moisture | 1 year |
| Water in swamps | 5 years |
| Water storages in lakes | 17 years |
| Groundwater | 1400 years |
| Mountain glaciers | 1600 years |
| World ocean | 2500 years |
| Polar ice floes | 9700 years |

Source: *World Water Balance and Water Resources of the Earth*, Gidrometeoizdat, Leningrad, 1974 (in Russian)

Over the past century, population growth and human activity have caused this precious resource to dwindle. Between 1900 and 1995, world demand for water increased more than sixfold—compared with a threefold increase in world population. The ratio between the stock of fresh water and world population seems to show that in overall terms there is enough water to go round. But in the most vulnerable regions, an estimated 460 million people (8 per cent of the world's population) are short of water, and another quarter of the planet's inhabitants are heading for the same fate. Experts say that if nothing is done, two-thirds of humanity will suffer from a moderate to severe lack of water by the year 2025.

*The water from the fountain glides,
flows and dreams as,
almost dumb, it
licks the mossy stone.*

Antonio Machado
(1875–1939), Spain

Inequalities in the availability of water—sometimes even within a single country—are reflected in huge differences in consumption levels. A person living in rural Madagascar uses 10 litres a day, the minimum for survival, while a French person uses 150 litres and an American as many as 425.

Scarcity is just one part of the problem. Water quality is also declining alarmingly. In some areas, contamination levels are so high that water can no longer be used even for industrial purposes. There are many reasons for this—untreated sewage, chemical waste, fuel leakages, dumped garbage, contamination of soil by chemicals used by farmers. The worldwide extent of such pollution is hard to assess because data are lacking for several countries. But some figures give an idea of the problem. It is thought for example that 90 per cent of waste water in developing countries is released without any kind of treatment.

Things are especially bad in cities, where water demand is exploding. For the first time in human history, there will soon be more people living in cities than in the

# A thirsty planet

We now have less than half the amount of water available per capita than we did 50 years ago. In 1950, world reserves, (after accounting for agricultural, industrial and domestic uses) amounted to 16.8 thousand cubic metres per person. Today, global reserves have dropped to 7.3 thousand cubic metres and are expected to fall to 4.8 thousand in just 25 years.

Scientists have developed many ways of measuring supplies and evaluating water scarcity. In the maps at right, "catastrophic" levels mean that reserves are unlikely to sustain a population in the event of a crisis like drought. Low supplies refer to levels which put in danger industrial development or ability to feed a population.

Just 50 years ago, not a country in the world faced catastrophic water supply levels. Today, about 35 per cent of the population lives under these conditions. By 2025, about two-thirds will have to cope with low if not catastrophic reserves. In contrast, "water rich" regions and countries—such as northern Europe, Canada, almost everywhere in South America, Central Africa, the Far East and Oceania—will continue to enjoy ample reserves.

The sharp declines reflect the soaring water demands of growing populations, agricultural needs and industrialization. In addition, nature has been far from even-handed. More than 40 per cent of the water in rivers, reservoirs and lakes is concentrated in just six countries: Brazil, Russia, Canada, the United States, China and India. Meanwhile just two per cent of river, reservoir and lake water is found in about 40 per cent of the world's land mass.

As a result, in 2025 Europe and the United States will have half the per capita reserves they did in 1950, while Asia and Latin America will have just a quarter of what they previously enjoyed. But the real drama is likely to hit Africa and the Middle East, where available supplies by 2025 may be only an eighth of what they were in 1950.

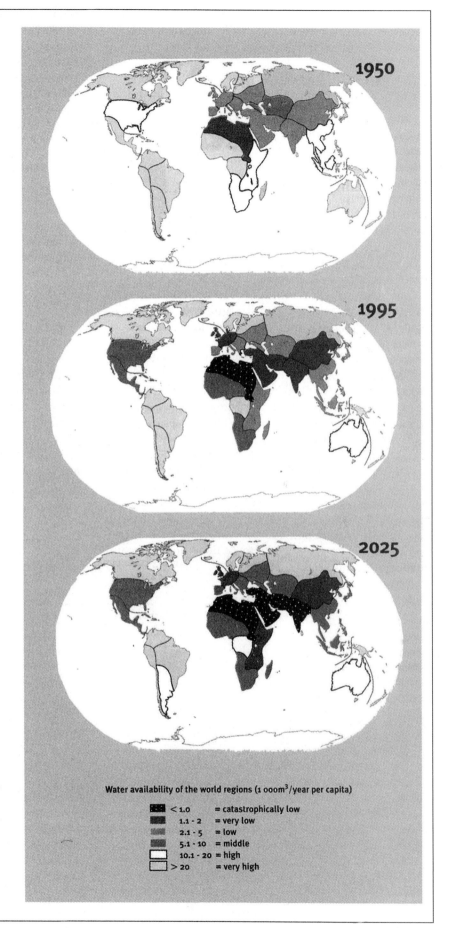

Water availability of the world regions (1 000m³/year per capita)

| | |
|---|---|
| < 1.0 | = catastrophically low |
| 1.1 - 2 | = very low |
| 2.1 - 5 | = low |
| 5.1 - 10 | = middle |
| 10.1 - 20 | = high |
| > 20 | = very high |

countryside and so water consumption will continue to increase. Soaring urbanization will sharpen the rivalry between the different kinds of water users.

## Curbing the explosion in demand

Today, farming uses 69 per cent of the water consumed in the world, industry 23 per cent and households 8 per cent. In developing countries, agriculture uses as much as 80 per cent. The needs of city-dwellers, industry and tourists are expected to increase rapidly, at least as much as the need to produce more farm products to feed the planet. The problem of increasing water supply has long been seen as a technical one, calling for technical solutions such as building more dams and desalination plants. Wild ideas like towing chunks of icebergs from the poles have even been mooted.

But today, technical solutions are reaching their limits. Economic and socio-ecological arguments are levelled against building new dams, for example: dams are costing more and more because the best sites have already been used, and they take millions of people out of their environment and upset ecosystems. As a result, twice as many dams were built on average between 1951 and 1977 than during the past decade, according to the US environmental research body Worldwatch Institute.

Hydrologists and engineers have less and less room for manoeuvre, but a new consensus with new actors is taking shape. Since supply can no longer be expanded—or only at prohibitive cost for many countries—the explosion in demand must be curbed along with wasteful practices. An estimated 60 per cent of the water used in irrigation is lost through inefficient systems, for example.

Economists have plunged into the debate on water and made quite a few waves. To obtain "rational use" of water, i.e. avoiding waste and maintaining quality, they say consumers must be made to pay for it. Out of the question, reply those in favour of free water, which some cultures regard as "a gift from heaven." And what about the poor, ask the champions of human rights and the right to water? Other important and prickly questions being asked by decisionmakers are how to calculate the "real price" of water and who should organize its sale.

## The state as mediator

The principle of free water is being challenged. For many people, water has become a commodity to be bought and sold. But management of this shared resource cannot be left exclusively to market forces. Many elements of civil society—NGOs, researchers, community groups—are campaigning for the cultural and social aspects of water management to be taken into account.

Even the World Bank, the main advocate of water privatization, is cautious on this point. It recognizes the value of the partnerships between the public and private sectors which have sprung up in recent years. Only the state seems to be in a position to ensure that practices are fair and to mediate between the parties involved—consumer groups, private firms and public bodies. At any rate, water regulation and management systems need to be based on other than purely financial criteria. If they aren't, hundreds of millions of people will have no access to it.

---

*Moroccan jurist, president of the Maghreb–Machrek Water Union, vice-president of the International Water Secretariat

---

Reprinted with permission from *The UNESCO Courier*, February 1999, pp. 18–21.

# Article 44

# In Race to Tap the Euphrates, the Upper Hand Is Upstream

By DOUGLAS JEHL

TELL AL-SAMEN, Syria—The Euphrates River is close by, but the water does not reach Abdelrazak al-Aween. Here at the heart of the fertile crescent, he stares at dry fields.

The Syrian government has promised water for Mr. Aween's tiny village. But upstream, in Turkey, and downstream, in Iraq, similar promises are being made. They add up to more water than the Euphrates holds.

So instead of irrigating his cotton and sugar beets, Mr. Aween must siphon drinking and washing water from a ditch 40 minutes away by tractor ride. Just across the border, meanwhile, Ahmet Demir, a Turkish farmer, stands ankle deep in mud, his crops soaking up all the water they need.

It was here in ancient Mesopotamia, thousands of years ago, that the last all-out war over water was fought, between rival city-states in what is now southern Iraq. Now, across a widening swath of the world, more and more people are vying for less and less water, in conflicts more rancorous by the day.

From the searing plains of Mesopotamia to the steadily expanding deserts of northern China to the cotton fields of northwest Texas, the struggle for water is igniting social, economic and political tensions.

The World Bank has said dwindling water supplies will be a major factor inhibiting economic growth, a subject being discussed at a weeklong international conference in South Africa starting Monday [August 26, 2002] about balancing use of the world's resources against its economic needs.

Global warming, some experts suspect, may be adding to the strain. Droughts may be extended in already dry regions, including parts of the United States, even as wetter areas tend toward calamitous downpours and floods like those ravaging Europe and Asia this summer. In general, the world's climate may be more prone to extremes, with too much water in some areas and far too little in others.

Both the United Nations and the National Intelligence Council, an advisory group to the Central Intelligence Agency, have warned that the competition for water is likely to worsen. "As countries press against the limits of available water between now and 2015, the possibility of conflict will increase," the National Intelligence Council warned in a report last year.

By 2015, according to estimates from the United Nations and the United States government, at least 40 percent of the world's population, or about three billion people, will live in countries where it is difficult or impossible to get enough water to satisfy basic needs.

"The signs of unsustainability are widespread and spreading," said Sandra Postel, director of the Global Water Policy Project in Amherst, Mass. "If we're to have any hope of satisfying the food and water needs of the world's people in the years ahead, we will need a fundamental shift in how we use and manage water."

An inescapable fact about the world's water supply is that it is finite. Less than 1 percent of it is fresh water that can be used for drinking or agriculture, and demand for that water is rising.

Over the last 70 years, the world's population has tripled while water demand has increased sixfold, causing increasing strain especially in heavily populated areas where water is distant, is being depleted or is simply too polluted to use.

Already, a little more than half of the world's available fresh water is being used each year, according to one rough but generally accepted estimate. That fraction could climb to 74 percent by 2025 based on population growth alone, and would hit 90 percent if people everywhere used as much water as the average American, one of the world's most gluttonous water consumers.

Water tables are falling on every continent, and experts warn that the situation is expected to worsen significantly in years to come. On top of the shortages that already exist, the outlook adds to the tensions and uncertainty for countries that share water sources, like Turkey and Syria, where Mr. Aween is among those still waiting and hoping for the Euphrates to be brought to his door.

# A Few Miles' Difference

The stories of Mr. Aween and Mr. Demir illustrate how the growing fight for water can make or ruin lives.

Until last year, Mr. Demir, 42, a father of nine in Turkey, was living an itinerant life as a smuggler and a migrant laborer. But on a recent scorching afternoon, he stood sunburned and content, his striped pants rolled above his knees, bare feet squishing in Euphrates mud.

"It seems like we have all the water we need," Mr. Demir said, leaning on his shovel and running a hand through his close-cropped hair. What has changed in this swath of southern Turkey is the arrival of irrigation. It is part of one of the world's largest water projects, an audacious $30 billion plan by Turkey's government to spread the Euphrates' gifts across a vast and impoverished region of the country.

By now, Mr. Aween, the Syrian, might have been celebrating, too. Under Syria's irrigation plan, ambitious in its own right, water from the Euphrates should have reached Mr. Aween's door, less than 50 miles from Mr. Demir's.

But strong doubts are emerging about whether the vast scope of Turkey's project will leave enough water for its neighbors downstream—so much so that Syria appears to have put the brakes on its development plan.

"We're still waiting," Mr. Aween, 40, said on behalf of his 2 wives, 3 children, and 17 brothers and sisters, who all live in a hamlet that bears the family name. He wore a loose, Arab-style outer garment and cheap plastic sandals as he hitched a rusty tanker trailer to a sputtering tractor, his water bearers. "But the water hasn't come."

The trouble over the Euphrates can be expressed in a simple, untenable equation.

As best as anyone can determine, the river, in an average year, holds 35 billion cubic meters of water. But the separate plans drawn up by Turkey, Syria and Iraq for building dams and irrigating fields would, taken together, consume nearly half again more water than the river holds.

Each country has acknowledged the impossibility of marrying their schemes. But none has shown any willingness to scale back. Trying to accommodate fast-growing populations and to head off a migration to the cities, each country is still clinging to its irrigation dreams.

On both sides of the Turkish-Syrian border, snapshots of those dreams still unfold on summer dawns, in fields that used to be good for little but grazing, but where new irrigation canals are now delivering Euphrates water in regular supply.

Thirsty crops like cotton and sugar beets have begun to thrive. Farm incomes have tripled. Young women who turn out in sparkling dresses tend prized plants with special care, shepherding the water down each muddy row. Young boys cavort in irrigation ditches that provide relief from the intense midday heat.

Now that there is water, Mr. Demir said, it would not be such a bad thing if his four sons, ranging in age from 4 months to 22 years, decided to stay and work the land—something he could not have imagined only a year ago when farming was far harder.

But in a world in which so much depends on having water, he bristled at the idea of sharing it. "If I used any less, the others would use more," he said. "I use what I need, and as for the rest, it's their business."

That kind of thinking has not helped Mr. Aween's village, not far from Tell al-Samen, where the absence of water is almost equally on display. With his family, Mr. Aween raises scraggly goats and grows whatever barley and wheat he can coax from dry, unirrigated land. It is a far cry, he said, from the lush green crops he would grow with irrigation—the difference between sustenance and comfort.

Syria and Turkey have been at odds since the late 1980's, when Turkey decided to proceed with its development project without consulting its neighbors.

To Turkey, the plan was an essential step in the country's development, a means of transforming an area that is home to six million people, including many restive Kurds, into a zone of greater economic and political stability. But to Syria, which depends on the Euphrates for half its fresh water, it continues to be seen as a major threat. Iraq, the far downstream neighbor, has been mostly a bystander because of its international isolation.

The basic disputes between Turkey and Syria—over water rights and allocations—would be familiar to any landowner who has struggled over competing claims to a stream or a well. But the fact that the parties are nations, with large armies at their disposal and large populations at stake, has added to the weight of the clash.

Less powerful militarily than Turkey, Syria resorted through much of the 1990's to indirect pressure, by giving support to the terrorist leader Abdullah Ocalan, the No. 1 enemy of the Turkish government. In response, Turkish officials sometimes went so far as to warn that the flow of the Euphrates into Syria could be cut.

Such talk has cooled in the last three years since the arrest of Mr. Ocalan. But the potential for trouble lingers.

Over the last year, as Turkey has completed many dams and has begun to extend its irrigation, the flow of Euphrates water into Syria has grown consistently smaller. The flow has fallen below levels allowed under Turkey's only existing commitment to Syria, an interim 1987 agreement intended for the period of dam construction.

Turkey has blamed drought and the overuse of the new, cheap water by farmers for the shortfall. The government says much of that can be reversed. But Turkish officials have also said they no longer see the 1987 deal as binding. The trend has raised deep concerns in Syria, which is already facing water shortages in Damascus, the capital.

Both countries say they are ready to strike a deal, but cannot imagine scaling back their development plans.

"For half of Syria, the Euphrates is life," said Abdel Aziz al-Masri, a top official at Syria's Ministry of Irrigation.

Mumtaz Turfan, the director of Turkey's department of hydraulic works, echoed the sentiment. "Without our dams, life in Turkey would be impossible," he said.

## Rising Tension, Dwindling Water

It has been a decade since Syria, Turkey and Iraq sat down to formal negotiations, and their positions remain as starkly opposed now as they were then.

Syria and Iraq want the water to be divided roughly in thirds. Turkey, on the other hand, claims more than half for itself. It says any sharing must take into account Turkey's status as the source of most Euphrates water and the home to a population that is half again as big as Iraq's and Syria's put together.

"I am sure that if we forget about borders, we can solve the problem," Mr. Turfan said. Over the last 30 years, he has presided over the construction of 700 dams in Turkey, and a map that stretches across his wall details plans for 500 more.

"I am providing water for 22 million people who take drinking water from our dams, and for 35 percent of our agriculture," he said. It would be a mistake, he suggested, for Turkey to accept less than the lion's share of the Euphrates. "Without irrigation, we can't do anything," he said. "We can't give it up and destroy our people."

Around the world, the question of water ownership has never seemed so divisive.

Despite efforts by the United Nations and others, the world has yet to come up with an accepted formula on how shared waters should be divided. That situation applies to nearly 300 rivers, including the Nile, the Danube, the Colorado and the Rio Grande, all subject to major disputes.

In 1997, a United Nations convention declared that international waterways should be divided reasonably and equitably, without causing unnecessary harm. But Turkey, along with China and Burundi, refused to sign the agreement, a sign of reluctance on the part of major upstream countries to cede their dominant positions.

Beyond current water shortages, one reason for the tension is a lack of optimism that pressures on the world's water supply might be eased.

Of course, some new technologies, including advanced irrigation techniques, innovative desalination methods and bold water-moving schemes, offer some hope.

But experts say the task of making fresh water available where there is none—from seawater, from icebergs or by moving surface water long distances—remains too costly to be widely adopted. Most easily accessible fresh water sources have already been tapped.

To those who would bend nature, there are many cautionary tales, including the Soviet-orchestrated drying up of the Aral Sea, an environmental catastrophe. Even for its project on the Euphrates, Turkey could not win World Bank support because of the plan's significant costs, including the submerging of an important archaeological site.

But sometimes, as in Turkey's case, the thirst for water is greater than any such opposition. Along the Euphrates, Turkey is paying its own way, despite the financial hardship and despite the uncertainty over how its plans will affect others.

In the dry heat of a Mesopotamian summer, how that tenacity is regarded depends on which way the water is flowing.

In an office in Al Raqqa, a Syrian city on the banks of the Euphrates, Iliaz Dakal, a consultant to Syria's land development agency, was studying a map of his own. It depicted, in vivid green, the land that Syria is already irrigating along the river. Much more land, though, like the fields that are home to Mr. Aween, was crosshatched to show uncertainty—its water and its fate now perhaps in Turkey's hands.

"For now, thank God, we have enough to grow the wheat we need," Mr. Dakal said. "But what about the future?"

Mr. Demir, in his muddy Turkish field, had an answer, saying he had come to see water as a byproduct of power.

"If Turkey is stronger," he said, "then I get to keep my water."

---

From the *New York Times.com*, August 25, 2002. © 2002 by The New York Times Company. Reprinted by permission.

# The Coming Water Crisis

Many billions of dollars will be needed to quench America's thirst, but is private business the answer?

By Marianne Lavelle and Joshua Kurlantzick

The tap water was so dark in Atlanta some days this summer that Meg Evans couldn't see the bottom of the tub when she filled the bath. Elsewhere in her neighborhood, Gregg Goldenberg puts his infant daughter, Kasey, to bed unbathed rather than lower her into a brew "the color of iced tea." Tom Crowley is gratified that the Publix supermarket seems to be keeping extra bottled water on hand; his housekeeper frequently leaves notes saying, "Don't drink from the faucet today." All try to keep tuned to local radio, TV, or the neighborhood Web site to catch "boil water" advisories, four of which have been issued in the city since May to protect against pathogens. "We've gotten to the point where I'm thinking this is just normal," Evans says. "It's normal to wake up and take a bath in dirty water."

In a nation where abundant, clear, and cheap drinking water has been taken for granted for generations, it is hard to imagine residents of a major city adjusting to life without it. But Atlanta's water woes won't seem so unusual in the years ahead. Across the country, long-neglected mains and pipes, many more than a century old, are reaching the end of their life span. When pipes fail, pressure drops and sucks dirt, debris, and often bacteria and other pathogens into the huge underground arteries that deliver water. Officials handle each isolated incident by flushing out contaminants and upping the chlorine dose (Atlanta says its water meets health standards despite its sometimes unappetizing appearance), but no one sees this as a long-term solution. America's aging water infrastructure needs huge new investment, and soon.

Decayed pipes alone would be a serious challenge. Now, add these: Providing water free of disease and toxins is ever more difficult, as old methods prove inadequate and new hazards emerge. Shortages have become endemic to many regions, as record drought and population sprawl sap rivers and aquifers. Then there's the threat, unthinkable a year ago, that now seems to trump all others: terrorism. Put it all together, and it's easy to see why concern over clean drinking water might someday make the energy crisis look like small potatoes.

"The idea of water as an economic and social good, and who controls this water, and whether it is clean enough to drink, are going to be major issues in the country," says economist Gary Wolff, at Oakland's Pacific Institute for Studies in Development, Environment, and Security. In March, Environmental Protection Agency Administrator Christie Whitman called water quantity and quality "the biggest environmental issue that we face in the 21st century."

Water providers say that Americans can still trust the product on tap. "People should feel good about their water. Water is safe and we're working hard to keep it that way," says Thomas Curtis, deputy executive director of the American Water Works Association. But the Natural Resources Defense Council's Erik Olson detects a "schizophrenic" element in industry assurances. "They say we need hundreds of billions of dollars to fix the system, but when people ask, 'Is there a public-health issue?' they say, 'No, no.' But clearly, there's a public-health problem."

Both the sanguine and the worried agree on one thing: High costs will force the nation's water delivery system to evolve into something quite different. Citizens will be asked to pay more and use less. And big business, still a minor player in this country's water scene, is seeking a leading role. Private industry promises needed new capital and greater efficiency, but the jury is still out on whether it can deliver. Witness, for instance, the plight of Atlanta, which in 1999 became the largest U.S. city to privatize its water system. Already the city is weighing whether to nullify its 20-year contract with United Water, a subsidiary of the French company Suez.

**Buried troubles.** For now, issues of ownership, infrastructure, and health have been back-burnered while governments grapple with the threat of water system terrorism (box, "Security First"). Terrorism, however, cannot long postpone action on the fissures spreading in the 700,000 miles of pipes that deliver water to U.S. homes and businesses. Three generations of water mains are at risk: cast-iron pipe of the 1880s, thinner conduits of the 1920s, and even less sturdy post-World War II tubes. While refusing to call it a crisis, Curtis says, "We are at the dawn of an era where utilities will need to make significant investments in rebuilding, repairing, or replacing their underground assets."

Cost estimates range from EPA's $151 billion figure to a $1 trillion tally by a coalition of water industry, engineering, and environmental groups. The AWWA projects costs as high as $6,900 per household in some small towns.

---

### Pathogens

**SOURCE:** sewage discharges and farm runoff can introduce E. coli bacteria, cryptosporidium, and other harmful microorganisms

**PROBLEMS:** gastrointestinal illness, severe in people with weak immune systems

**HOT SPOTS:**
- New Haven, Mich.
- San Antonio
- anyplace with treatment or pipe system breakdowns

---

Health is at risk if nothing is done. Already, water mains break 237,600 times each year in the United States. An industry study last year found pathogens and "fecal indicator" bacteria at significant levels in soil and trench water at repair sites. Of the 619 waterborne disease outbreaks the Centers for Disease Control and Prevention tracked between 1971 and 1998, 18 percent were due to germs in the distribution system. Researchers also question whether Americans are getting sick from their drinking water far more often than is recognized. "Is this happening below the radar screen, with low-level [gastrointestinal] things, where people will stay home from work, or be miserable at work, and not ever go to the doctor?" asks Jack Colford of the University of California–Berkeley. He is leading a major EPA-CDC-funded study comparing disease rates between participants who drink tap water through a sophisticated filter and those using a fake look-alike filter. Harvard researchers reported in 1997 that emergency-room visits for gastrointestinal illness rose after spikes in dirt levels that still remained well within federal standards.

**Quality concerns.** Just keeping up with federal regulations is increasingly difficult. The next five years will see more new rules than have been adopted in all the years since enactment of the Safe Drinking Water Act in 1974. Environmental advocates blame the logjam on delays in addressing many health hazards. The arsenic standard, which produced an uproar early in the Bush administration, was years in the making. The EPA ultimately approved the same standard President Bill Clinton chose in his last days in office—reducing the arsenic limit from 50 to 10 parts per billion. The change of heart coincided with a National Academy of Sciences report, released to little notice the week of September 11. It indicated that even the Clinton standard was weak: As little as 3 ppb arsenic carries a far higher bladder and lung cancer risk than do other substances EPA regulates.

---

### PRICEY FENCES

## Security First

Poisoning of drinking water is unlikely, experts agree, but they expect terrorism to consume money and attention. Large water systems will spend $450 million on "vulnerability assessments" Congress has ordered, says the American Water Works Association. An additional $1.6 billion is needed for basic security at pumping stations and treatment plants: alarms, cameras, fences. Better walls and backup pumps will cost billions more. "The threat is imponderable, but… it could be catastrophic," says G. Tracy Mehan, Environmental Protection Agency assistant administrator. "It doesn't fit any cost-benefit calculation."

Only great quantities of chemicals or biological contaminants could cause sickness, and water treatment kills many pathogens. Far more threatening, say system managers, are explosions that would cut off water to firefighters while other sites are set ablaze. Plans of U.S. water systems have been found in al Qaeda hideouts, industry officials warn.

After September 11, Congress approved $90 million in anti-terrorism grants to water systems. But customers will have to pay the far larger bill to come, says AWWA official Thomas Curtis. Utilities won't neglect pollution and delivery problems, he says, but, "certainly other investments will be deferred to security needs." —M.L.

---

New science has also undermined confidence in older methods of purifying water. Chlorination has been one of the 20th century's great public-health achievements, smiting the deadliest waterborne diseases, cholera and typhoid. But this sword has developed a double edge. Nearly 200 women in Chesapeake, Va., sued their water system, claiming that miscarriages they suffered in the 1980s and 1990s are traceable to trihalomethanes, chemicals produced when chlorine reacted with their region's murky river water. While pregnancy-risk research is hotly debated, the EPA decided that cancer risk from chlorine by-products is high enough that it ordered water system reductions earlier this year. Localities have already spent millions of dollars converting to another disinfectant, chloramine (a chlorine and ammonia mix), which curbs some byproducts.

Cities and towns are finding that they must deal with new science on contaminants at a much faster pace than the EPA can regulate them. This summer, Bourne, Mass., the southern gateway to Cape Cod, had to close three of its six drinking water wells, having discovered they were contaminated with perchlorate, a rocket fuel component that leaked from a nearby military reservation. Across the country, the Metropolitan Water District of Southern California, serving 17 million people, announced in April that its new treatment system "will remove a large portion of perchlorate" leaking into a major regional reservoir, Lake

Mead. But *U.S. News* has obtained material distributed at a June 11 MWD board meeting showing the treatment was not working as hoped.

> ## Arsenic
>
> **SOURCE:** occurs naturally in groundwater and sometimes as a residue of mining and other industrial operations
>
> **PROBLEMS:** a strong poison at high doses; at low doses linked to cancer, diabetes, and other diseases
>
> **HOT SPOTS:**
> - Albuquerque, N.M.
> - Norman, Okla.
> - towns throughout the Southwest

The EPA is still studying possible drinking water limits for perchlorate as well as for MTBE, a gasoline additive meant to reduce air pollution that proved to be a frighteningly efficient groundwater pollutant. (South Tahoe and Santa Monica, Calif., last month obtained big settlements from oil and chemical companies to help restore MTBE-poisoned water supplies.) And in April, a U.S. Geological Survey report revealed that streams nationwide are laced with prescription and over-the-counter drugs and even caffeine.

Pollution is shrinking water supplies for communities at the same time that burgeoning population and weather are causing severe shortages. Norman, Okla., with 95,700 people the largest system currently afoul of arsenic standards, very likely will shut down some wells even though it expects average daily water demand to more than double in the next 40 years. "We don't want to be a poster child" for arsenic contamination, says utilities director Brad Gambill. This summer, more than 40 percent of the nation—over twice the normal rate—has suffered drought conditions. "Normally, we get tons of flowers, but now we have nothing growing," says Donna Charpied, a farmer in Riverside County, Calif., pointing to withered plants on her small homestead. Some ecologists believe global warming will make drought the norm in much of the West. Drought breeds anger: The CIA predicts that by 2015, drinking-water access could be a major source of world conflict.

Some cities have already instituted drastic conservation programs. Santa Fe has restricted lawn watering, leading New Mexicans to decorate yards with spray-painted artificial flowers. In parched Denver, a conservation campaign encourages residents to shower in groups. Omaha has an odd-even residential address lawn-watering program.

One spring Saturday morning this April, Chuck Maurer of San Antonio realized while brushing his teeth that he and his neighbors had become victims of a water conservation program gone awry. "It was grotesque," he recalls. "The water was brown in color and cloudy with particulates, and a really bad odor. It was sewer water." Precisely. The San Antonio Water System had accidentally cross-connected his neighborhood's drinking water lines with pipes delivering treated sewage water to a public golf course. Watering fairways and greens with "reclaimed water" has become popular in water-short tourist areas, especially Florida. But experts say such systems require extra care to keep sewage from entering potable systems.

**Big business to the rescue.** With immense challenges ahead, U.S. drinking water systems are considering something never tried here on a large scale: privatization. In March, Indianapolis announced a $1.5 billion agreement with USFilter, the largest U.S. privatization to date, and in May, San Jose, Calif., voted to consider privatizing. Private firms helped supply water to Boston as early as 1796, and utilities have long hired outside contractors to build, but not operate, plants and distribution systems. But over the past five years, an IRS ruling that helped firms obtain longer-term tax-free water contracts, combined with politicians' push for deregulation and municipal-system breakdowns, opened the door for firms to actually manage systems. Only 15 percent of utilities are investor-owned, but in recent years, a handful of big water corporations, mostly foreign owned, have moved onto the U.S. scene: from France, Suez and the media-water conglomerate, Vivendi; from Germany, the utility RWE. (One domestic player with giant ambitions was Enron's water subsidiary, Azurix, which had touted a plan to plumb the Everglades and manage the water.)

> ## MTBE
>
> **SOURCE:** a fuel additive designed to reduce air pollution that has turned into a swift, efficient groundwater polluter through spills and storage tank leaks
>
> **PROBLEMS:** stomach, liver, and nervous system effects, possible cancer risk
>
> **HOT SPOTS:**
> - Pascoag, R.I.
> - Santa Monica, Calif.
> - New Hampshire

Congress is considering hiking federal funding for infrastructure, but the Bush administration encourages the privatization trend, saying that water systems cannot expect to get all the dollars they need from Washington. Says G. Tracy Mehan, EPA assistant administrator for water: "I think the needs are so great especially when you see the demands of homeland security and the federal budget. Private capital is one of several options that are going to have to be considered much more than they have been."

One private-sector success story is Leominster, Mass., a town of 40,000, which signed a 20-year deal with USFilter in 1996. Before then, "our treatment plant was totally corroded. We fixed leaks by putting out old coffee cans to catch the water," says Mayor Dean Mazzarella. USFilter saved the city

## What do Peoria and Paris have in common?

*In the United States, consumers pay above-average rates in many water systems run by private companies, a sampling of quarterly bills shows.*

| PRIVATE U.S. SYSTEMS | QUARTERLY BILL |
|---|---|
| Peoria, Ill. | $100.17 |
| Bloomsburg, Pa. | $94.69 |
| Hoboken, N.J. | $88.50 |
| Camden, N.J. | $74.42 |
| Atlanta | $51.00 |
| Jersey City, N.J. | $49.80 |
| **U.S. average (public and private)** | **$47.50** |
| Leominster, Mass. | $44.70 |
| Corvina, Calif. | $35.80 |

Sources: Raftelis Financial Consulting, 2002 Water and Wastewater Rate Survey, *U.S. News* research

*U.S. water rates are low compared with those of other countries, but that could change if localities begin to address water system problems.*

| CITY | QUARTERLY BILL |
|---|---|
| Paris | $171.80 |
| Osaka, Japan | $115.39 |
| Vienna | $97.02 |
| Hong Kong | $88.73 |
| **United States** | **$47.50** |
| Riga, Latvia | $27.05 |
| Sofia, Bulgaria | $16.49 |
| Buenos Aires | $10.72 |
| Palmerston North, New Zealand | $5.48 |

Source: Raftelis Financial Consulting, 2002 Water and Wastewater Rate Survey. Rates are for 22,450 gallons, typical quarterly U.S. household water use.

money it then used to upgrade a 60-year-old filtration plant that was "held together by wire and chewing gum," says city environmental inspector Matthew Marro.

Experience in other countries suggests that privatization can, indeed, pour needed capital into drinking water. Investment in the United Kingdom increased more than 80 percent after it turned to total privatization. "Public-private partnerships are going to sweep the U.S," says Andrew Seidel, president of US-Filter. "The country has 50,000 different water systems, and those will consolidate into bigger systems aligned with private companies and able to handle the growing number of water-treatment issues."

But in Atlanta, the experience has not been so positive. This summer, Mayor Shirley Franklin sent a formal notice to United Water that the city was dissatisfied with its performance under the 20-year contract signed with the city's previous administration. Problems cited by Franklin included the firm's staffing levels, bill collection, and meter installation. Atlanta had hoped to halve the $49 million annual cost of running its water system by privatizing; one city official says savings are less than $3 million. "You have to keep in mind that a public-private partnership is an ongoing dialogue between the customer and its private partner," says United Water spokesman Rich Henning. "We certainly have struggled. But within the last six to nine months we have dedicated more resources, and we've been listening more to the client." He calculates Atlanta's savings to be about $15 million a year but says the city should be using that money to address the infrastructure problems that United Water inherited.

Gordon Certain, president of the civic association of North Buckhead, the neighborhood hardest hit with water-quality problems, says United Water is unresponsive to complaints. "They're acting kind of like they have a 20-year contract," he says, wryly. (Of course, they do.) The company's response to complaints has ranged "from polite to totally inappropriate," he says. "They told one woman who wanted her water tested that she should get it tested herself." But resident Jacques Davignon thinks privatization "has only made the finger-pointing much more complex." He says the company and the city should share responsibility. "Let's not get on TV and beat United Water up," he says. "Let's do a little forward thinking, come up with a strategic plan."

Private enterprise also has rushed in with water-shortage solutions. The agribusiness firm Cadiz Inc. wants to store water in the barren Mojave Desert, where tidal waves of dust sweep across salt-rimmed dry lakes. The water, taken from the Colorado River and from an indigenous underground aquifer, would flow to thirsty Los Angeles during droughts. "Storing and selling aquifer water will be the key to California's future," says Mark Liggett, Cadiz's senior vice president.

Jim André, a desert biologist working in the Mojave, says Cadiz has no impartial scientific study of the potential impact. Environmental groups warn that drawing groundwater from the Mojave will create a dust bowl similar to California's Owens Lake region, a water grab that inspired the film *Chinatown*. But Cadiz says it has a monitoring system to prevent overpumping. "We have solicited tons of input from all groups for our environmental assessment," Liggett says.

**Creative solutions.** Other ideas seem somewhat fanciful. Ric Davidge, a former Reagan administration official, wants to siphon 10 billion gallons of water each winter from northern California rivers, pump it into 850-foot-long plastic bladders, and ship it downstate. Other entrepreneurs suggest melting Alaska icebergs. Oilman T. Boone Pickens hopes to pipeline water from Texas's Ogallala aquifer to water-short cities like San Antonio and Dallas.

Privatization projects are also dogged by accountability concerns. Industry sources worry that the terrorism vulnerability assessments U.S. water systems are developing will wind up in

## DO IT YOURSELF

### If there's trouble at the tap

Consumers are embarking on their own efforts to ensure safe drinking water at home. But choices about testing water, filtering it, or switching to bottled are far from clear.

**Testing.** Many tests measure data that make no difference to health: hardness, acidity and alkalinity, and color. But given the health hazards, testing for lead and arsenic is probably worthwhile. Lead in drinking water often leaches from old pipes. "It's a national health tragedy, but people can empower themselves to address it," says Richard Maas of the University of North Carolina's Environmental Quality Institute. Usually, running the water for a minute before use eliminates lead. Maas's institute (www.leadtesting.org) does lead and arsenic testing for about $15 each.

Tests for other harmful contaminants like chlorine byproducts and pesticides usually are more expensive and may involve sending samples by overnight mail. The EPA (www.epa.gov/safewater/faq/sco.html) provides a list of state agencies to contact to find an approved laboratory. California's Silver Lake Research sells tests that give consumers immediate results. Its WaterSafe all-in-one package, which tests for lead, pesticides, and E. coli bacteria, costs $14.99. These yes-or-no tests can signal a problem but do not tell consumers contaminant levels.

**Filters.** As a rule, filters remove lead, chlorine, and many byproducts from tap water. Consumers should read labels to ensure that the models they choose—whether inserted into carafes, installed on faucets, or in refrigerator water systems—act on these contaminants. To guard against parasites such as cryptosporidium, look for filters that meet "NSF Standard 53 for cyst reduction." NSF International (www.nsf.org) tests and certifies water-treatment equipment. Regular replacement of filters, often costly, is crucial to prevent bacteria from colonizing. Beware the much touted boiling of water. While cheap and effective at killing some contaminants, it can also release harmful chlorine byproducts into the air.

**Bottled.** U.S. consumers gulped down $6.5 billion in bottled water last year, up more than 50 percent from five years ago. The Food and Drug Administration prohibits firms from labeling as "spring water" anything not from an underground formation that flows naturally to the Earth's surface. In theory, bottled water must meet the same federal standards as public water, but there are exemptions. For example, the FDA does not regulate the plasticizer DEHP, despite evidence that water long stored in plastic bottles could exceed tap water limits.

Most bottled water is clean. But in 1999, a Natural Resources Defense Council study showed that four of 103 tested brands of bottled water violated federal standards for chemicals or coliform bacteria, while one quarter fell short of stricter California standards for other contaminants. The bottled-water industry maintains that its product is well regulated and clean.

Much bottled water is not from mountain springs but from city taps. Kansas City, Mo., bottles its own "City of Fountains" water, and Oak Creek, Wis., sells its "Claire Baie" brand. New Orleans Mayor C. Ray Nagin wants to sell the product of his financially strapped water system as "Crescent City Clear."

But by far, the behémoths in this business are Pepsi and Coke, which in only five years have gained the top two spots in the market. Pepsi's No. 1 Aquafina is municipal water from spots like Wichita, Kan., while Coke's Dasani (with minerals added) is taken from the taps of Queens, N.Y., Jacksonville, Fla., and elsewhere. Both firms use high-tech reverse-osmosis filtering to obliterate all contaminants. A 1-liter bottle of Aquafina last month cost $1.09 in one suburban Washington, D.C., supermarket. That is 1,741 times the average U.S. public drinking water price of $17.70 per 1,000 cubic feet. Drink up!

—M.L.

---

corporate parent offices overseas, possibly unprotected from disclosure. In New Orleans, an official highly familiar with its water system told *U.S News* that the Big Easy's move toward privatization lacks oversight. "The whole approach to having companies bid for the water system was 'public, catch us if you can,' since after bids were taken the public had only 10 days to examine the proposals," she says.

Privatization worries have even made it to Broadway: In the comedy *Urinetown*, a firm privatizes toilets and raises toilet fees. Residents caught urinating in other places are arrested. "With private control, who guarantees that the less well off will get affordable water, and who picks up the cost if the private company fails?" asks Sandra Postel, director of the Global Water Policy Project, a research institute in Amherst, Mass.

**Progress report.** Indeed, the financial viability of some leading water companies has been called into question recently. Cadiz lost $2.5 million in the most recent quarter; the firm recently tried to reduce its debt through a deal with Saudi Prince Al Waleed ibn Talal, but in July the effort collapsed. Suez's water arm saw revenues grow by just 1 percent. Vivendi, though experiencing revenue growth of 12 percent, made major missteps in its media division that have left it laden with debt and is divesting its stake in one water investment, Philadelphia Suburban.

Nor have private companies, by and large, delivered savings to consumers. In fact, most private water providers surveyed by *U.S. News* charged higher-than-average rates (table). George Raftelis, a Charlotte, N.C., industry consultant, points out that unlike public utilities, private firms do not enjoy tax-exempt financing, are subject to income taxes, and must return profits to shareholders. Moreover, "privatization does not equal competition," says Janice Beecher, director of the Institute of Public Utilities at Michigan State University. "After bidding, you're transferring the monopoly powers of a public utility to a private company." She suggests cities and towns award shorter contracts and make public utilities and private firms compete.

Citizen outcry over the water rates private firms charge has boiled over into riots in countries such as Bolivia. But so far in the United States disputes have been hashed out in the political process. Peoria and Pekin, Ill., both are moving to deprivatize their water systems, having determined that if private ownership continued, future rate increases would be as much as 60 percent higher than if the systems were publicly run. Because other communities have done the same, Curtis of AWWA does not see a mass movement to privatize: "Some are looking at it, and some are trying to move in the other direction."

---

**Perchlorate**

**SOURCE:** a component of solid rocket fuel, munitions, and fireworks; has leaked from at least 58 U.S. military bases and manufacturing plants

**PROBLEMS:** interferes with functioning of the thyroid gland

**HOT SPOTS:**
- Riverside, Calif.
- Bourne, Mass.
- contamination confirmed in 20 states

---

But the harsh reality is that the price of drinking water will most likely rise whether private industry or government manages the system. The EPA estimates that the water bill consumes only seven tenths of 1 percent of U.S. household median income; Americans spend more than triple that on bottled water and filters. A recent Harvard School of Public Health analysis pointed out that rates in many developed countries are significantly higher. "[W]ater rates have been insufficient to cover long-run costs," such as maintenance of pipes and plants, let alone larger issues such as preserving clean rivers and surrounding watershed, the report said.

"People think water is free because it falls from the sky," says Seidel of USFilter. "Well, it is—but treated, filtered, and piped water isn't." Privatization advocates contend that only market-oriented pricing can force $H_2O$-hogging Americans to conserve. "Unless you put a market-determined price on something, it is not respected," says Clay Landry, a research associate at Bozeman, Mont.'s Political Economy Research Center. "Right now, who even thinks about the cost of water coming out of their tap?"

But public officials are loath to hike rates for fear of burdening lower-income families. That's certainly a problem in big cities, but even more so in small towns, where, lacking economies of scale, water treatment and distribution is more expensive. Consultant Raftelis found that water bills in small systems average 25 percent higher than in large ones he has surveyed. The new arsenic rule is projected to cost households under $1 annually in the largest systems but over $300 in those serving fewer than 100 customers.

Economist Wallace Oates of the think tank Resources for the Future says arsenic's economic realities make a case for abandoning national standards and letting localities weigh costs and benefits on their own. Congress and the EPA already let small water systems operate with less regulation and enforcement—some will have 14 years, instead of four years, to meet the new arsenic rule. The Bush administration is studying whether to relax small-system standards even more. Yet all but a fraction of health violations occur in small systems, which serve some 50 million citizens. "What you have is a two-tier drinking water system, and that's pretty troubling," says NRDC's Olson. He argues that health and efficiency require a major consolidation among the 54,000 U.S. water suppliers. Says EPA's Mehan, "Citizens and systems are going to have to look at this option."

**Turning off the tap.** Citizens are certainly looking at other options, but less with an eye to changing the system than to just protecting themselves and their families. "We're looking at having a plumber put a filter on our entire house," said Atlanta resident Davignon. In the meantime, he buys bags of ice and water from the supermarket, adding, "I hate to pay for water, but if it's undrinkable, or the kids can't bathe, you do it." Already, 76 percent of Californians rely on bottled or filtered water. "We have reached a breaking point beyond which central treatment can no longer go," says Peter Censky, executive director of the Water Quality Association, which represents filter makers. Joseph Cotruvo, a former EPA water administrator, agrees: "You wouldn't think of drinking orange juice out of a pipe, would you? I wouldn't be surprised if 25 years from now the thought of drinking water as a beverage rather than a commodity will dominate."

The drive toward bottled water and filters will, however, widen the gap between haves and have-nots, a result some hope technology can prevent. "[G]oing into the 21st century, you can't get the kind of long-term improvements in water quality that are needed without the next generation of technology," says Olson. A few U.S. water systems are trying disinfectants used in Europe: ozone, ultraviolet light, and perhaps the best purifier (used by bottlers Pepsi and Coke), reverse-osmosis membrane technology. "It removes just about everything," says Olson, "so you don't have this contaminant-of-the-month approach."

---

**THMs**

**SOURCE:** trihalomethanes form when chlorine reacts with organic material, from decayed leaves to feces, in water; extremely common contaminant

**PROBLEMS:** linked to bladder cancer, with some evidence of miscarriage risk

**HOT SPOTS:**
- Waco, Texas
- Washington, D.C., suburbs

And yesterday's clean water may not be clean enough for the future. L. D. McMullen, chief executive officer of the Des Moines water system, believes as the population ages and more people have compromised immune systems, cities and towns will have to provide water much lower in contaminants than they do today. "We will totally have to deliver water to customers in a totally different way," he says. "You may see what I like to call 'neighborhood polishing units,' that develop ultrapure water in the neighborhoods and deliver it to homes" through much smaller pipe systems. Households need relatively little superclean water, McMullen points out, since less than 15 percent of "drinking water" is drunk or bathed in. Most goes to flushing toilets and watering lawns.

Des Moines has learned from experience that its citizens will pay for such improvements: In 1992, the city raised water rates 25 percent to build the world's largest removal plant for nitrate, an agricultural runoff that can reduce infants' oxygen uptake (blue-baby syndrome) and cause other ills in adults. But whether public water systems tackle their challenges on their own or turn the job over to private enterprise, or some combination, the changes ahead will require a revolution in how Americans think about drinking water. "People's knowledge of water comes from beer commercials, focused on the land of sky-blue waters, or mountain springs and aquifers underlying some Wisconsin hillside," says Censky of the Water Quality Association. "The public thinks water in these sources is pure, but it's not. Really, pure water is a man-made product."

*With David D'Addio*

# Index

## A

Abbey, Edward, 65
acid precipitation, GIS and, 151
adaptive management, restoring ecosystems disrupted by dams and, 19–21
Africa, AIDS in, 100, 102
African superswell, 22
African Union (AU), 13–14
aging: economic burden of, 197; in Florida, 197; global crisis of, 197; pensions and, 199, 201; politics of, 201–202
agriculture: Euphrates River and, 207–209; European Union policy on, 95; precision, 165
Agriculture Department, 75
AIDS (acquired immune deficiency syndrome), 100, 102–108, 187–190; blood selling in China resulting in, 105–106, 192; in China, 191–193; public education about, 107–108; vaccine for, 104, 190; women and, 105
Airlift Deployment Analysis System (ADANS), use of GIS during Persian Gulf War and, 141, 155–156
airport development, use of web-based GIS in, and public involvement, 137–139
Alaska, glacial movement in, and GIS, 152–153
American Association of Retired Persons (AARP), 201
American Steel Barge Company, 125
aquifer, Ogallala, 60
ArcIMS, 138
area studies tradition, 16
Armenia, oil in, 195
Army Corps of Engineers, 64–65
arsenic, in drinking water, 211, 212
Association of American Geographers (AAG), 2
Atlanta, Georgia, water treatment in, 45, 210, 213
automobiles, fuel-cell powered, 84
Aween, Abdelrazak al-, 207, 208, 209
Azerbaijan, oil in, 194, 195, 196

## B

Babbitt, Bruce, 62–66
Baltimore, Maryland, water treatment in, 45
Bangladesh, population growth in, 180–186
Betts, Roland, 35
birth control. *See* contraceptives
Bisbee, North Dakota, 116–117
blood selling, in China, and AIDS, 105–106, 192
blue-baby syndrome, drinking water and, 216
BMW, 113–114
*bosque*, 118–123
bottled water, 214
Bourne, Massachusetts, 211
bridges, effects of human modifications on Salt River and, 50, 51–53
Bright, Ed, 140
Brooks, Al, 150
Brower, David, 65
brown trout, 20
Bureau of Land Management (BLM), 71
Bureau of Reclamation, 66
Burnally, Bonnie, 97
Bush administration: energy policy of, 29; global warming and, 84

## C

Cadiz Inc., 213
California, pollution-control initiatives by, 43
Canada, 92
cancer, chlorination and, 211
carbon dioxide, 22–27, 67, 68–69; global warming and, 84
carbon sequestration, 67, 68–69
carbon tax, 31, 67
Central Asia, oil in, 194–196
central business districts (CBDs), 131, 133–135
channel scour, effects of human modifications on Salt River and, 54–55
channelization, effects of human modifications on Salt River and, 48–59
*Charles W. Whetmore*, 125
Cheney, Dick, 30–32
Chernobyl, 29
ChevronTexaco, 194
Chicago World's Fair, whaleback ships at, 125
China, 172–177, 196; AIDS in, 102–108, 191–193; Great Wall of, 97–99; Hong Kong and, 96
chlorination, of water, 211, 214
*Christopher Columbus*, 125
*City of Everett*, 125
Clean Air Act, 67
Cleveland, Ohio, water treatment in, 45
climate change, 75–76; carbon sinks and, 82
coal, 67, 68–69
coal reservoirs, dumping carbon dioxide in, 69
coal seams, dumping carbon dioxide in, 69
coastal change analysis, GIS and, 151–153
Coastal Change Analysis Program (C–CAP), 151–152
Coleman, Phil, 151
Columbian Exposition, whaleback ships at, 125
Committee on Geography of the Board of Earth Sciences, 2–3
computer-based field mapping, 158–159
*Congo* (Crichton), 140
conservationism, 70
continents, vertical motion of, 24–26
contraceptives, population growth and, 180, 182–183, 185–186
convection, in mantle of Earth, 22–28
CO2e.com, 69
Cretaceous period, 25, 27
cropland, 71–73, 75
crytosporidium, 211
Curtis, Thomas, 210

## D

dams, 62–66; construction of, 62–63; destruction of, 62, 64; economic issues of, 63–64; effects of human modifications on Salt River and, 48, 49–50, 52; hydroelectric, 63–69, 172, 175; restoring ecosystems disrupted by, 19–21; social and economic issues around, 63–64; Three Gorges, 172–177
Demir, Ahmet, 207, 208, 209
Denver, Colorado, 25
Des Moines, Iowa, drinking water in, 216
Donghu, China, AIDS and, 191–193
droughts, 83
Durfee, Richard, 142, 149, 151, 154

## E

E. coli, 211
Earth First!, 63, 65
earth science tradition, 15, 16–17
Economic and Monetary Union (EMU), 199–200
El Nino, 122, 163
emissions trading, 83–84
EnCana, 69
Environmental Protection Agency (EPA), 43, 44, 45, 46–47, 72, 151
environmental research, 160
environmental restoration, GIS and, 152–154
environmentalism, 70
Erie Municipal Airport Authority, use of GIS to expand, 137–139
ESRI Inc., 138
Euphrates River, water use conflicts and, 207–209
euro, 199–200
European Union (EU), 199–200; expansion of, 94–95
expressways, 134
ExxonMobil, 194

## F

Fairfax County, Virginia, water treatment in, 45
family planning services: population growth and, 180–184; in Thailand, 183
Farallon plate, 26
Federal Energy Regulatory Commission (FERC), 66
fertility rates, 180, 201
field geology, 158–159
field mapping, computer-based, 158–159
filtering, of drinking water, 214

# Index

fish populations: ecosystem disruption by Glen Canyon Dam and, 19–20, 21; GIS and, 151
flooding, 83; advantages of, 120; in China, 172; effects of human modifications on Salt River and, 48–49, 50, 51–52; seasonal, 118; of Yangtze, 172
forests, as resource, 71, 72, 74
fossil fuels: carbon sequestration and, 68–69; climate change and, 82–85; coal burning and, 67; government subsidies for, 85; new oil fields in Central Asia and, 194–196
*Frank Rockefeller,* 125
fresh water, 203–206; contamination levels in, 204
fuel cells, 67, 84

## G

geophysics, 23–28
Georgia, oil in, 195, 196
Gezhouba Dam, 175
GIS and Spatial Technologies (GISST), 153, 154
GIS (Geographic Information Systems) technology, 158–159, 162–164; ORNL and revolution in, 140–157; public involvement in airport development and, 137–139
Glen Canyon Dam, 19–21, 65
Global Positioning System (GPS), 159, 165
global warming. *See* climate change
Gold, Barry, 19–20, 21
Gondwana, supercontinent of, 27
Gordon, Slade, 64–65
Gore, Al, 167
Granite Reef Dam, 48, 49, 50, 52
grassland, 71, 74
gravel mining, effects of human modifications on Salt River and, 54, 56
gravity, 23, 28
Great Lakes, whaleback ships on, 124–125
Great Wall of China, 97–99
greenhouse gases, 75
Greenville County, South Carolina, 109–115
Ground Zero, plans for rebuilding, 33–35
Guangdong province, in China, 90
guns, oil in Central Asia and, 194

## H

habitat studies, satellite imaging and, 162
heat generation, carbon sequestration and, 68
HEC-RAS, 57
HIGHWAY modeling, GIS and, 155
Hinze, Bill, 150
Hokokam people, 49–50
homosexuality, AIDS and, 106, 108
Honea, Bob, 149
Hong Kong, 96
Hoover Dam, 62
Hubbard Glacier, GIS and movement of, 152
humpback chad, 20
Huntington, Samuel, 201

hurricanes, 77
hydroelectric dams, 63–64, 172, 175; licenses for, 66; Three Gorges, 172–177. *See also* dams

## I

Imagine New York, 33
immigrant workers, 198
"impact fee," for real estate development, 79–80
income, by county, 168–170
India: AIDS in, 112–118; homosexual life in, 116
Indonesia, 160–161
in-stream gravel mining, effects of human modifications on Salt River and, 54, 56
integrated gasifier combined cycle (IGCC) approach, to carbon sequestration, 68
Intergovernmental Panel on Climate Change (IPCC), 82–83
INTERLINE modeling, GIS and, 155
International Conference on Population and Development (ICPD): in Cairo, 183, 185; in Mexico City, 185
International Energy Agency (IEA), 30
intravenous drug use, AIDS and, 114
Iraq, Euphrates River and, 207, 208, 209

## J

Japan, 91–92
Jetty Jacks, 120–121
Joint Flow and Analysis System for Transportation (JFAST), GIS and, 155, 156

## K

Kazakhstan, oil in, 194, 195
King, Amy, 153
Krugman, Paul, 11
Kyoto Protocol, 83
Kyrgyzstan, oil in, 194, 195, 196

## L

lake acidification, GIS and, 151
Lake Mead, 79
Lake Powell, 65–66
land cover, 70–77
land transformation, 70–77
land-use modeling, GIS and, 149
Las Vegas, Nevada, 78–81; Development Services Center of, 78
Leominster, Massachusetts, water supply in, 212–213
Lewiston, Idaho, 63
Lindesay, William, 97, 98
locks, 176
Lower Manhattan Development Corporation (LMDC), plans for rebuilding at Ground Zero and, 33, 35

## M

Madagascar, 13–14
Managua, Nicaragua, 171

Manchester, New Hampshire, water treatment in, 45
man-land tradition, 16–17
mantle, 22–28
mapping: art of, 16; computer-based field, 158–159; lack of, in Managua, Nicaragua, 171
*maquiladoras,* 86–87
Margle, Steve, 148
Massachusetts, pollution-control initiatives in, 46
McDougall, Alexander, 124–125
Medicare reform, 187
*Meteor,* 125
metropolis, American, 128–136
Mexico, 86–87
Miami/Dade County, Florida, 133
Minnesota, 44
Mojave Desert, storage of water in, 213
Montana, 44
MTBE, in drinking water, 211
mutation rates, in AIDS virus, 104

## N

NASA (National Aeronautics and Space Administration), 160
National Academy of Public Administration (NAPA) panel, 44, 46
National Acid Precipitation Assessment Program (NAPAP), 151
National Center for Atmospheric Research, 83
National Climatic Data Center, 82
National Energy Plan, GIS and, 150
National Environmental Policy Act (NEPA), 149
National Marine Fisheries Service (NMFS), 64–65, 151
National Oceanic and Atmospheric Administration (NOAA), 151
National Research Council (NRC), 2–3
National Resources Inventory, 71
National Science Foundation (NSF), 149
National Uranium Resource Evaluation (NURE), 150
Nebraska, 44
New Hampshire, pollution-control initiatives in, 43–44
New Jersey, pollution-control initiatives in, 44, 46
New Mexico, 118–123
New Partnership for Africa's Development (NEPAD), 14
New York City, New York, rebuilding Ground Zero and, 33–35
Nicaragua, 171
nongovernmental organizations (NGOs), 165–166
Norman, Oklahoma, water supply in, 211
North American Free Trade Agreement (NAFTA), 86–87; costs and benefits of, 30–31; deregulation in, 30; liability insurance in, 31; nuclear energy industry and, 29–32; subsidies for, 31–32
North Carolina, 44
Nuclear Regulatory Commission (NRC), 30

# Index

## O

Oak Ridge National Laboratory. *See* ORNL
Oak Ridge Regional Modeling Information System (ORRMIS), 145, 149
oceans: dumping carbon dioxide in, 68–69; heat content of, 83
Office of Surface Mining, 150
Ogallala aquifer, 60
oil, in Central Asia, 194–196
oil reservoirs, dumping carbon dioxide in, 69
Oklahoma City, Oklahoma, water treatment in, 45
O'Malley, Martin, 45
*101*, 124
Ontario Power Generation, 69
Oregon, 46
Organization of African Unity (OAU), 13
ORNL (Oak Ridge National Laboratory), GIS revolution and, 140–157
orphans, due to AIDS, 178, 191–193

## P

Pace, Peter, 141–142
Pakistan, population growth in, 180–183
Pattison, William, 15
"pebble-bed" technology, 29
perchlorate, in drinking water, 211–212, 215
Persian Gulf War, use of GIS during, 141, 155–156
Peterson, Randall, 20, 21
Phoenix, Arizona, effect of human modification on Salt River in, 48–59
plate tectonics, 22–28
Population and Community Development Association, in Thailand, 183
population growth, 180–186; AIDS and, 189–190; global social emergency of, 186; loss of fresh water and, 204; momentum in, 184; U.S. Congress and, 185; women and, 182, 184
Port Authority (PA), plans for rebuilding at Ground Zero and, 33, 35
preservationism, 70
privatization, of water treatment operations, 210–216
Putin, Vladimir, 196

## R

rainbow trout, 20
region states, 90–93
regional studies tradition, 16
Remote Sensing and Special Surveys (RSSS) program, 153–154
retail trade, shopping centers and, 134
Rio Grande, 60, 118–123
rivers, 62–66; Colorado, 177; endangered, 118; Euphrates, 207–209; Rio Grande, 60, 118–123; Salt, 48–59; Snake, 63; Yangtze, 172–177
Rumsfeld, Donald H., 194, 196
rural poverty, decline of small towns and, 116–117
Russia, 196

## S

Safe Water Drinking Act, 211
saline aquifers, dumping carbon dioxide in, 69
salmon, 62–66
Salt River, effects of human modification on, 48–59
San Antonio, Texas, water supply in, 211
Santa Ana Pueblo, 121
satellite technology, 160–167; meteorological, 161; navigational, 165; surveillance, 160, 162–163
Savage, John L., 172
scour, effects of human modifications on Salt River and, 54–55
scrubbing, power plants and, 68
sea level, fluctuation of, 24
seismology, 22–28
Senegal, AIDS in, 188–189
sequestration, carbon, 67, 68–69
sexually transmitted diseases (STDs), 103, 188–189; women and, 103–104
ships, whaleback, 124–125
Sierra Club, 65
Silicon Valley, 92
Sky Harbor Airport, 56
small towns, decline of, 116–117
Snake River, 63
Social Security, 197
*South Park*, 125
Spatial Data Transfer Standard (SDTS), 143
spatial tradition, 16
states, pollution-control initiatives by, 43–47
Stauber, Karl, 116
steam reformation, carbon sequestration and, 68
street maps, lack of, in Managua, Nicaragua, 171
subduction zones, 23–28
suburbs, 130–136
Syria, Euphrates River and, 207–209

## T

Tajikistan, oil in, 195
territoriality, 4–5
terrorism, water supply safety and, 210, 211
testing, of drinking water, 214
textile industry, in Greenville County, South Carolina, 111–112
Theodore Roosevelt Dam, 50, 52
THMs (trihalomethanes), 215
*Thomas Wilson*, 125
Three Gorges Dam, 172–177
Three Mile Island, 29
Tinnel, Ed, 148, 150
topography: maps and, 159; undersea, 162
tourism, Great Wall of China and, 97–98
toxic waste, 86
transportation, 5; automobile, 131–132, 134; growth of, 128–136; mass, 131; railroad, 129; trolley, 130; whaleback ships and, 124–126
transportation modeling, GIS and, 154–156
tree, cottonwood, 118. *See also* forests
Tsang, Donald, 96
Turkey, Euphrates River and, 207–209
Turkmenistan, oil in, 195

## U

Uganda, AIDS in, 188–189
United States Transportation Command (USTRANSCOM), use of GIS by, 155–156
United Water, 213
urban growth, 128
*Urinetown*, 214
USFilter, 212–213
Uzbekistan, oil in, 195

## V

vaccines, for AIDS, 104, 190
Varney, Robert W., 43–44

## W

waste-water treatment plants, 45
water: as commodity, 206; conflicts over, and Euphrates River, 207–209; conservation districts, 61; consumption of, 60; extraction of, 61; pipelines, 60; quality of, 204; rights, 60, 120; supply, 79, 181, 205
water-treatment facilities, 45; privatization and, 210–216
wealth, personal, by county, 168–170
Weinberg, Alvin, 149
Weltin, Bob, 116
wetlands, 71, 74
whaleback ships, 124–125
Whitman, Christine Todd, 44, 47
Wilford, John Noble, 2
Wisconsin, pollution-control initiatives in, 44, 46
women: AIDS and, 105; population growth and, 182, 184; STDs and, 103, 104
World Commission of Dams, 177
World Economic Forum, at Davos, 82
World Trade Center, plans for rebuilding on site of, 33–35

## Y

Yangtze River, 172–177

## Z

Zedong, Mao, 172
Zimbabwe, 13–14
zoning, 134

# Test Your Knowledge Form

We encourage you to photocopy and use this page as a tool to assess how the articles in *Annual Editions* expand on the information in your textbook. By reflecting on the articles you will gain enhanced text information. You can also access this useful form on a product's book support Web site at *http://www.dushkin.com/online/*.

NAME: DATE:

TITLE AND NUMBER OF ARTICLE:

BRIEFLY STATE THE MAIN IDEA OF THIS ARTICLE:

LIST THREE IMPORTANT FACTS THAT THE AUTHOR USES TO SUPPORT THE MAIN IDEA:

WHAT INFORMATION OR IDEAS DISCUSSED IN THIS ARTICLE ARE ALSO DISCUSSED IN YOUR TEXTBOOK OR OTHER READINGS THAT YOU HAVE DONE? LIST THE TEXTBOOK CHAPTERS AND PAGE NUMBERS:

LIST ANY EXAMPLES OF BIAS OR FAULTY REASONING THAT YOU FOUND IN THE ARTICLE:

LIST ANY NEW TERMS/CONCEPTS THAT WERE DISCUSSED IN THE ARTICLE, AND WRITE A SHORT DEFINITION:

# We Want Your Advice

ANNUAL EDITIONS revisions depend on two major opinion sources: one is our Advisory Board, listed in the front of this volume, which works with us in scanning the thousands of articles published in the public press each year; the other is you—the person actually using the book. Please help us and the users of the next edition by completing the prepaid article rating form on this page and returning it to us. Thank you for your help!

## ANNUAL EDITIONS: Geography 03/04

### ARTICLE RATING FORM

Here is an opportunity for you to have direct input into the next revision of this volume.
We would like you to rate each of the articles listed below, using the following scale:

1. **Excellent: should definitely be retained**
2. **Above average: should probably be retained**
3. **Below average: should probably be deleted**
4. **Poor: should definitely be deleted**

Your ratings will play a vital part in the next revision.
Please mail this prepaid form to us as soon as possible.
Thanks for your help!

| RATING | ARTICLE |
|---|---|
| | 1. The Big Questions in Geography |
| | 2. Rediscovering the Importance of Geography |
| | 3. Birth of the African Union |
| | 4. The Four Traditions of Geography |
| | 5. Restoring an Ecosystem Torn Asunder by a Dam |
| | 6. Sculpting the Earth From Inside Out |
| | 7. Nuclear Power: A Renaissance That May Not Come |
| | 8. Filling the Void |
| | 9. Global Warming: The Contrarian View |
| | 10. The Pollution Puzzle |
| | 11. Human Modification of the Geomorphically Unstable Salt River in Metropolitan Phoenix |
| | 12. Texas and Water: Pay Up or Dry Up |
| | 13. Beyond the Valley of the Dammed |
| | 14. Environmental Enemy No. 1 |
| | 15. Carbon Sequestration: Fired Up With Ideas |
| | 16. Past and Present Land Use and Land Cover in the USA |
| | 17. Operation Desert Sprawl |
| | 18. A Modest Proposal to Stop Global Warming |
| | 19. A Greener, or Browner, Mexico? |
| | 20. The Rise of the Region State |
| | 21. Continental Divide |
| | 22. A Dragon With Core Values |
| | 23. The Late Great Wall |
| | 24. A Continent in Peril |
| | 25. AIDS Has Arrived in India and China |
| | 26. Greenville: From Back Country to Forefront |
| | 27. Death of a Small Town |
| | 28. The Rio Grande: Beloved River Faces Rough Waters Ahead |
| | 29. "You Call That Damn Thing a Boat?" |
| | 30. Transportation and Urban Growth: The Shaping of the American Metropolis |
| | 31. Internet GIS: Power to the People! |
| | 32. ORNL and the Geographic Information Systems Revolution |
| | 33. Mapping the Outcrop |
| | 34. Gaining Perspective |

| RATING | ARTICLE |
|---|---|
| | 35. Counties With Cash |
| | 36. A City of 2 Million Without a Map |
| | 37. China Journal I |
| | 38. Before the Next Doubling |
| | 39. A Turning-Point for AIDS? |
| | 40. AIDS Scourge in Rural China Leaves a Generation of Orphans |
| | 41. The Next Oil Frontier |
| | 42. Gray Dawn: The Global Aging Crisis |
| | 43. A Rare and Precious Resource |
| | 44. In Race to Tap the Euphrates, the Upper Hand Is Upstream |
| | 45. The Coming Water Crisis |

*(Continued on next page)*

ANNUAL EDITIONS: GEOGRAPHY 03/04

**BUSINESS REPLY MAIL**
FIRST-CLASS MAIL  PERMIT NO. 84  GUILFORD CT
POSTAGE WILL BE PAID BY ADDRESSEE

**McGraw-Hill/Dushkin**
**530 Old Whitfield Street**
**Guilford, Ct 06437-9989**

NO POSTAGE
NECESSARY
IF MAILED
IN THE
UNITED STATES

## ABOUT YOU

Name _____  Date _____

Are you a teacher? ❏   A student? ❏
Your school's name _____

Department _____

Address _____ City _____ State _____ Zip _____

School telephone # _____

## YOUR COMMENTS ARE IMPORTANT TO US!

Please fill in the following information:
For which course did you use this book?
_____

Did you use a text with this ANNUAL EDITION?  ❏ yes  ❏ no
What was the title of the text?
_____

What are your general reactions to the *Annual Editions* concept?
_____

Have you read any pertinent articles recently that you think should be included in the next edition? Explain.
_____

Are there any articles that you feel should be replaced in the next edition? Why?
_____

Are there any World Wide Web sites that you feel should be included in the next edition? Please annotate.
_____

May we contact you for editorial input?  ❏ yes  ❏ no
May we quote your comments?  ❏ yes  ❏ no